Earth and Environmental Sciences Library

The *Earth and Environmental Sciences Library* (EESL) is a multidisciplinary book series that showcases innovative approaches and in-depth reviews, aimed at enhancing the role of the Earth and Environmental Sciences communities. The series also offers valuable guidance for stakeholders, decision-makers, policymakers, international organizations, and NGOs.

Topics of interest include oceanography, the marine environment, atmospheric sciences, hydrology, soil sciences, geophysics, geology, agriculture, environmental pollution, remote sensing, climate change, water resources, and natural resource management.

To further advance these critical areas, the *Earth and Environmental Sciences Library* invites contributions from the Earth and Environmental Sciences communities in the form of edited volumes, monographs, textbooks, conference proceedings, and more. The series fosters knowledge exchange and collaboration across disciplines, encouraging comprehensive discussions and solutions to pressing environmental challenges.

Emmanuel Daanoba Sunkari · Shakir Ali
Editors

Geology and Gold Mineralization in Ghana

Geological Provinces and Gold Mineralization

Springer

Editors
Emmanuel Daanoba Sunkari
Mining Engineering
Faculty of Integrated and Advanced Technology
Sir Padampat Singhania University
Udaipur, Rajasthan, India

Shakir Ali
Department of Geology
University of Delhi
New Delhi, Delhi, India

ISSN 2730-6674 ISSN 2730-6682 (electronic)
Earth and Environmental Sciences Library
ISBN 978-3-031-97162-4 ISBN 978-3-031-97163-1 (eBook)
https://doi.org/10.1007/978-3-031-97163-1

This Springer imprint is published by the registered company Springer Nature Switzerland AG
The registered company address is: Gewerbestrasse 11, 6330 Cham, Switzerland

Preface

Despite Ghana's rich history and significant contributions to the global gold mining industry, there is a notable absence of comprehensive literature specifically addressing the geology and gold mineralization of the country. This gap in academic and practical resources hinders the ability of geologists, mining professionals, and students to fully understand and leverage the unique geological characteristics that make Ghana a leading gold producer. A dedicated book on this subject would not only fill this critical knowledge void but also support the advancement of local and international mining practices, foster educational growth, and promote sustainable development within the sector. The creation of such a resource is essential for enhancing the scientific community's understanding and for driving innovation in gold exploration and extraction in Ghana. Hence, this book primarily aims to provide an in-depth exploration of Ghana's geology and its link to gold mineralization. It incorporates the latest advancements in studying gold deposits through field studies, petrology, structural geology, geochemistry, geostatistics, isotope geology, economic geology, and other related disciplines.

The book *Geology and Gold Mineralization in Ghana* is divided into five comprehensive parts with eight chapters, which deeply delve into the intricate geology and gold mineralization processes that have made Ghana one of the world's leading gold producers.

In the first part **"Introduction to the Geology and Gold Mineralization in Ghana"**, readers are introduced to the diverse geology and gold mineralization styles found across Ghana. This part provides a foundational understanding of the occurrence and general geological features of the main gold deposit types in the country. The first chapter of the book entitled "Geology and Gold Mineralization Styles in Ghana" by *Sunkari et al.* is an introductory chapter that provides a comprehensive review and synthesis of the geological characteristics and principal forms of gold mineralization in Ghana, as documented in existing literature. The authors provide a concise overview of the distribution and fundamental geological attributes of the primary gold deposit types found in Ghana.

Structural controls play a pivotal role in gold mineralization in Ghana, as discussed in the second part **"Structural Controls on Gold Deposits in Ghana"**. This part highlights the significant influence of geological structures on the distribution and concentration of gold deposits, providing a detailed analysis of the structural framework that controls gold mineralization. The chapter "Structural Controls on Orogenic Gold Mineralization: Evidence from Major Operating Mines in Ghana" presented in this part by *Atta* demonstrates that orogenic gold mineralization in Ghana is predominantly influenced by geological structures. The author drew conclusions solely from data and information gathered from gold deposits currently being mined by major mining companies in Ghana. Additionally, the chapter investigates the key structures controlling gold mineralization at the mines and their implications for improving exploration success.

The third part **"Petrogenesis of Auriferous Host Rocks in Major Gold Belts in Ghana"** focuses on the petrogenesis of lithologies within major gold belts in Ghana. Understanding the origin and evolution of these lithologies is crucial for comprehending the geological context of gold mineralization. Two chapters are presented in this part. The first chapter "Petrographic and Geochemical Characteristics of Ayanfuri Gold-Bearing Granitoids (Abenabena Gap and Fetish Pits) in the Edikan Gold Mine, Kumasi Basin, Ghana: Implications for Petrogenesis" by *Fiadonu et al.* gives petrographic and geochemical insights, critical to understanding gold-bearing granitoid formation processes in the Ayanfuri area within the Edikan Gold Mine of the Kumasi Basin. The second chapter "Petrographical Studies of the Subika Deposit, Ahafo, Ghana: Insights into the Litho-Characterization, Ore Textures and Role of Host Rocks During Hydrothermal Gold Mineralization" by *Adomako-Ansah et al.* also characterizes host rocks in the Subika deposit of the Ahafo South Camp, within the Sefwi Granite-Greenstone Belt of Ghana. This provides critical information on the role of ore textures on hydrothermal gold mineralization.

In the fourth part **"Geology, Mineralogy, Geochemistry, Fluid Evolution and Genesis of Hydrothermal Gold Deposits in Ghana"**, the book delves into the unique characteristics of specific hydrothermal gold deposits as well as their genesis. This comprehensive analysis sheds light on the complex processes that lead to the formation of these economically significant deposits. Three chapters are presented in this part featuring some gold prospects at Iduapriem Mine, Damang Mine, and Birimian gold deposits in general. The first chapter in the fourth part entitled "Footprint of Hydrothermal Gold Mineralization in the Kawere Sandstone, Tarkwaian Group, Ghana: Findings from Two Drillhole Intersections at the Badukrom Prospect, Iduapriem" by *Adomako-Ansah et al.* outlines the mineralogical and geochemical characteristics of the hydrothermal gold potential at the Badukrom prospect, highlighting its implications for exploration in the area. The second chapter "Trace Element Characteristics in Ore Sulfides and Their Records of Fluid Conditions During Overprinting Hydrothermal Gold Mineralization in the Damang Placer Gold Deposit, SW Ghana" also compiled by *Adomako-Ansah et al.* documented trace element geochemical data of hydrothermal sulfides linked to gold-bearing quartz $\pm$ carbonate veins overprinting a paleoplacer gold deposit in Damang, SW Ghana. By examining the texture and trace elements of these sulfides, the authors gained insights into the ore fluid conditions and processes that led to hydrothermal gold mineralization. The third chapter "Constraints on Fluid Composition and Physico-Chemical Conditions for the Birimian Gold Deposits in Ghana" compiled by *Manu* evaluates the characteristics of ore-forming fluids and proposes a model for the development of gold mineralization in the Birimian of West Africa, particularly in Ghana. The author contrasts this new model with earlier ones that emphasized comparing the early Proterozoic Birimian greenstone belts to the Archaean greenstone belts. All these chapters employ a multi-method approach to unravel the formation processes, sources, and genesis of notable and recently discovered hydrothermal gold deposits in Ghana.

Finally, the fifth and last part **"Application of Machine Learning in Predicting Gold-Bearing Lithologies in Ghana"** introduces readers to the cutting-edge application of machine learning in predicting gold-bearing lithologies. This emerging field holds great promise for the future of gold exploration, offering new tools and techniques for identifying potential gold deposits with greater accuracy and efficiency. Therefore, the chapter in this part "Application of Machine Learning Algorithms to Predict Rock Types Using Geochemical Data: A Case Study from the Obuasi Gold District, Ghana" by *Sunkari et al.* features the use of machine learning algorithms to predict gold-bearing host rocks in the Obuasi Gold District in Ghana. This district is found in the prolific Ashanti Gold Belt, and thus, the authors employ machine learning algorithms such as adaptive boosting (AdaBoost), random forest (RF), gradient boosting (GBoost), and support vector machine (SVM) to produce a rock prediction model, which will augment rock classification in the area.

Overall, this book aims to provide a thorough and insightful investigation of the geology and gold mineralization in Ghana, serving as a valuable resource for geologists, students, researchers, and anyone interested in the fascinating world of gold exploration. We hope that readers will find this book both informative and inspiring.

Lastly, we thank all contributing authors who tirelessly prepared their chapters and revised them multiple times according to our suggestions. This was done just to make the book handy for the readers. Furthermore, the book could not be completed without the support of the Editor-in-Chief of the book series, **Prof. Abdelazim Negm**, and the timely help from Springer Nature.

The editors and authors believe that more books on gold deposits in Ghana should be developed in the future so that we can continue to uncover gold and other geological treasures of Ghana.

Tarkwa, Ghana/Udaipur, India

Dr. Emmanuel Daanoba Sunkari
emmanueldaanoba.sunkari@spsu.ac.in
edsunkari@umat.edu.gh

New Delhi, India/Johannesburg, South Africa

Dr. Shakir Ali

April 2025

Contents

Introduction to the Geology and Gold Mineralization in Ghana

Geology and Gold Mineralization Styles in Ghana

Emmanuel Daanoba Sunkari, Abdurrahman Lermi, Kofi Adomako-Ansah, Etornam Bani Fiadonu, Samuel Edem Kodzo Tetteh, Obed Oppong, Salaam Jansbaka Adams, Yılmaz Demir, and Shakir Ali

1 Introduction

Approximately 66% of Ghana's geology consists of Birimian rocks that were formed during the Paleoproterozoic period and have five NE-SW trending metavolcanic belts with equal spacing between them, and one N-S trending belt [1]. Sediments and metasedimentary rocks dominate in the basins occupying spaces between the belts [2]. The remaining geology of Ghana consists of post-Birimian rocks [3]. The Birimian supra-crustal rocks have undergone extensive deformation during the Eburnean Orogeny with a characteristic pronounced folding within the metasedimentary rocks [4]. Although, the lavas within the metavolcanic belts may be rhyolites, dacites, and andesites, they are mostly basalts [5]. According to Kuma et al. [6], rocks within the metasedimentary basins show some configuration of facies, which changes

E. D. Sunkari (✉) · K. Adomako-Ansah · E. B. Fiadonu
Department of Geological Engineering, Faculty of Geosciences and Environmental Studies, University of Mines and Technology, Tarkwa, Ghana
e-mail: edsunkari@umat.edu.gh; emmanueldaanobasunkari@gmail.com

K. Adomako-Ansah
e-mail: kadomakoansah@umat.edu.gh

E. B. Fiadonu
e-mail: ebfiadonu@umat.edu.gh

E. D. Sunkari
Department of Chemical Sciences, Faculty of Science, University of Johannesburg, Johannesburg, South Africa

E. D. Sunkari · A. Lermi
Department of Geological Engineering, Faculty of Engineering, Niğde Ömer Halisdemir University, Niğde, Türkiye
e-mail: alermi@ohu.edu.tr

E. D. Sunkari
Mining Engineering, Faculty of Integrated and Advanced Technology, Sir Padampat Singhania University, Rajasthan State, Udaipur, India

E. D. Sunkari and S. Ali (eds.), *Geology and Gold Mineralization in Ghana*, Earth and Environmental Sciences Library, https://doi.org/10.1007/978-3-031-97163-1_1

from the nuclei of the basins to the fringes. Opoku-Ntim et al. [1] noted that distinct chemical facies characterize the precinct of transition between the metasedimentary basins and the metavolcanic belts. This zone has been identified as an important spot for most gold mineralization in Ghana [6].

The belts and basins have intrusions of granitoids with different ages, chemical composition, and mineralogical composition [7]. The major types of granitoids found include the mica-rich Cape Coast (basin) granitoids, the hornblende-rich Dixcove (belt) granitoids, and the late K-dominated (post-Tarkwaian) granites [8]. Lying unconformably over the belts is the Tarkwaian Group [9]. According to Koffi et al. [10], the rocks of the Tarkwaian Group, which were derived from the Birimian Supergroup are mostly composed of conglomerates, phyllites, sandstones and slates with a characteristic upward fining. The sediments enclosed in the Tarkwaian are arenaceous and argillaceous with clastic origins. Tay et al. [11] stated that there are about two identifiable zones with beds of conglomerates and pebbles within the Banket Formation and the Kawere Conglomerates (the lower members of the Tarkwaian Group). Epidioritic sills, dykes or laccoliths and folds that trend NE are found within the Tarkwaian Group [12].

In southeast Ghana, four alternating felsic and mafic belts form a system known as the Dahomeyan Supergroup with SSW–NNE trend [13]. Most parts of the

S. E. K. Tetteh
Mineral Exploration Research Centre, Harquail School of Earth Sciences, Laurentian University, Sudbury, ON, Canada
e-mail: stetteh@laurentian.ca

O. Oppong
Ore Deposit and Exploration Center, School of Resources and Environmental Engineering, Hefei University of Technology, Hefei, Anhui Province, China

Anhui Province Engineering Research Center for Mineral Resources and Mine Environments, Hefei, Anhui Province, China

O. Oppong
e-mail: obedoppong918@gmail.com

S. J. Adams
School of Natural and Environmental Sciences, University of Environment and Sustainable Development, Eastern Region, Somanya, Ghana
e-mail: sajansbaka@uesd.edu.gh

School of Nuclear and Allied Sciences, College of Basic and Applied Sciences, University of Ghana, Accra, Ghana

Y. Demir
Department of Geological Engineering, Faculty of Engineering and Architecture, Recep Tayyip Erdoğan University, Rize, Türkiye
e-mail: yilmaz.demir@erdogan.edu.tr

S. Ali
Department of Geology, University of Delhi, Delhi, India
e-mail: shakiriitb@gmail.com

Dahomeyan terrane consist of an undulating and unvaried low-lying plain with inselbergs and ridges. According to Nuamah Daniel et al. [14], the topography of places underlain by felsic gneisses is undulating, whereas the topography of places underlain by mafic gneisses is flat, in general. Also, the mobile belt within which the Dahomeyan Supergroup was emplaced contains the Togo Structural Unit and the Buem Structural Unit. While the Togo Structural Unit consists of metamorphosed sediments, the Buem Structural Unit consists of both metasedimentary and metavolcanic rocks. Jones [15] noted that both form a range of highlands in Ghana. One-third of Ghana's landmass is largely underlain by the Voltaian Supergroup, which hosts undeformed clastic sediments formed during the Neoproterozoic [16]. Along the coast, sedimentary rocks ranging in age from the Early Cretaceous to the Tertiary period are found. These include the Amisian and Appollonian Groups, Sekondian Group, and the Accraian Group.

Ghana is located in the Leo-Man Shield of the West African Craton (WAC; Fig. 1) and is enriched in several mineral deposits, especially gold, hence was once called "Gold Coast" prior to independence in 1957. Apart from gold, some of the major ore deposits currently being mined are diamonds, bauxite and manganese. Today, Ghana is the largest gold producer in Africa and sixth largest globally. Gold production in the country constitutes about one-third of total revenues. For example, in 2021, total gold production was 138.7 tonnes. The deposits have been exploited over centuries. The bulk of the gold deposits in Ghana is hosted in the Paleoproterozoic rocks belonging to the Birimian Supergroup and the overlying Tarkwaian Group. The belts that host the gold deposits are generally known as the Birimian gold belts of Ghana and lie from southeast to northwest as Kibi-Winneba Belt, Ashanti Belt, Asankrangwa Belt, Sefwi Belt, Bui Belt, Bole-Nangodi Belt, and Wa-Lawra Belt (Fig. 2). Two main types of mineralization occur in these belts: mesothermal quartz-vein type/intrusive gold mineralization and paleoplacer gold mineralization [18, 19]. Paleoplacer gold mineralization is only restricted to the Ashanti and Bui belts.

Since minerals are depletable resources, conscious efforts have been made to find new deposits as well as identify other minerals that have the potential to support national development and for export. However, except for few isolated and probably, old materials, modern-day literature, which puts the geology and gold mineralization of Ghana in a summary context is lacking. Also, studies related to the genesis of the major types of gold mineralization in Ghana is inadequate. Hence, this chapter reviews and summarizes the geology and major types of gold mineralization in Ghana. Specifically, the study presents a brief description of the occurrence and general geological features of the main gold deposit types. The study provides a conceptual framework for identifying gaps for future research in Ghana.

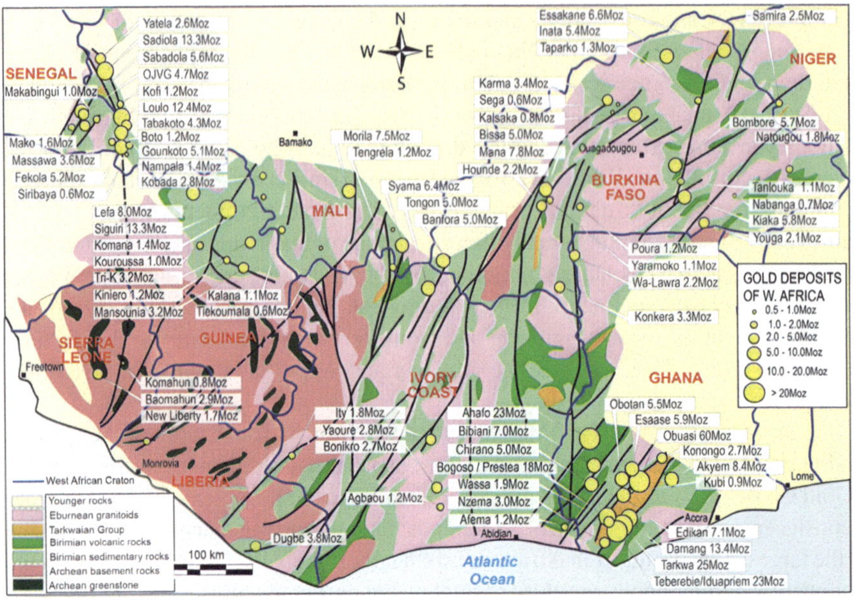

Fig. 1 Geological map of the West African Craton showing the major gold deposits (after Le Mignot et al. [17])

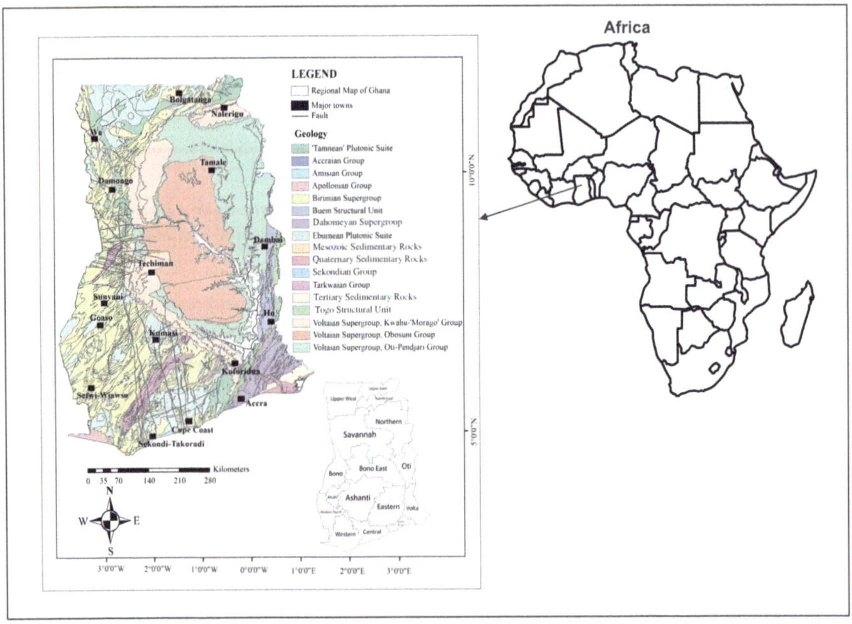

Fig. 2 Geological map of Ghana showing the major geological provinces and faults

2 Geology of Ghana

The West African Geological Setting

The dominant geological domains in West Africa include the West African Craton (WAC), the Pan-African Mobile Belts, and the Neoproterozoic-Palaeozoic sedimentary basins [20]. The WAC is made up of an Archaean Basement complex and the Birimian Supergroup, which are exposed in two locations, namely, the Reguibat rise in Mauritania and the Leo-Man rise stretching from Ghana to Senegal [21] (Fig. 1; [22]). These two prominent geological features are remnants of ancient cratonic nuclei and are separated by the Proterozoic-Palaeozoic Taoudeni Basin. According to Arhin [23], the Leo-Man rise dominates the southern part of the WAC. This rise is composed of Liberian-aged rocks and Birimian supracrustals.

Ghana is found in the southeastern portion of the WAC and the southwestern portion of the Nigeria-Benin Shield. Graham et al. [24] stated that "the geology of Ghana can be divided into three main geologic provinces:

- an early Proterozoic Birimian Supergroup and Tarkwaian Group (Fig. 2) of the main West African shield occupying the west and northern parts of the country;
- a Pan-African province covering the Dahomeyan Supergroup, Togo and Buem Structural Units (Fig. 2) in the southeast and eastern parts of the country; and
- an Infracambrian/Palaeozoic sedimentary basin situated in the central and eastern parts of the country (Fig. 2)".

Birimian Supergroup

In Ghana, the Birimian Supergroup is made up of the metasedimentary Birimian basins and the metavolcanic Birimian belts (Fig. 3). It is commonly known that the metasedimentary basins are of a marine sedimentary origin while the metavolcanic belts are composed of metalavas and pyroclastics. According to Eisenlohr and Hirdes [25], a suite of fine-grained rocks makes up the metasedimentary Birimian basins together with a significant volcanoclastic constituent. Phyllites, tuffaceous sediments and meta-tuffs, shale, siltstones, and greywackes are characteristic rock types within these basins. On the other hand, basalts and volcanic interflows are the main rock types found within the metavolcanic Birimian belts [24]. Although the metavolcanic Birimian belts are considered younger than the metasedimentary Birimian basins in Ghana, current investigations point to a contemporaneous formation as evidenced by lateral facies equivalents [26].

The metasedimentary Birimian basins are predominantly composed of rhythmic laying of meta-greywackes, meta-siltstones, tuffaceous phyllites and meta-sandstones. They are divided into six stratigraphic units from bottom to top as: Lower Arenaceous sub-basin, Lower Argillaceous sub-basin, Middle Arenaceous sub-basin, Middle Argillaceous sub-basin, Upper Argillaceous sub-basin, and Upper

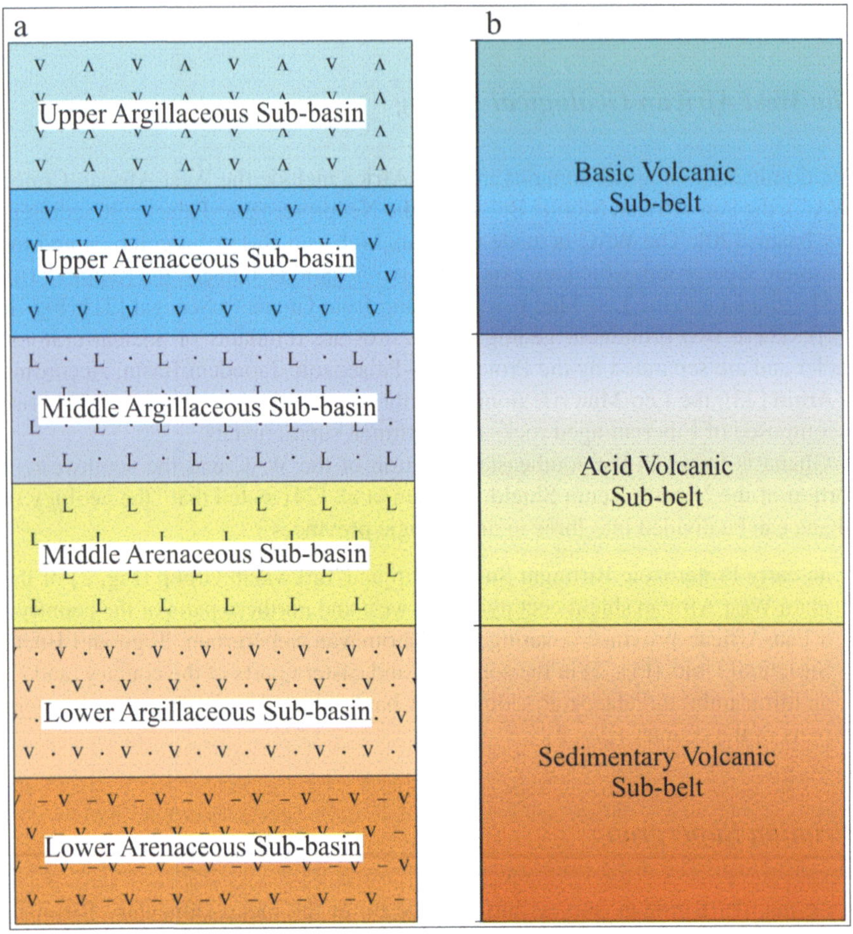

Fig. 3 Stratigraphy of the Birimian Supergroup **a** Metasedimentary basins and **b** Metavolcanic belts

Arenaceous sub-basin (Fig. 3). However, the metavolcanic Birimian belts are stratigraphically divided into three sub-belts from bottom to top as: sedimentary-volcanic sub-belt, acid volcanic sub-belt, and basic volcanic sub-belt (Fig. 3).

Smith et al. [27] reported the presence of six greenstone (metavolcanic) belts in Ghana including Ashanti, Sefwi, Kibi-Winneba, Bole-Navrongo, Wa-Lawra, and Bui belts. An additional belt identified is the Manso Nkwanta/Asankrangwa Belt. In the exception of the Wa-Lawra belt, which trends N–S, the greenstone belts in Ghana have a NE–SW trend. Kesse [28] noted that the "keel-shaped" morphology of the belts might be due to plutonic and diapiric intrusions within them. The belts are wider in southern Ghana (between 20 and 70 km) than in northern Ghana (from 10 to 20 km) and are separated by metasedimentary rocks [27]. Tholeiitic lavas, basalts, interbedded dacites, rhyodacites, and pyroclastics, together with calcareous chloritic schists and amphibolites are the major rocks found within the belts [29]. Metamorphism, as recorded within these rocks, is in the greenschist facies, although rocks of amphibolite facies may also be found, especially along the margins of granitic intrusions [22]. While researchers such as Dampare et al. [29] stated that the belts were formed in an oceanic island environment, others such as Feybesse et al. [30] argue that a combination of plutonic and depositional events on a continental margin led to the formation of the belts. Age constraints derived from zircons in the belts vary between 2162 ± 6 and 2266 ± 2 Ma [18].

The metasedimentary Birimian basins in Ghana include the Maluwe, Cape Coast, Kumasi, and Sunyani Basins. These basins are about 60–70 km wide and host metamorphosed and tightly folded sedimentary rocks [27]. The basins are, however, narrower in the north of the country. Graham et al. [24] revealed that rocks of the metasedimentary Birimian basins have been subdivided into volcaniclastic rocks, turbidite-related wackes, argillitic rocks and chemical sediments with gradational boundaries. "The volcaniclastic sediments comprise chiefly of sand-to silt-sized, partly reworked pyroclastics that is shown by the presence of quartz, idiomorphic plagioclase crystals, chloritized glass fragments, and the absence of heavy minerals" [31]. Griffis et al. [18] stated that rocks of the metasedimentary Birimian basins were obtained from the greenstone belts and deposited in basins within which they occupy presently. Age constraints from detrital zircon grains from the metasedimentary rocks yield ages of between 2180 and 2130 Ma [18]. Petersson et al. [32] also reported similar ages (2220–2130 Ma) using detrital zircon grains. These studies both used Laser Ablation Inductively Coupled Plasma Mass Spectrometry (LA-ICP-MS) and the U–Pb geochronology technique to determine these ages.

Post-Birimian Granitoids

About 20% of the lithologies in Ghana is composed of granitoids (mostly granites, granodiorites, and granite gneisses), which are located in the southern, western, and northern fringes. Karikari et al. [33] stated that these granitoids are considered as "post-Birimian and pre-Tarkwaian" and have granites, granitic gneisses, pegmatites, aplites, syenites, granodiorites, and diorites as the major rock types. These granitoids are rich in hornblende, either with or without biotite or muscovite and have been categorized into three types, namely, the Basin-type (Cape Coast) Granitoids, Belt-type (Dixcove) Granitoids, and Bongo (late K-rich) Granitoids [34].

Ahenkorah et al. [35] noted that the Basin-type Granitoids are older than the Belt-type Granitoids. These granitoids are biotite- and muscovite-rich granodiorites, have gneissic, and often migmatitic character. The Basin-type Granitoids are huge granitic batholiths, which are foliated and found to be in intrusions within the metasedimentary Birimian basins [35]. On the other hand, the Belt-type Granitoids are unfoliated intrusions within the metavolcanic Birimian belts. The Belt-type Granitoids are hornblende-rich granodiorites and tonalites and are usually not found as intrusions in the older Basin-type Granitoids [32]. Nevertheless, in several parts of the country, there are reports of the presence of rocks similar to the Belt-type Granitoids belt in biotite-rich gneisses of the Basin-type Granitoids [28].

It is also worth noting that Belt-type Granitoids have a lower Na/K ratio, more severely altered feldspars, and less wide metamorphic aureoles when compared with the Basin-type Granitoids [8]. According to Ganyaglo et al. [36], "the Bongo granite (reddish microline granites) found in isolated patches in the central part of the study area are generally foliated, undifferentiated and outcrops are exposed as boulders of varying sizes". Their relationship with the other granitoid suites in Ghana remains unclear.

Tarkwaian Group

The Tarkwaian Group is located within the greenstone belts in Ghana (Fig. 2) and is predominantly composed of conglomerates, sandstones, phyllites, and argillaceous rocks [37]. The Ashanti and Bui belts in Ghana host the best-developed Tarkwaian Group in Ghana. According to Abdul-Rashid [38], the system has a vertical extent of about 2.5 km in the Ashanti belt and about 9 km thick in the Bui belt. Nunoo et al. [39] pointed to the presence of the Tarkwaian Group in other belts such as the Kibi-Winneba and Sefwi belts, though not well developed as those of the Ashanti and Bui belts. In the Ashanti belt, the Tarkwaian Group extends from the Kwahu area north of Konongo to Tarkwa District, with nearly horizontal Neoproterozoic clastic sediments concealing it. Two sets of sedimentary cycles make up the Tarkwaian Group within the Bui belt, occupying synclines that have an overturning history [18]. These cycles are the Bui cycle and the Sabiyi cycle. Smith et al. [27] observed that while the Mundale, Nuapo, and Kane are the formations making up the about 2 km thick Bui cycle, the approximately 7 km thick Sabiyi cycle comprises the Tombe and Nyanchulo Formations.

The general stratigraphy of the Tarkwaian Group from the lowermost to the topmost units includes the Kawere Conglomerates, Banket Formation, Tarkwa Phyllites, and Huni Sandstones (and Dompim phyllites) (Fig. 4). The Kawere Conglomerates, with a total vertical extent 200–700 m, are weakly worked (polymictic) conglomerates with alternating phyllites, sandstones, and quartzites [40]. The Banket Formation is economically the most significant member of the Tarkwaian Group because of its gold mining history. The thickness of the formation is between 120 and 160 m in the Tarkwa area. According to Nyame [41], the formation includes interbedded sandstones and quartzites, and gold-rich quartz conglomerates. More so, Griffis et al. [18] stated that extensive paleoplacer gold deposits like that mined in Tarkwa are found in the Tarkwaian conglomerates of the Bui Belt. Above the Banket Formation are the Tarkwa Phyllites, which are about 120–400 m thick grading upwards into the Huni Sandstones. The Huni Sandstones are about 1.4 km thick and are composed of quartzites and phyllites with an interbedding relationship [27].

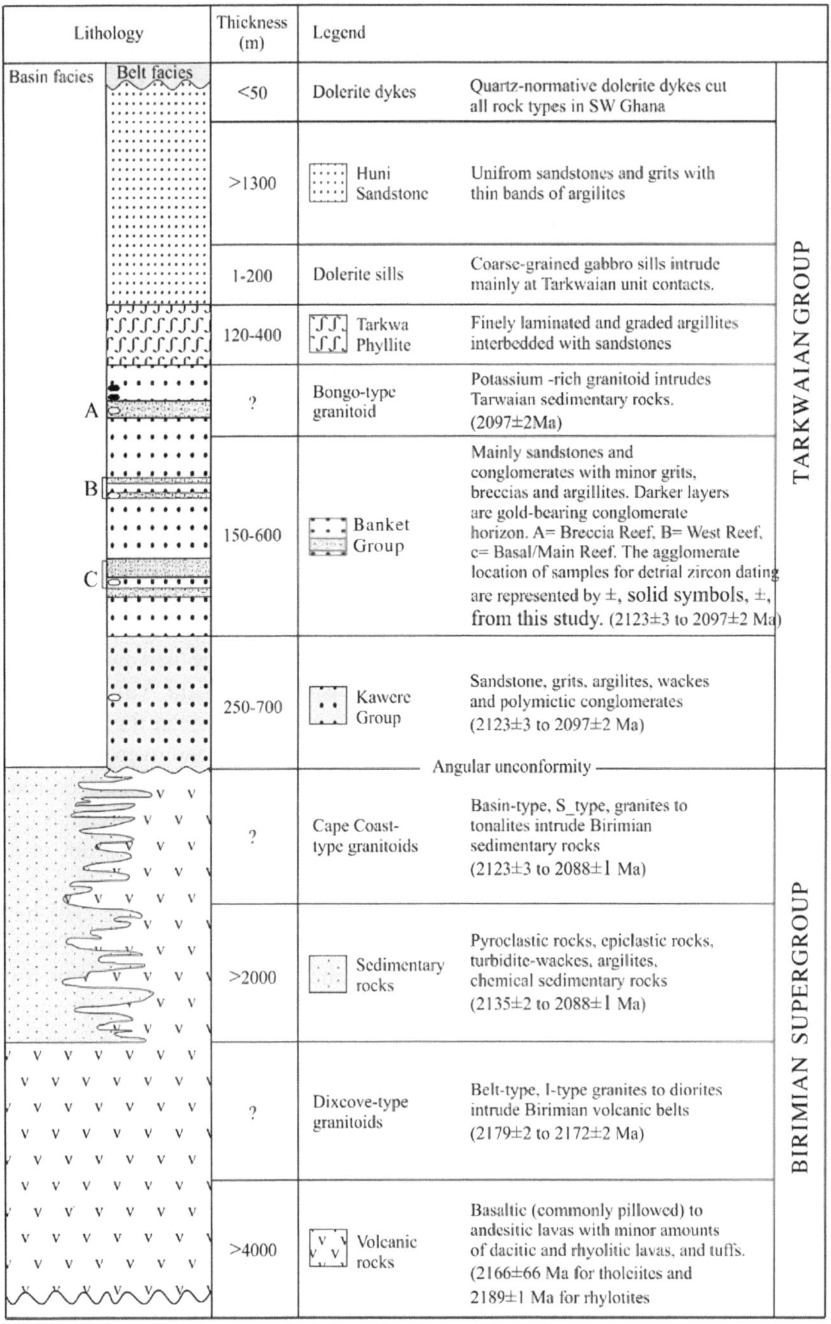

Fig. 4 Stratigraphy of the Tarkwaian Group and its relationship with the Birimian Supergroup (after Pigois et al. [19])

The Tarkwaian Group has been inferred to have been deposited in alluvial fans and reworked by braided channels [18]. This reworking has locked gold in fine particles within the conglomerates of the Banket Formation. Griffis et al. [18] pegged the depositional age of the Tarkwaian Group at 2133–2132 Ma using detrital zircons from the Banket Formation and the Kawere Conglomerates.

Voltaian Supergroup

The Voltaian Supergroup consists of a 5–7 km thick sequence of flat-lying Neoproterozoic-early Palaeozoic sandstones, siltstones, mudstones, and limestone (Fig. 2) [16, 42]. The Voltaian Supergroup could also be found to occupy small areas in Togo, Benin, Niger, and Burkina Faso (Fig. 5). However, in Ghana, the Voltaian Supergroup occupies about an area of 115,000 km^2, dominating the eastern half of the country [16]. According to Boher et al. [44], the Voltaian Supergroup lies unconformably above the basement rocks of the Leo-Man Block (Birimian metasedimentary basins and metavolcanic belts, post-Birimian Granitoids, and Tarkwaian Group) along its western boundaries [42]. Generally, the rocks of the Voltaian Supergroup have not undergone significant deformation or reworking. This is evidenced by the low-lying, E-SE pointing dips they have, which range narrowly from 1–2° [42]. However, there has been considerable disturbance of the Voltaian Supergroup along and near its eastern boundary. Here, the Pan-African thermotectonic event has largely deformed the rocks resulting in steeper dips and over-thrusting of the Pan-African Dahomeyide orogenic or mobile belt onto the Voltaian Supergroup. In Ghana, the Voltaian Supergroup is thicker close to the Oti River (in Yendi/Sabari areas) and becomes shallower towards the west, where it finally thins out in the Larabanga area [45].

Numerous researchers including Anani et al. [46] and Abu et al. [16] have concluded that the Voltaian Supergroup consists of three primary groups of rocks, namely the Kwahu-'Morago'/Bombouaka, Oti-Pendjari, and Obosum Groups from bottom to top (Fig. 5). Rocks of the Kwahu-'Morago'/Bombouaka Group are believed to have been deposited in fluviatile, flood plain, and shallow marine environments [16]. Some aeolian materials could also be found in the upper members of this group. The major lithologies in this group include sandstones, shales, and interbedded sandy shales. The Oti-Pendjari Group, according to Affaton et al. [47] was deposited in an ancient valley between Gambaga and Kwahu Massifs. The Oti-Pendjari Group is made up of glacial and marine deposits together with limestones. The northern areas of this group comprise of some minor conglomerates and tuffs [48]. Anani et al. [49] stated that the Obosum Group was deposited during the final uplift event of the Pan-African Dahomeyide orogenic or mobile belt. According to Carney et al. [48], this group fills a syncline within the Voltaian Supergroup and is made up of "terrestrial molasse sediments". Lithologies found in this group include conglomerates, arenites, and certain argillaceous sediments [16].

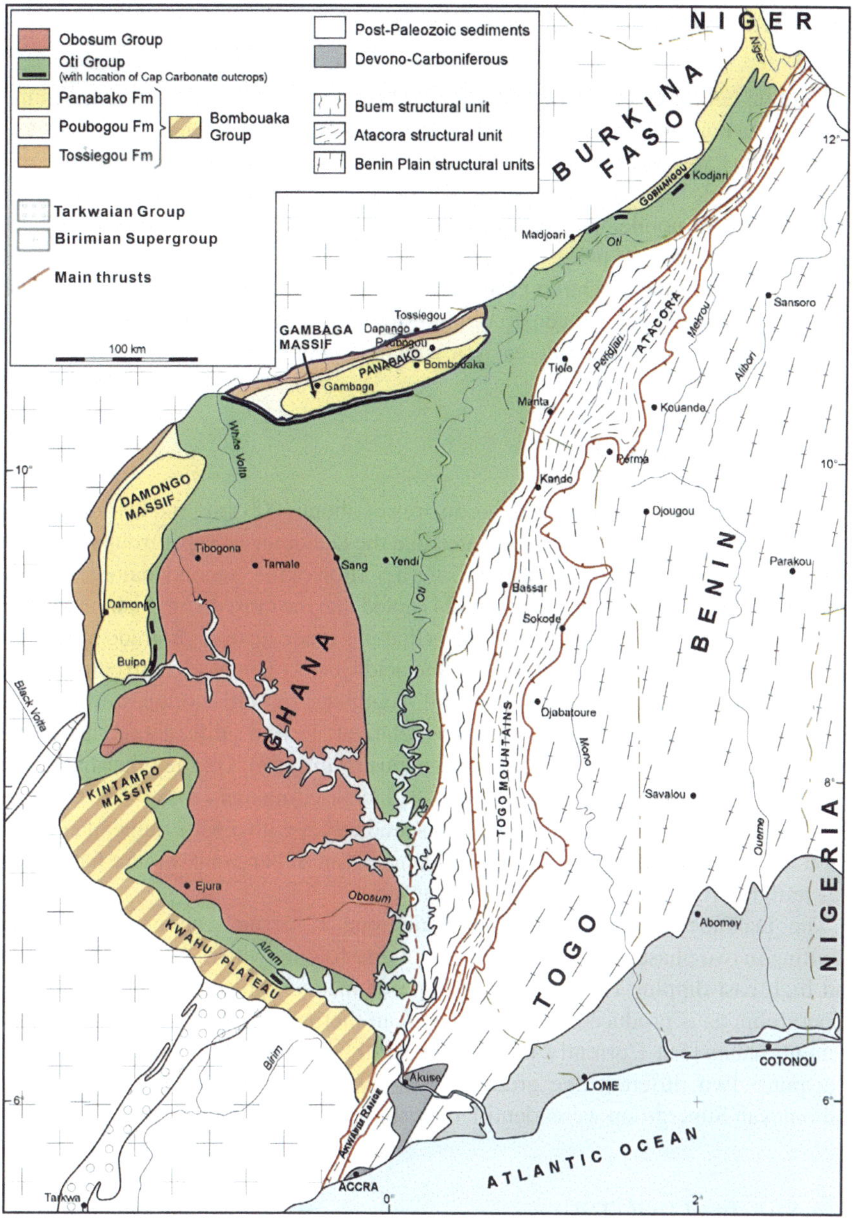

Fig. 5 Geological map of the Voltaian Supergroup (after Deynoux et al. [43])

The Metamorphic and Mobile Belt

The mobile belt in Ghana is part of a larger Trans-Saharan belt and exposed in the southeastern part of Ghana, along with Togo and Benin. This southeastern portion is known as the Pan-African Dahomeyide orogenic or mobile belt and is part of the Pan-African orogenic belt. Attoh and Nude [50] mentioned that the Pan-African Dahomeyide orogenic or mobile belt was created during the breaking-up of a supercontinent named Rodinia. This event led to the congregation of cratonic rocks on Gondwana's northwest. According to Darko et al. [51], the Pan-African Dahomeyide orogenic or mobile belt in Ghana is made up of the Dahomeyan Supergroup, Togo Structural Unit, and Buem Structural Unit, each of which is separated by thrust faults.

Dahomeyan Supergroup

The Dahomeyan Supergroup occupies an area of about 7000 km^2 and is found in the SE corner of Ghana. Yidana [52] indicated that the Dahomeyan Supergroup forms the eastern most rock group in Ghana and forms a portion of the "second major tectonostratigraphic terrain in Ghana". Graham [24] noted that the most distinguishing feature of the Dahomeyan Supergroup in Ghana is that it is made up of high-grade metamorphic rocks and consists of alternate basic and acidic belts of gneiss. These belts extend from the Accra Plains to Togo with a south-southwest to north-northeast trend. The basic belts have a tholeiitic nature and are relatively rich in salite, garnet, andesine, oligoclase, and hornblende gneisses of igneous origin [53]. On the other hand, the acid belt gneisses are meta-volcaniclastics and metasedimentary rocks [53]. Graham [24] stated that granites, nepheline syenites, and dykes with a wide range of composition are found as intrusions within the Dahomeyan Supergroup. Figure 6 presents the stratigraphy of the Dahomeyan Supergroup.

The Dahomeyan Supergroup has undergone a thermo-tectonic deformation resulting in two phases of metamorphism, intense folding with SSW-NNE fold axes, and high east-dipping rocks [55]. The type of metamorphism in the Dahomeyan Supergroup, as is produced by the Pan-African thermo-tectonic event, is cataclasis metamorphism [56]. Currently, the exact age of the Dahomeyan Supergroup remains a dispute. Two different age groups, 1700–2050 Ma and 450–580 Ma, for the Dahomeyan Supergroup were identified by studies in Benin and Togo [57].

Togo Structural Unit (TSU)

The Togo Structural Unit (TSU) is delimited to the east by younger Accraian and older Dahomeyan rocks and to the west by the Buem Structural Unit, Voltaian Supergroup, and Cape Coast Granitoids (Fig. 2; [49]). While the rocks in the unit generally have a 135° dip direction, they strike at ~ 45° and contain "hornstones, jaspers and hematite quartz-schists as post-depositional rocks" [49]. According to Yidana [52],

Stratigraphic Unit		Lithology
Acid Dahomeyan		Quartz veins, granitic gneiss and migmatite, granite, gneiss
Alkalic Gneiss		Kpong conglomerate, nepheline gneiss
Basic Dahomeyan	Basic intrusives	Dolerite, norite, chromitiferous pyroxenite
	Metabasics	Garnet-hornblende-gneiss, garnet-hornblende (pyroxene) -gneiss, hornblende and biotite schist

Fig. 6 Stratigraphy of the Dahomeyan Supergroup in Ghana (after Mani [54])

the TSU is made up of chlorite and quartz sericite schists, phyllites, and quartzites. Nyavor [58] found that the quartzites and phyllites have undergone considerable deformation through recumbent folding and near-horizontal foliation, with the degree of deformation and metamorphism increasing towards the southeast. Nyavor [58] also revealed that while recumbent folds are the first events of folding in Togo, open and isoclinal folds developed later.

According to Bonsu [56], the TSU is composed of quartzites interbedding with phyllites, with a history of cataclastic metamorphism. The quartzites are generally grey, fine to medium-grained, and have thick foliations and extensive micro-fractures. The dominant minerals found in the phyllites underlying the quartzites are micas and chlorites. Bonsu [56] described these phyllites as fine-grained and thinly foliated with joints and micro-fractures. Isoclinal to chevron folds are common within the quartzites and the axial planes inclined at 30–60° in the southeastern direction. On the other hand, minor recumbent folds have an average dip of $< 30°$.

Buem Structural Unit (BSU)

The Buem Structural Unit (BSU) was formed around 512 Ma, which conforms to the metamorphism and metasomatism recorded by the rocks [15]. However, according to Affaton et al. [47], the BSU was formed between 620 and 640 Ma and are lateral facies of the Voltaian Supergroup in terms of both tectonism and metamorphism. Affaton et al. [59] identified an earlier weak metamorphic signature older than the Pan-African collision and is marked by prehnite–pumpellyite facies metamorphism under temperatures of 200–300 °C. According to Affaton et al. [59], "the alteration products of this metasomatic event include: (1) alteration of the volcanics to sericite, chlorite and carbonates; (2) formation of epidote/quartz veins in the volcanics; (3) intrusion of quartz veins into the sandstones, and (4) development of jasper from a possible limestone precursor".

Stratigraphic succession of the BSU from the lowest bed to the topmost member is given as:

- B1 Akroso Conglomerate (mostly green arkosic sandstone and quartzite);
- B2 Lower Shale 5500 m (brown or purple shale and sandstone);
- B3 Volcanic Group (basal pillow lavas followed by basalts, few rhyolite flows, agglomerate and tuff, and some intercalated sandstone);
- B4 Limestone/Jasper/Tillite;
- B5 Lower Massive Sandstone (massive, structureless feldspathic sandstone);
- B6 Upper Shales (red brown shale with sandstone horizons); and
- B7 Upper Massive Sandstone/Ironstone/Jasper.

The BSU occupies the northern and northwestern portions of the TSU [60]. The main lithologies found in the BSU include arkose, greywacke, sandstones, conglomerates, and shales [24]. These shales are characteristically sandy, calcareous, ferruginous, and argillaceous. There are also tuffs, beds of agglomerates, jaspars, and lavas that have solidified as trachyte, basalt, or andesite [24]. High degrees of folding with N–S trends are seen within the BSU. The beds also have easterly dip directions and dips between 10 and 90° (60–65° are most common) [61]. Metamorphism is not common in the BSU; however, shear zones and schistosity are frequent in the locality of the Buem-Togo contact [62]. This same contact serves as sites of basic intrusions within the BSU, with the intrusive bodies being either gabbros or dolerites.

The Coastal Sedimentary Basins

Sekondian Group

The Sekondian Group was formed during the late Ordovician to early Cretaceous along the coasts of Western and Central Regions in Ghana, as indicated by micropalaeontological data [63] and unconformably overlies the Birimian Supergroup [64]. Averagely, this shale and sandstone-dominated group, which has been well faulted has a vertical extent of 1200 m [63]. Asiedu et al. [65] suggested that the Sekondian Group was sourced from the post-Birimian Granitoids and were deposited in a mixture of non-marine to coastal marine setting as revealed by sedimentary textures, structures, and palaeontology. Bär and Riegel [66] noted that the Sekondian Group correlates with the Maranhao Basin formations found in present day Brazil.

Six formations make up the stratigraphy of this group namely: Adjua Shales, Elmina Sandstone, Takoradi Sandstone, Takoradi Shale, Efia-Nkwanta Beds, and Sekondi Sandstones [67]. The Adjua Shales lie directly on the Birimian Supergroup and are part of the lowest Sekondian Group formation [68]. They are dominated by shales, which may be black or grey and are thinly bedded. However, there are grit beds and laminations of arenaceous material interbedded with these shales [69]. Well-rounded to angular boulders and other clastic fragments of different rock types exist across the Adjua Shales. One characteristic of the Adjua Shales is their varved and ripple marked nature [70]. The Elmina Sandstone is a chocolate-chocolate purple formation inferred to be directly above the Adjua Shales [65]. The coloration is due to the presence of feldspars and limonitic cement material. This formation is a uniformly massive, hard, poorly bedded but well jointed, strongly cross-laminated, and medium-grained sandstone member of the Sekondian Group [65]. The Elmina Sandstone grades from coarse grains at its base to fine shale-bearing material at its apex.

The Takoradi Sandstone and Takoradi Shales are collectively known as the Takoradi Beds. The Takoradi Sandstone is usually cream to dark rusty brown in hand specimen with dominant angular and well-rounded quartz grains. This formation is massive, bedded, cross-laminated, and fossiliferous. Some fossils found in the Takoradi Sandstone, lower part of the Takoradi Beds, include remains of fish, brachiopods, and lamellibranches [71]. The Takoradi Shales that also make up the upper part of the Takoradi Beds have gastropods in addition to these fossils [71]. Unlike the Takoradi Sandstones, the Takoradi Shales contain some traces of hydrocarbons and bitumen [72]. These shales are compact and hard with dark grey or black coloration. They have been found to contain unusual sulfate minerals such as jarosite and halotrichite. They also contain discoidal nodules of compact finely granular, grey siderite or clay ironstone. Calcite and gypsum occur as veinlets traversing the shale and siderite nodules and along the bedding planes.

The Effia Nkwanta Beds consist of a range of rock types with bright colours and have been divided into three beds: lower, middle and upper Effia Nkwanta Beds [65]. The Lower Effia Nkwanta bed is made of sandstones, which are cross-laminated and fine grained with ripple marks and floury textures [65]. The Middle Effia Nkwanta Bed is made up of quartzose sandstone interbedded with mudstone and shale. The sandstone is friable, well bedded and somewhat massive [65]. This bed is generally brighter, pink or orange. The Upper Effia-Nkwanta beds are made up of shales and siltstones with minor mudstone and sandstone [65]. The shales and siltstones are usually purple, pink, grey or green and are well bedded. One key characteristic of this bed is the presence of chert in thin bands or as rounded nodules along bedding planes. The entire Effia Nkwanta Beds rest directly below the topmost formation of the Sekondian Group and the Sekondi Sandstones. The Sekondi Sandstones are divided into two distinct beds. The Lower Sekondi Sandstone is a thick and massive sandstone unit with fragments of irregular and angular chert [65]. It is cross-laminated and orange pink in colour. The Upper Sekondi Sandstone is a soft and argillaceous sandstone, which is somewhat resistant to weathering. It is feldspathic, pebbly, and chocolate pink in colour. The well-rounded pebbles are largely quartz, and minor chert, phyllite, or greenstone.

Amisian Group

Dapaah-Siakwan and Gyau-Boakye [73] described the Amisian Group as outcropping along the Amisa River estuary between Winneba and Saltpond. Major lithologies in the Amisian Group include grits, conglomerates, sandstones, clay, and arkose, which are interbedded with dips ranging from 5 to 20° to the north-northwest. Birimian phyllites, granites, gneisses, Togo quartzites and pebbles are the source of the materials from which the Amisian Group was formed. One characteristic of the Amisian Group is its poor sorting and intense cross-stratification. The cross-stratification shows that the source material may be from the sea. Machens [74] noted that the Amisian Group was formed during the Late Jurassic through to the Early Cretaceous.

Apollonian Group

The Apollonian Group is found in the southwestern corner of Ghana and in south-eastern Ivory Coast. In Ghana, the Apollonian Group is synonymous with the 1165 km^2 Tano Basin located in the area drained by the Tano and Ankobra rivers. According to Annan [75], the Apollonian Group was formed during the Cretaceous-Eocene Epoch in a crescent-like basin along the Atlantic Coast. Kesse [76] stated that the rocks of the group are mostly limestones with alternating sands and clays. Where these sands and clays are compacted, sandstones and shales are respectively found. Nodules of pyrite or marcasite are common in clays and shales [76]. According to Nazifi [77], thin and fossil-laden limestone is the sole stratigraphic marker in the area. The rocks of the group dip at 1–3° and have an S-SW dip direction [75].

Accraian Group

The Lower-Middle Devonian Accraian Group is directly and unconformably above the Dahomeyan Supergroup, covering an area of about 11.7 km^2 around the capital city of Ghana, Accra [78]. According to Anani et al. [79], the group is made up of quartz grits, sandstone, shale, and mudstones. Fossils found in the group were attributed to Middle Devonian due to their resemblance with North American fauna of that age [80]. The Accraian Group is subdivided into the Korle Lagoon Formation, Korle Bu Formation, and Kaneshie Formation. The Korle Lagoon Formation extends from the headland near Osu Fishery in the east to 300 m west of Christiansborg Castle. Rocks of this formation are mostly composed of sandstones with minor grits, breccias, pebble beds, and shales [81]. Beds of the sandstone formation are conspicuously current bedded, and some bedding surfaces are either massive or thinly bedded. The sandstones dip S–SW at 30° and their contact with the Dahomeyan Supergroup is an unconformity. The Korle Bu Formation contains fossils. The fossils are ichnofossils of trilobites and lamellibranch as represented by cast and impressions. The Kaneshie Formation is made up of interbedded quartzitic sandstones and argillaceous shales. Individual beds are generally not thicker than 30 cm.

Keta Basin

The Keta Basin is located in the southeastern corner of Ghana and is part of the Dahomey coastal basin extending from the Niger delta in Nigeria [82]. According to Kumapley [83], the major lithologies in this basin include siltstones, shales, clays, and limestone with some sands and gravels. Kesse [76] noted that the rocks gently dip towards the southeast at about 2° and were deposited on an unknown basement, though the basement is assumed to be the Dahomeyan Supergroup. The Keta Basin has an overall tectonic block structure bounded by a fault or fault system on its northern flank. The trend is generally NE–SW. However, in the western half of the area, the trend is NNE–SSW. Also, at the southern part, the trend is E–W [76]. "The Keta basin is a fault-controlled Mesozoic/Tertiary sedimentary basin along the coast of the Gulf of Guinea" [84].

The Early to Middle Devonian basal sedimentary sequence of the Keta Basin consists of marine shale, sandstone, and siltstone overlain by dolerites and sills formed during the Jurassic [85]. Near Ghana's border with Togo, a suite of marine limestone, shale, and sandstones formed within the Cretaceous-Eocene Epochs are found. The central part of the basin is dominated by Quaternary unconsolidated coastal sediments, marine sands, and gravels, with an average thickness of approximately 30 m around Keta. These sediments become thicker toward the Volta River estuary [86].

History of Gold Exploration in Ghana

It is widely known in history that the old Ghana Empire, whose name was associated to its king, was sited about 800 km north of the present-day city, Accra, and occupied regions within present-day Mali and Mauritania [87]. For over seven centuries, Arab traders traded in gold before the Europeans first came to Sub-Saharan Africa. The name "Ghana" was subsequently utilized by these Arab traders to refer to the more extensive empire within which they carried out their trade [88]. Until today, the old Ghana Empire, the pre-independence Gold Coast, and present-day Ghana have produced a chunk of the world's gold [89].

Before the Europeans came to mechanize gold mining in Ghana, locals used very simple methods to identify areas or spots rich in gold, mostly by looking for nuggets of the metal after torrential downpours and digging up those spots. This is corroborated by Ofosu-Mensah [90] who indicated that senior citizens in some parts of the Central Region of Ghana testified that traditional mining was lucrative in the wet month of April and reduced during the dry season. The testimonies also revealed that people mostly looked for nuggets and started digging with their families or close friends near rivers, streams, or old river channels. They looked out for certain types of plants, such as ferns or grasses, as a sign of gold rich soils, as well as the instructions of the gods through dreams of the land since they believed that the gold was controlled by these deities. Upon discovery of a relatively rich gold accumulation in the soil, the traditional miner usually mined this gold in three simple ways, mostly with rudimentary tools [91]. The methods range from panning along stream channels, shallow pitting (most lucrative), and deep-shaft mining for gold reefs.

Boys and women usually pan for placer gold after downpours, especially in gravel deposits near estuaries along the central and western coast of Ghana. According to Bosman [92], "they used large troughs and trays, which they filled with sand and gravels and repeatedly washed with fresh water till all the lighter materials were removed", leaving behind concentrates of gold. According to Arhin [93], the bowls used for the panning were named *sika kuduo* among the Asante and Akyem. Shallow pits between 3 and 10 ft (usually small holes or trenches) were dug into the regolith. Kumah and Nyarko [94] indicated that this method known as *mmoaboa* was the most lucrative method of gold extraction among the Akans. Hence, entire families and partnerships were involved. Where rich reef gold sources were identified at deeper levels within the ground, shafts of varying sizes and depths were dug to access them. Though shafts were 10–20 ft deep, there were deeper shafts extending about 100 or more ft below the earth. According to Arhin [93] and Ofosu-Mensah [91], the deep shafts were locally called *amena pea nkron* and were usually chimney-shaped. This reef gold was mostly within hard rocks. Thus, to be able to mine it, the miners heated the rock for at least 3 days with fire and then attacked it with cold water to crack it so that they can easily chip it out of the shaft [91]. Arhin [93] mentioned that gunpowder was also used to carry out some form of blasting, during the early years of the coming of the Europeans. According to Ofosu-Mensah [91], "the entry of Europeans into a previously exclusively African field of activity revolutionized the mining industry in the Gold Coast. The industry became highly mechanized". This mechanization made modern mining labour-intensive, capital intensive, and highly technical unlike traditional mining, which was done with rudimentary skills and tools. Some of the traditional mining tools used in ancient times in Ghana are given in Fig. 7.

Between 1400 and 1500, Europe was confronted by monetary security, underlined by gold scarcity, which drove their search for resources and distant lands. It was within this period that the Portuguese arrived at the shores of the Gold Coast, the first Europeans to do so. As recorded by history, Europe was confronting a shortage of gold, which was subverting monetary security [87]. The Portuguese arrived in 1471, and according to Miller et al. [96], they realized that there were numerous African kingdoms who possessed rich deposits of gold, even in the soil. In other words, they

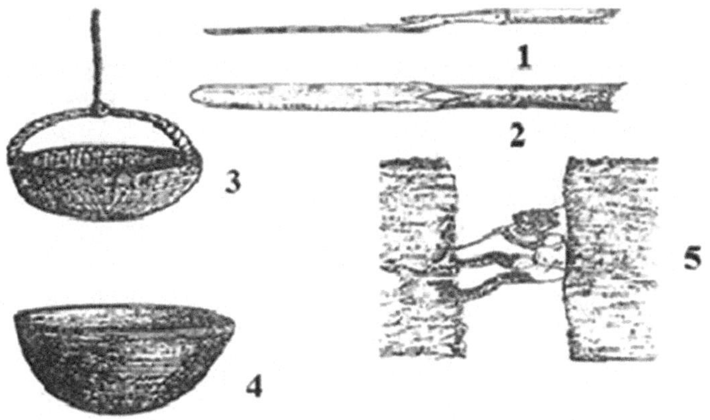

Fig. 7 Traditional mining implements *1. Soso toa, 2. Soso tupre, 3. Small can with rope attached, 4. Koro, and 5. Amena pea nkron* [95]

saw the Gold Coast as a wealthy region they could exploit via trade to stabilize the finances of then Europe [97]. Hence, they constructed the famous Elmina Castle and began trading in "gold, beads, knives, rum, guns", and subsequently slaves [96]. They presented new mining methods to bolster production levels sharply. According to Addy [98], the Gold Coast produced about 35% of the global gold supply because of improved mining technology. Four centuries later, the British issued about 70 mining concessions in the Gold Coast after officially colonizing the Gold Coast in 1874 [99]. However, very little increase in gold production was seen. A major reason for this situation as stated by Hilson [88] is the risky and arduous nature of transporting raw materials and mined ore relative to South Africa's Transvaal gold region towards the end of the nineteenth century. However, an appreciation in production happened when numerous mines were commissioned in the Obuasi area in 1895, including Ashanti Goldfields [87]. Jedwab and Moradi [100] and Kesse [28] stated that railway lines began to be laid between Kumasi and Accra in 1901 to enhance access to thousands of concessions in just 1 year. This corresponded to what is known as the "first gold boom of the Gold Coast" because gold production increased by four times. Ofosu-Mensah [101] stated that the Boer Wars in South Africa within the same period drove investors toward Ghana.

Events of the First World War led to an increase in demand for metals to perpetuate the industrial revolution, gold inclusive. Thus, the second gold boom of the Gold Coast happened just after the war ended, with the colony producing about 882,300 oz in 1941 [102]. However, production levels plummeted during the Second World War and remained below 550,000 oz annually for the rest of the 1950s. "The decline in production was worsened by the uncertainty facing foreign investors as the colony's independence movement gained momentum, allowing mines to greatly deteriorate" [102]. Ghana eventually gained independence in 1957 and many mining firms relocated their operations offshore after the government denied Ghana of firms that had underinvested their mines grants to rehabilitate their mines [87]. Thus, the then government took full control of the mining space in response. Jackson [103] revealed that the only mines that were not nationalized by 1961 were the Obuasi and Konongo mines and that by 1966, the Konongo mine was also in the government's hands. According to Hilson [88], private firms owned 45% of Ashanti Goldfields while the government controlled the rest. The nationalization drive of the government led to a decrease in gold production. Akabzaa and Darimani [102] provided evidence to that effect by stating that by 1976, production was about 33% and 80% of what was realized in 1941 and 1960, respectively. The same study stated the reasons for this decline. The major reason was that the Ghanaian government had "neither the required capital nor the management skills to rejuvenate the country's gold mines".

Elbra [87] further discussed that after the independence of Ghana, political instability was characteristic of the country's political space, with coup d'états being common. The resulting economic declines dealt further blows to the production levels of gold. To escape the economic woes that presented themselves, the military regime in 1981 developed what is known as "Economic Recovery Plan" with the International Monetary Fund (IMF). According to Hilson [104], an increase in private ownership of the country's mines was one key agreement under the IMF program. This privatization and liberalization of mining became formal under the Minerals and Mining law in 1986. This was done to introduce diversity into the country's economy through tax waivers and royalties [87]. Opoku-Dapaah and Boko [105] stated that the privatization and liberalization of the mining sector succeeded in attracting investments into the gold mining industry from multinational mining firms across America, Europe, and Australia.

The modern mining sector in Ghana has about 13 large-scale mines, numerous small-scale operations and an undocumented number of illegal mining operations popularly known as "galamsey" [106]. Bloch and Owusu [107] stated that the illegal small-scale miners might be about 50,000–200,000. Table 1 presents ownership structure and production data of the country's large-scale gold mines as of 2014.

The government of Ghana developed the Mining and Minerals Act 703 [108] in a bid to attract more foreign investment and generate more revenue from the mining sector. Several authors, including Bloch and Owusu [107] and Elbra [87] agreed that the changes introduced by this Act were effective. Evidence to this was that about US$3 billion was invested in the industry in the 3 years that followed [108]. The mining sector contributed about 43% of foreign exports and 5.8% of GDP by 2009 [87] and became the 9th largest producer of gold globally [108]. As expected, these improvements, coupled with a steady rise in global gold prices between 2003 and 2013 led to rapid economic growth [87].

3 Gold Occurrences in Ghana

The Birimian terrane of Ghana has been a potential location for gold mining, mostly on the NE-trending volcanic belts and the only NS-trending volcanic belt (Fig. 2). These belts include from southeast to northwest: Kibi-Winneba Belt, Ashanti Belt,

Table 1 Ghanaian gold mine ownership and production data as of 2014 [108]

Mine	Majority owner	Government stake	Output (Oz)	% of total production
Tarkwa	Gold fields	10%	558,222	17.7
Akyem	Newmont Ghana gold	Nil	471,658	14.9
Ahafo	Newmont Ghana gold	Nil	442,020	14.0
Chirano	Kinross	10%	286,326	90.1
Various	PMMC (Purchases from various small-scale mines)	100%	265,350	8.4
Obuasi	AngloGold Ashanti	Nil	243,223	7.7
Edikan	Perseus Mining	10%	187,363	50.9
Damang	Gold fields (Through Abosso Goldfields Limited)	10%	177,741	5.6
Iduapriem	AngloGold Ashanti	Nil	176,930	5.6
Prestea/Bogoso	Golden Star	10%	147,957	4.7
Wassa	Golden Star	10%	112,835	3.6
Nzema	Endeavour Mining Corporation	10%	88,476	2.8
Total output			**3,158,101**	**100**

Asankrangwa Belt, Sefwi Belt, Bui Belt, Bole-Nangodi Belt, and Wa-Lawra Belt (the only north–south trending belt) (Fig. 2).

Ghana's gold occurrences can be divided into:

(i) Mesothermal quartz vein systems and intrusives;
(ii) Paleoplacers; and
(iii) Alluvial.

Mesothermal Quartz Veins and Intrusives

The quartz vein systems almost invariably appear to be related to NNE to NE trending regional structures (tectonic corridors), which are typically concentrated along the margins of various Birimian greenstone belts and adjacent metasedimentary basins [115] (Fig. 2). The metasedimentary units on the margins of the volcanic belts and in the broad intervening basins are tightly folded (Fig. 8). Interbedded argillites, greywackes, and volcaniclastic units frequently deposited in the transitional zones between the belts and basins are the host rocks for this type of deposit. These transitional zones also commonly feature a variety of fine-grained chemical sediments (siliceous, graphitic, manganese, carbonate, fine-grained sulfides, etc.), which have drawn special attention in recent decades [53, 109–117]. It is also quite clear that in some areas, especially in the Ashanti Belt (Konongo and the coastal sector), mafic volcanics also host important gold occurrences [116]. Elsewhere in the northwestern part of Ghana within the Paleoproterozoic greenstone Birimian Wa-Lawra belt (the only N–S trending belt), is an emerging opportunity for shear-hosted gold deposits in the Julie and Bekpong prospects [53]. These deposits are lithologically hosted in a series of granitoid intrusives of TTG composition and include a network of deformed, boudinaged quartz veins [53]. Towards the northeastern part of the country occurs vein-type mineralization in the Dokrupe area along the Bole-Nangodi belt [18, 118].

All these gold deposits feature extensive, steeply dipping, vein systems, usually with multiple stages of vein development. In some deposits, the veins are relatively narrow (1–2 m) and discrete and there may be several parallel sets within broad structural zones such as in Prestea. In other areas, such as Bogosu and Obuasi, the multiple vein systems form broad stockwork structures [119] and appear to have a variety of combinations. On one of the deeper levels at Bibiani, one vein complex was essentially solid quartz over a width of 55 m. Not all vein systems are created

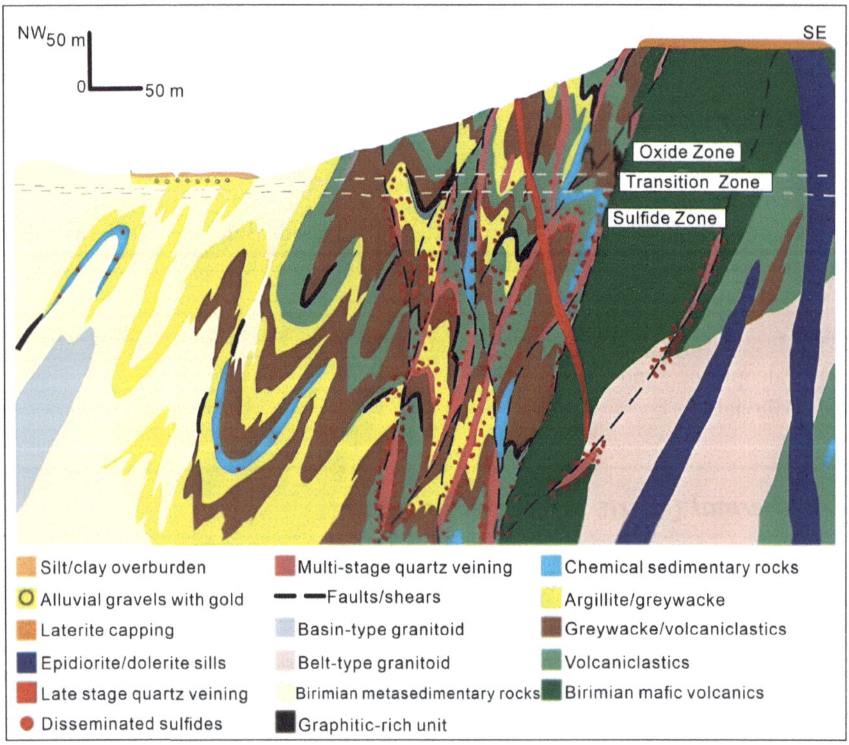

Fig. 8 Schematic cross-section of the northwest to the southeast margin of a typical Ashanti-type, Birimian mesothermal quartz-vein and associated sulfide gold deposit (after Griffis et al. [18])

equal; some are relatively barren in gold, whereas others can be extremely high-grade. Individual vein systems can be traced for long distances along strike (1000 + m) and be up to 10 + m wide (Obuasi, Bibiani), although most are narrower (Prestea, Konongo). The veins also can extend to a considerable depth (300 + m). An individual vein system can pinch out rapidly, but a separate one can be discovered along strike, down dip, or anywhere within the confines of the broader, favourable structural zone.

The mineralized veins often feature medium to dark grey quartz and commonly contain visible gold along with relatively abundant granular pyrite, arsenopyrite, chlorite, tourmaline, ankerite, and calcite, usually in relatively fine-grained, acicular form (Fig. 9). However, in many of the mineralized vein systems, such as at Bogosu, visible gold is quite uncommon. In the more fortunate vein occurrences (Obuasi, Prestea), much of the gold (40–60%) occurs in a free-milling form, whereas the remainder is usually tied up within grains and crystals of pyrite and/or arsenopyrite. Unfortunately, in most of the deposits within the Bogosu concession (more than 12 individual bodies), the vast majority (80+%) of primary gold is intimately tied up with fine-grained arsenopyrite and, to a lesser degree, pyrite grains. A variety of minor accessory polymetallic sulfides are reported in Prestea [120] and Obuasi [121] within quartz veins. Other gangue minerals typically include chlorite, sericite, carbonate, and carbonaceous matter, which give the vein the grey color seen in many of the major and minor deposits. The quartz vein systems are intimately tied up with wide zones of disseminated sulfides in the host rocks [122]. This feature has long been recognized at virtually all the major gold producers, but the zones were too low grade to be mined in the past. However, in recent years, they have become very important sources of new ore reserves, especially at Obuasi. At Prestea, the disseminated sulfides are reported to extend for at least 100 m in the host rocks of some of the vein systems [117], but these appear to be too low grade to be of commercial interest now.

At Bogosu, the mineralized zones feature extensively disseminated sulfides over widths commonly exceeding 50 m [112]. The disseminated sulfides consist mainly of pyrite and arsenopyrite in variable amounts [123]. The gold in these zones is frequently locked up in the sulfide minerals on a submicroscopic scale and a lot of the mineralization is highly refractory. However, some of the new gold discoveries (Nkran Hill, Chirano, Ntotoroso, Ayanfuri), which include vein systems and disseminated sulfides, appear to contain substantial free-milling gold within the sulfide zones. The geometry of the disseminated zones is not well understood, but certainly, at Bogosu they appear to be very large to small pods, and this appears to be the case at Prestea and Obuasi. The disseminated sulfide zones often appear to be related to relatively early-stage hydrothermal activity and are frequently cut by later stages of quartz veins, some of which are rich in gold, whereas others may be quite barren [112].

There appears to be a relatively common alteration pattern at most of these occurrences, which consists of intense silicification, chlorite and carbonate alteration, sericite, and minor feldspar. Albitization has become increasingly recognized in many deposits in recent years. Some of the vein systems have relatively narrow

Fig. 9 Mesothermal quartz vein mineralization in the Julie deposit of NW Ghana **a** A-type vein in drill core, showing vein quartz including chlorite, tourmaline, calcite, and pyrite, **b** Drill core showing strongly deformed TTG containing small, boudinaged A-type veins. Sulfides and tourmaline are abundant in the groundmass, and **c** A sample of C-type vuggy quartz veins with native gold mineralization (arrow). Cal-calcite, Py-pyrite, Chl-chlorite, Tur-tourmaline, Qz-quartz, Ank-ankerite, Au gold (taken from Amponsah et al. [53])

alteration envelopes, whereas others are much more extensive. The broad disseminated sulfide zones are generally highly silicified and form part of the alteration envelopes of the mineralized systems. Tourmaline and rutile are present in some of the vein systems, and minor green mica (chromium or vanadium?) is reported in a few deposits. In recent years, there have been quite a few studies on the geochemistry and fluid inclusions of the vein systems in the Ashanti Belt [121–128]. While these studies have added considerable useful information on the nature and timing of hydrothermal activity in these complex mineralized systems, results and conclusions from different studies show a wide range of divergent views. However, a variety of data confirm that the hydrothermal temperatures cover quite a broad range (approximately 250–450 °C) but, in general, are within a typical range for mesothermal deposits.

Deposits Associated with Intrusives

A few important gold occurrences were discovered where granitoids have turned out to be very favorable host rocks [27]. The first notable occurrences were at Ayanfuri where several of the deposits had intermediate granitoids that hosted significant mineralized quartz stockwork systems. The granitoids have intruded Birimian metasedimentary rocks and the mineralized systems occur in both the metasedimentary rocks and granitoids [124]. Shortly after the Ayanfuri discoveries, work in the Manso Nkwanta district also revealed a very similar pattern where comparatively small intermediate granitoids and the host metasedimentary rocks contain numerous significant gold deposits. For example, within the Obotan concession, the Nkran Hill open-pit operation features a broad stockwork system mainly in Birimian metasedimentary rocks but also some intermediate granitoids. Just north of the Obotan concession, an extensive low-grade quartz stockwork is hosted mainly in an intermediate granitoid at Abori. At the very north end of the Manso Nkwanta district, a small gold deposit consisting of a quartz stockwork is largely confined to a granitoid at Mpesetia and numerous other prospects in the same district consistently show a similar granitoid and metasedimentary rock affiliation.

Most of the mineralized stockworks in the Manso Nkwanta district have features similar to typical Ashanti-type deposits with free gold in the veins and vein selvages, which also contain abundant pyrite and subordinate arsenopyrite although the latter does not appear to contain substantial gold. The alteration also features silica, carbonate, sericite, and chlorite alteration in both the metasedimentary rocks and granitoids. Although only limited petrographic work is available on these intrusions, it appears that many are plagiogranites with considerable Na and relatively minor K. In the Sefwi Belt, there are a series of significant gold prospects on the Chirano concession immediately south of Bibiani [116]. Several of these occurrences feature vein systems hosted in intermediate granitoids. On the north side of the Sefwi Belt, prominent occurrences appear in the Yamfo district, which is starting to emerge as a major new gold district. One of the most important prospects appears to be at Ntotoroso where shallow dipping vein systems are hosted in an intermediate belt granitoid, and similar host rocks are observed in several of the prospects around Kenyase. The general characteristics of the vein systems and alteration features in these areas appear to be similar to those at Ayanfuri and in the Manso Nkwanta area.

Unusual gold occurrences are located on the Dabokrom concession immediately NW of Takoradi, where the Mpohor mafic intrusive complex hosts significant shallow to moderate dipping quartz vein systems [27]. The preferred host rock is a quartz diorite phase, which contains abundant disseminated titaniferous magnetite. However, close to the vein systems, the host rock is completely altered to silica, chlorite, epidote, carbonate, and abundant pyrite. The veins are generally quite high-grade, and visible gold is common; there appears to be no arsenopyrite, and the highly silicified and pyritic host rocks contain substantial gold, most of which appears to be free milling. The mafic intrusive is pre-metamorphic but appears to post-date nearby belt granitoids and may be related to the extensive mafic sills very common in the Tarkwaian sedimentary rocks further north [27]. The vein systems are related to regional structures and have features similar to Ashanti-type vein systems, although they do not appear to display as much multiple-stage veining. The veins in the Mpohor mafic intrusive appear to be quite late-stage features, post-dating peak Eburnean metamorphism, and appear to be akin to the Poura-type vein systems, although there are no conspicuous polymetallic sulfide accessory minerals. The Mpohor intrusive may have been particularly favoured because of abundant magnetite that reacted with sulfur in the hydrothermal fluids to produce pyrite and gold, which were precipitated at the same time.

Most of the above occurrences are largely within the confines of the volcanic belts, although some of the prospects at the southern end of the Manso Nkwanta district may be more within the basin domain. Some of the intrusives in the Manso Nkwanta district may have affinities to the younger basin granitoids than the generally older belt intrusives. Certainly, yet, there do not appear to be any major known prospects obviously within major basin batholiths, although, in principle, if favourable structures extend into the batholiths, they could host significant gold mineralization. This may be the case at Mpesetia where mineralization is hosted in a granitoid, which could be a phase of the large Kumasi batholithic complex. Much further north in the Maluwe basin, the extensive mineralized vein systems in the vicinities of Tinga and Wasipe as well as at Kuri, are hosted in intermediate granitoids that are well into the basin domain. It has long been known that some of the pegmatites and aplite dikes in the Winneba district [125] contain minor gold mineralization along with cassiterite, wolframite, and molybdenite [129]. These are late phases of the Winneba granitoids, and some of the pegmatites in the Cape Coast-Saltpond also have traces of gold. However, none of these occurrences appear to be of any commercial interest and are mostly geological curiosities.

Sources and Timing of the Intrusive Deposits

Yao and Robb [130], after detailed geological investigations of the granitoid intrusion in Obuasi, realized that for the granitoid-hosted gold deposits at Obuasi, fluid inclusion data revealed mineralizing fluids characterized by low-salinity (< 6 wt.% NaCl equivalent) H_2O-CO_2-rich fluids. The alteration mineral assemblages typically reflect mesothermal *P–T* conditions and the addition of Si, K, S, As, and CO_2 components to the fluids. Metal zonation and fluid compositional variation away from the granitoids were not observed, and "granophile" components such as W-Sn-Bi-Mo were not found either from field investigations or from chemical analysis of the ores. Furthermore, the high-precision U–Pb rutile isotopic dating indicates that the mineralization ages significantly post-date the intrusion of host granitoids by about 4–12 million years. Thus, it is considered that the granitoids provided preferential conduits for fluid flow due to their brittle rheology. Under the *P–T* and fO_2 conditions relevant to the gold deposits hosted within the granitoids at Obuasi, it is suggested that the gold species in the fluids were dominated by bisulphide complexes of gold such as $Au(HS)_2^-$ (Shenberger and Barnes [131]; Hayashi and Ohmoto [132]). In this intrusion, the physical contact between gold grains and fluid inclusions was not observed. However, gold is closely associated with sulfides (pyrite and arsenopyrite) in both quartz veins and alteration zones. Hence, they suggested that the gold deposition from the fluids may result mainly from fluid immiscibility and/or sulfidization of wall rocks. These processes are a common phenomenon in lode gold deposits (e.g., Bowers [133]; Craw et al. [134]).

Considering the geochemical characteristics, fluid inclusions, and radiogenic data of ores and granitoids, together with the geological setting (brittle regime) and geochronological sequences of the Birimian rocks, it is suggested that the granitoid-hosted gold deposit at Obuasi was formed in a brittle structural environment by metamorphic low-salinity H_2O–CO_2-rich fluids related to the waning stages of the Eburnean orogeny approximately 2100 Ma. As concluded by John et al. [135], the metamorphism in the entire Ashanti area reached amphibolite facies and the greenschist-facies metamorphism in the region is the result of retrograde metamorphism. It was also suggested by John et al. [135] that gold mineralization in the Birimian terrane was formed during retrograde (post-peak) greenschist facies metamorphism.

Paleoplacers (Tarkwaian-Hosted Deposits)

The Banket conglomerates of the Tarkwaian Group have now produced about 10 million ounces of gold over the past 100 years and although the last remaining underground mine closed in 1999, the district now has very substantial open-pit gold production, and the regional resource potential is excellent. As the general geology of the Tarkwaian Group has been discussed earlier the following section will focus mainly on the general features of the gold-bearing conglomerates in the Tarkwaian Group.

The Tarkwaian Group [136] is 2000–2500 m thick and consists of four main formations; the lower Kawere Formation, which is represented by poorly sorted polymictic conglomerates of variable thickness and overlain by an upper sequence of mainly sandstones [18]. This is overlain by up to about 600 m of mainly sericitic quartzites that include a distinctive zone of conglomerates, which contain very significant gold and are commonly referred to as Banket conglomerates. There are four conglomeratic units with coarse quartz clasts (fine to coarse pebble size) in a sandy matrix with considerable sericite, hematite, magnetite, and a variety of minor heavy minerals (Fig. 10). The Banket Formation is succeeded by the relatively thin Tarkwa Phyllite (approximately 150 m) followed by the very thick (up to about 1500 m) Huni Sandstone units [136]. The 'basal' or B conglomerate in the Banket Formation usually contains the best gold values, but the overlying conglomerates also contain substantial gold, and the intervening cross-bedded quartzites frequently contain low gold values. On the eastern margin of the Tarkwa Syncline, where most of the underground mining has been concentrated, the Banket conglomerates are comparatively thin (often less than 1 m), whereas they thicken towards the west. The high-grade zones (15–45 g/t) are largely confined to the thin conglomerate beds that contain a high proportion of

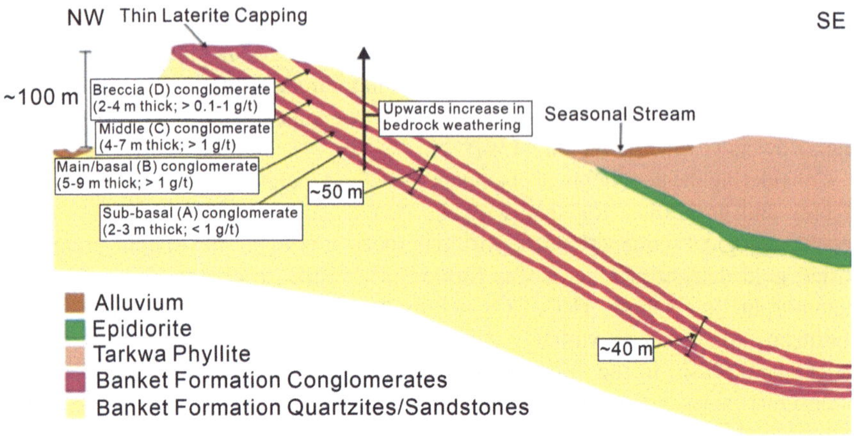

Fig. 10 Northwest to the southeast cross-section of a typical Banket Formation-hosted Tarkwaian palaeoplacer gold deposit (after Griffis et al. [18])

well-rounded, coarse quartz pebbles. The gold is virtually restricted to the matrix of the conglomerates where it occurs as quite fine (10–15 μm), equidimensional grains in clusters, or as individual grains, often in zones with abundant hematite (up to 60% hematite in the matrix of some high-grade shoots).

The Tarkwaian gold is typically very pure and fine. The conglomeratic sequence appears to be about 30–45 m thick on the eastern margin of the basin but thickens towards the west to about 60–75 m. In the western area of Kotraverchy, the Banket Formation is about 60 m but it thickens to about 100 m towards the north [137]. Individual lenses of Banket conglomerates vary in length from 600 to 1000 m and vary in width from 100 to 150 m. Quite importantly, the general current directions consistently indicate a source area for the conglomerates and sandstones to be from the east, southeast, and south. Sestini [137] concluded that the sedimentary features of the Banket Formation indicate that the sediments developed as a series of alluvial fans on a piedmont surface. Braided streams originating largely in the east during a wet climatic environment dispersed the sedimentary debris. The braided channels joined up in the central part of the basin and then flowed to the north. Strogen's [138] analysis of the sedimentary and structural features places more emphasis on the importance of a series of braided tributary channels also originating in the east and joining up with the main trunk channel flowing towards the north. The basin itself appears to be a half-graben structure bounded to the west by steep boundary faults. Strogen [138] cites similarities with the graben basins of the East African Rift System and the Salton Sea basin in southern California.

Although few geologists have speculated on an epigenetic origin for the Tarkwaian gold occurrences, most opinions by those most familiar with the area, are essentially unanimous that these deposits are true paleoplacers. However, they have been modified after deposition and the source of gold appears to be somewhat debatable.

Vein Systems in the Tarkwaian

Quartz veins are very common in the Tarkwaian Group [129] but historically, these have been of no commercial interest. The one exception was noted at Chiribroso [129], located approximately 18 km east of Obuasi, where a narrow (less than 1 m), WNW striking quartz vein was mined on a small scale in 1900–1904. This all changed in the late 1980s when a large mineralized stockwork system was discovered by Abosso Goldfields (Ranger Minerals) at Damang in the northern part of the Abosso concession. The Damang quartz stockwork system is hosted in the Banket Formation units on the east limb of a tight anticlinal structure with NNE trend. The fracture system has been traced for several kilometers along strike and features two dominant and generally NS trending vein systems; one dips shallow (20–35°) to the east, whereas the second set dips more steeply (60–75°) to the east. The fracture systems also appear to be related to the late-stage regional fault. Although the quartz veins contain some visible gold, most of the gold is located in the selvages of the veins which are closely associated with extensive silicification, pyrite, and pyrrhotite mineralization. The veins contain negligible sulfides but show minor but persistent indications of carbonate, sericite, tourmaline, and ilmenite.

Since the discovery of the Damang deposit, this type of occurrence has been the target of considerable exploration effort. At least one similar occurrence (Kayeya) has been outlined along with a regional structure in the Tarkwaian sediments just north of the Ofin River, about 20 km due east of Dunkwa. Also, the Chiribroso vein system has similar features. It has also become very apparent in recent years that some of the old mining districts along the western margin of the Ashanti belt also contain mineralized quartz veins hosted in Tarkwaian rocks. These are closely associated with the major structures that have concentrated gold mineralization in nearby Birimian metasedimentary and volcanic rocks. A similar situation is evident in the Konongo district where the Birimian and Tarkwaian rocks are intensely tectonized and there is some evidence of mineralized vein structures in the Tarkwaian metasedimentary rocks. The studies to date suggest the vein systems in general, are relatively low-grade but they may be quite wide (20–50 m) and, in the right circumstances, could be developed over considerable strike-length and thereby forming quite large potential resources for future possible interest.

Oxide and Laterite Occurrences

Virtually all the primary gold deposits in Ghana have been intensely weathered to produce oxide caps. These oxide caps (also called saprolite) have been a major target for exploration over the past 15 years because they are usually amenable to inexpensive surface mining and treatment schemes. Oxide deposits are gradational with underlying primary mineralization and the extent of oxidation is usually dependent on the climatic and geomorphological conditions. Oxide caps on the crest or flanks of hills may be 50–100 m thick, whereas oxidized zones in low-lying valleys may be less than 5 m thick. The lateral extent of oxide zones, in most cases, is largely controlled by the geometry and nature of the underlying primary mineralization and by the extent of the in-situ weathering. It is not generally typical for most of the oxide zones in the tropical climatic conditions of southern Ghana to develop pronounced mushroom-shaped vertical profiles and most of the oxide zones are only marginally larger than the underlying primary zones of gold mineralization. It also generally appears that the overall grades within the oxide zones are similar to underlying primary mineralization and that supergene enrichment within the oxide zone does not appear to have been very extensive, although, there may be important exceptions to this general trend. Within the oxide zones, the gold appears to be relatively evenly distributed in a very fine-grained form, which is easily recovered by cyanidation either in conventional milling systems or by heap leaching.

The effects of oxidation and near-surface weathering have had a particularly beneficial effect on those deposits with extensively disseminated sulfides in the primary zone and where a high proportion of the gold was originally contained within the sulfide grains. The liberation of gold from the sulfides has rendered many of the oxide caps suitable for mining whereas the underlying primary zones may be of too low a grade to justify more expensive refractory recovery systems. Of course, the physical effects resulting from the weathering have also resulted in many benefits, such as softer ores that require minimum blasting as well as substantially increasing the permeability of the host rocks, which may permit very low-cost heap-leach methods to recover the gold. These beneficial effects are not just confined to the mineralized vein systems and accompanying disseminated sulfides in Birimian units and belts intrusives. They have also had substantial benefits in the near-surface exposures of the Tarkwaian paleoplacers because the host rocks are softer and have a higher permeability, which permits very effective heap leaching of relatively low-grade mineralization. The lack of a near-surface mushroom effect in most of the oxide resources of southern Ghana may result from the fact that most of these occurrences have truncated weathering profiles with the removal of overlying lateritic and shallow saprolitic zones of weathering. Where laterites are preserved, a lateral dispersion of gold is often quite pronounced and is generally manifest in much broader geochemical soil anomalies and some gold enrichment in the near-surface environment. Where laterites overlie oxide and primary gold mineralization, it is common to see some depletion in the uppermost mottled saprolite zone so that shallow trenches within the mottled saprolite may yield gold values that are not representative of underlying mineralization. The southern part of the Ashanti Belt features many areas with thin but widespread laterites so high-grade trench results in these areas must be carefully assessed. It should also be noted that in some areas (i.e., part of the Yamfo-Kenyase district), there is an iron-laterite crust (duricrust), which has quite low background levels of gold above-buried mineralization. Also, there appear to be few examples in Ghana where gold in laterites has been formed on a large scale, but certainly, it appears that a considerable amount of artisanal mining has involved the washing of lateritic material for gold. Some of the early exploration efforts during the recent exploration boom were targeting possible bauxite targets similar to the Boddington-type of Western Australia. Areas in the Bibiani Range and at Awaso, where extensive bauxite occurrences are known, were examined and a reconnaissance survey was conducted in the Atewa Range, which is capped by large bauxite deposits.

It is also quite clear that lateritic caps overlying oxide and primary gold deposits in some of the major mining districts were enriched in gold. This was evident in the early days of open-pit mining at Bogosu where several of the major pits had lateritic caps carrying considerable gold. In addition, recent exploration in the Kanyankaw area south of Tarkwa has confirmed substantial gold in the iron laterites that cap the hills of this area, and which have been mined on an artisanal basis for generations. One can also reasonably conclude that at one stage, it is highly likely that there may have been numerous Boddington-type laterite gold occurrences in many areas of southern Ghana. The remnant deposits capping many of the highest ranges in southern Ghana indicate that bauxite and iron laterite deposits were probably very widespread throughout southern Ghana in early to middle Tertiary times. It is probably reasonable to assume that the major bauxite occurrences in Ghana formed at about the same time as other more extensive bauxite deposits further to the west in Guinea and these are believed to have formed mainly in the Palaeocene to early Miocene or about 65–20 Ma. It is therefore not difficult to conclude that any thick laterite/bauxite capping overlying most of the major mining districts must have contained very substantial concentrations of disseminated gold. These would have been subsequently eroded away into the rivers and over time, a large proportion of the gold may well have found its way into the coastal and offshore sediments. Moreover, in the Kunche area of northwestern Ghana, situated within the Wa-Lawra greenstone belt, laterite geochemistry has proven instrumental in delineating gold exploration targets within a complex lateritic environment, as demonstrated by Sunkari et al. [148]. The study introduced a geochemical methodology combining trace element analysis with statistical tools to identify pathfinder elements associated with gold in lateritic soils. The laterite samples were analyzed using ICP-MS and XRF, and their spatial distribution was mapped through GIS-based kriging. Key secondary minerals identified include goethite, hematite, and kaolinite. Multivariate analysis revealed that the element association of Pb, Cu, As, and Ag closely aligns with gold anomalies, making it a reliable indicator in such terrains. Threshold values derived using the median absolute deviation (MAD) method further highlighted zones of potential enrichment. These findings emphasize that residual laterites enriched in pathfinder elements represent promising targets for gold exploration in the Kunche area.

Alluvial Deposits

Most of the historical gold production from Ghana prior to the twentieth century came from a myriad of small streams and rivers draining areas with underlying oxide and primary gold deposits. In addition, several of the major rivers have been mined with dredges since the early 1900s and more recently, large alluvial operations have started on rivers in several districts. As noted above, the regional correlation of remnant bauxite/laterite capping on the highest ranges of southern Ghana gives a general indication of the extent of erosion during the middle to late Tertiary and Quaternary. Since then, approximately 400–600 m of bedrock has probably been eroded away and, along with it, a very large amount of gold. One can only speculate rather wildly on the amount of gold eroded since about the middle Tertiary, but it would be reasonable to estimate it to be at least in the order of the historical production to the beginning of the present century or about 60–80 Moz. If one goes back further in time and assumes that the mesothermal vein systems observed today extended to relatively shallow depths (shallow mesozonal to epizonal), then it is again obvious that more gold would have been eroded away.

About 1000 Ma time gap represents the extensive peneplain surface underlying the Voltaian Supergroup, and during this period, extensive erosion of Tarkwaian and Birimian rocks took place and probably included a very substantial amount of gold. There is very little detailed information on the ages of the alluvial terraces and river sediments observed in the present inland river system, but most appear to be relatively young and are probably of Late Quaternary age. Sestini [137] estimated that the present valley of the Ofin River is probably less than 30,000 years old and that the deepest incisions in the valley floor probably correlate with a lowering of sea levels up to 100 m below present levels. Hall et al. [139] estimate that in the Birim River valley, most of the sediments are younger than 15,000 BP and the basal gravels, containing abundant diamonds with accessory gold, were probably formed mainly in the period 7–14,000 BP. During this period, the area experienced high rainfall at a time when the vegetation cover initially was grassland and later became more heavily forested [139]. The same general pattern probably applies to most of the current inland river systems of Ghana. Alluvial gold was concentrated mainly in basal gravel overlying weathered Birimian units and occurred at a time when there was high rainfall, minimal vegetation cover and at least moderate relief. A significant drop (up to 100 m) of the sea level may have accentuated the relief. Evidence of earlier Quaternary erosion and depositional activity is present in the raised terraces, but, in general, the majority of the alluvials in the current river systems are Late Quaternary (Holocene) in age and, therefore, less than about 10,000 BP.

Rivers and Valleys

Several of the major rivers (Pra, Tano, Ankobra, Ofin) of southern Ghana were dredged in the early 1900s and the Ofin operations managed to struggle into the 1990s, although production in the last two decades has been very modest. Kesse [28] provides a very good summary of the alluvial dredging in the last century and reports on quite considerable alluvial exploration work carried out by the Ghana Geological Survey in the 1960s and 1970s. Sestini [137] summarized the results of Banka drilling programs on the Ofin River north of Dunkwa. The main valley is at least 1 km wide and features a lower gold-bearing gravel unit overlying weathered Birimian bedrock. The gravel is quite variable in thickness but usually in the range of 2–6 m, and it is overlain by barren silt and clay overburden of similar thickness. Within the gravels, there are broad zones (300–600 m wide) of higher-grade material (over 175 mg/

m^3), which show good continuity along the course of the valley. Sestini [137] cites regional evidence to suggest that the deepest valley incisions were probably related to a lowering of the sea level by about 100 m between 33,000–20,000 years ago. The gravel may have been related to a subsequent period of aggradation in the period 23,000–11,000 years BP. Reworking of the gravel has produced high-grade zones, which became the dredging targets. The Ankobra River has been dredged along various stretches of the current river channel in the early decades of the twentieth century.

The Ankobra River basin covers a very large area that includes numerous gold districts (Tarkwa, Prestea-Bogosu, Wassa, Akropong-Ayanfuri, central Asankrangwa belt, southern Bibiani Range, Axim district), and it has therefore been the conduit for very substantial amounts of alluvial gold. The margins of the existing main river valley contain significant alluvial gold occurrences in gravels that are 6–10 m thick, underlying barren overburden that is usually at least, if not slightly thicker than the gravels. The main valley gravels are relatively low grade (0.1–0.2 g/m^3) but contain higher grade sections and extensive terrace deposits along the margins of the valley, which also contain significant gold-bearing gravels.

Studies that are much more detailed were carried out on the central portion of the Birim River in the area where extensive alluvial diamond mining had taken place between Akwatia and Oda [139]. The Birim valley is from 200 to 1500 m wide and features a lower series of gravels, which overlie weathered Birimian bedrock. The gravels vary in thickness from 1–2 m to as much as 7 m, and these coarse clastics (with cobble and boulder clasts) are overlain by 1–2 m of sand and then 5–7 m of silt, sand, and clay. The gravels invariably carry alluvial diamonds, often in substantial concentrations (1–3 carats/m^3), and usually contain minor accessory gold.

Radiocarbon age dating of leaves and wood in the gravels indicate an age of 21,000–13,000 yrs BP for the lowermost gravels in scour channels and an age of 9000–7500 yrs BP (Holocene) for more extensive overlying gravels. The finer-grained overburden was largely formed in the last 7000 years and the current river channel, which has cut through overburden and the underlying gravels, was probably active over the past 3000 years. The major period of heavy denudation coincided with a sharp increase in rain at a time when the region was largely deforested; this occurred mainly in the period of about 13,500–10,500 years BP. The major tributaries of the Ankobra, the Fura (or Fure) and Mansi rivers were also tested extensively with Banka drilling in the early 1960s. Both these river valleys have a lower section of gravels 1–2 m thick overlain by 2–3 m of silt and clay overburden. The gravels usually carry widely disseminated alluvial gold (usually less than 100 mg/m^3) but contain higher grade zones (200–600 mg/m^3) as well and the current river channels have reworked portions of the underlying gravels and produced significant concentrations that are localized. Parts of the Fure, close to the confluence with the Ankobra, were dredged early in the 1900s. Areas dredged probably had in-situ grades in the range 200–400 mg/m^3 although recovered values were probably considerably lower due to inefficient recovery methods, which resulted in the loss of a high proportion of the fine-grained gold.

The Kibi District has long been known for very significant alluvial gold, and although some dredging was attempted in the upper reaches of the Birim River, most of the production has come from dozens of the smaller drainages all along the flanks of the Atewa Range. The most famous historical site was at Pusu Pusu where artisanal mining took place for generations and several sluicing operations were carried out in the early 1900s. Junner [129] indicates that the Pusu Pusu area features boulder conglomerates up to 2.5 m thick and these are overlain by clay/silt overburden of a similar thickness; very coarse nuggets (up to 8 oz) were apparently quite common in the gravels. In more recent years, the Goldenrae Mining Company on the north side of the Atewa Range attempted larger alluvial operations around Kwabeng. Their main resource on the Awusu River features a lower, poorly sorted gravel 2–3 m thick, which is overlain by silt/clay overburden of about the same thickness; the gravel resource can be up to 400–600 m wide and extends for at least 4 km downstream. The average grade of the gravel is about 600 mg/m^3 and the gold is quite coarse (90% + 2.0 mm) and pure (920–940 fineness).

Similar operations on the Jeni and Bonte rivers at the north end of the Manso Nkwanta district have been going on for about the past 10 years. In this area, the early work focused on the high-grade zones within the immediate confines of the present river channels where gold-bearing gravels and sands carry very good gold grades (800–1000 + mg/m^3). Along the margins of the channels, there are extensive deposits of gravels (1–2 m thick) with similar amounts of clay-rich overburden (1–3 m thick), also carrying quite good gold values (400–800 mg/m^3). The main gravel resource is 200–500 m wide and has been traced for about 14 km downstream. The area also features substantial resources in older terrace gravel deposits along the margin of the main river valley. The Manso Nkwanta district in general has many smaller alluvial gold deposits in areas where there are known bedrock occurrences of gold. The same general conditions are observed in virtually all the secondary and tertiary streams draining in most of the known gold districts, but the majority of the smaller streams usually have thinner gravels and overburden. These are the areas where much of the past historical small-scale production has come from and some of the original alluvial grades.

Beach and Offshore Sediments

Many of the earliest historical descriptions of the Gold Coast include stories of Cape Coast residents washing gold from sands in the town after rains. Also, early descriptions of the gold mining activities of the local people, referred to gold being recovered along the seashore, especially close to areas where streams entered the ocean [140]. There is absolutely no reason to question the early reports, and it seems quite certain that beach sands were at least a modest source of gold in past centuries. The coastal region of Ghana features a few raised beaches and old peneplain surfaces [141]. The raised beaches include three principal sets at current elevations of about 4.5, 9, and 12 m and older peneplain surfaces are mainly at about 22 and 55 m. These beaches would certainly be the most likely source of gold for the many stories of gold in the streets of the coastal towns. It is also somewhat curious that the name Elmina was given to a coastal town quite remote from any of the major gold producing centers and its reputation for gold must have been largely due to gold being brought down from the interior. However, as noted, there was a small gold operation at Abrobi Hill, immediately west of Kommenda and about 17 km west of Elmina. This small deposit was probably related to Devonian conglomerates (basal Amisian) but some gold in the area could have been from raised beaches. If there are minor occurrences of gold in raised beaches, then it would be reasonable to assume that there will be a similar occurrence in submerged beaches. In fact, some of the submerged beaches would probably contain considerably more gold than those now exposed. Certainly, that should be the case for those submerged beaches developed in the same period (approximately 7000–14,000 BP) as the extensive buried gravels now seen throughout southern Ghana.

There can be little doubt that a very considerable amount of alluvial gold would have been discharged along the coastal region where the major rivers enter the Gulf of Guinea. Although the current inland rivers still contain a very large resource of alluvial gold, and most of the historical production came from the inland rivers, it would seem unlikely that all the gold eroded from bedrock sources since the early Tertiary could have been mined or retained within the current inland river system. Therefore, a very significant amount of gold must have been flushed out into the near-shore marine environment. During high rainfall periods (approximately 7000–14,000 BP), some of the gold reaching the sea would have been modestly coarse and probably dropped out very close to the seashore of the time. However, the rivers must also have been carrying considerable fine-grained gold, which would have come from the Tarkwa district and from the many oxide deposits throughout the area. This fine-grained gold would have been carried further out to sea and deposited amongst the sand and silt. The prevailing long-shore currents (now towards the east) would be strong enough to move the gold some distance from their original discharge point and very likely, the persistent effects of the longshore currents would have resulted in a further concentration of the gold in the offshore environment. One may look at the offshore diamonds in Namibia or the rutile deposits along the coast of Sierra Leone to imagine what could occur in the coastal waters of Ghana to concentrate alluvial gold into deposits of potential economic interest.

Miscellaneous Occurrences

Junner [129] noted that "gold occurs locally in small amounts in the conglomerates and volcanic rocks of the Buem Structural Unit and in the Voltaian and Akwapimian conglomerates". Our knowledge of these occurrences has not improved since Junner's time and in the recent exploration boom, little attention has been paid to these indications. In principle, some of the basal conglomerates of the Voltaian Supergroup could well contain appreciable gold in areas where they overlie prospective greenstone belts such as the Ashanti and Sefwi belts. Some of these conglomerates may be exposed along the current margins of the basin but most will be deeply buried and therefore unlikely to be exploited as targets soon. There has been modest interest in alluvial and primary gold deposits in parts of Togo and Benin where the geology is similar to the eastern borderland area of Ghana and more systematic prospecting in this large area may yet reveal gold occurrences of more significance. The location, type of gold mineralization and estimated resources of known gold deposits in Ghana are summarized in Table 2.

4 Conclusion

The past 2500 years have witnessed the rich mining history of Ghana, the largest producer of gold in the West African sub-region. The ounces of gold mined and exported from the country are among the highest globally. This enviable position culminates in favourable economic development of the country, attracting significant foreign investment and fostering the growth of supporting industries. On a local community scale, the livelihoods of people, especially those in mining communities, have seen significant improvements through the activities of Corporate Social Responsibility of both multinational and indigenous mining firms. Challenges exist regardless, particularly related to environmental pollution and degradation. Significant efforts are being channelled into resolving these environmental challenges. The efforts include the reclamation and rehabilitation of mining sites and other responsible mining practices. Effective, fair, robust, and equitable regulatory systems are also being implemented to ensure a beneficial distribution of the revenue generated from mining in favour of mobilizing economic growth.

There are four geological provinces in Ghana: the shield (cratonic) domain, platform units or cratonic cover, metamorphic and mobile belt, and coastal sedimentary basins. The cratonic domain comprises the Paleoproterozoic Birimian Supergroup and the younger Tarkwaian Group. This cratonic domain, which has been intruded by granitic complexes, consists of volcanic and sedimentary assemblages deformed, folded, and metamorphosed during the Eburnean Orogeny (2.2–1.8 Ga). Unlike the cratonic domain, Neoproterozoic Voltaian sedimentation dominates the relatively precious metal-barren cratonic cover. The Pan-African Orogeny (950–550 Ma) deformed, and fault-separated rocks of the Dahomeyan Supergroup, Buem Structural Unit, and Togo Structural Unit, which make up the metamorphic and mobile belt. The youngest of Ghana's lithologies are within the coastal sedimentary basins restricted to the southern fringes of the country.

Two genetically distinct classes of gold mineralization are found in the cratonic domain, namely the metavolcanic/metasedimentary-hosted mesothermal quartz-vein orogenic gold deposits and the conglomerates-hosted paleoplacer gold deposits. The former class of gold deposits is interpreted to result from extensive hydrothermal processes related to the emplacement of basin-type granitoids within the Birimian Supergroup. The auriferous veins, formed through multiple stages, are consequently

Table 2 Summary of the known gold deposits in Ghana based on available records at the time of writing this chapter

Name of deposit	Location	Type of mineralization	Total resource estimates	References
Abore North	Asankrangwa Belt	Mesothermal vein type; Granitoid type	~ 0.3 Moz Au	[18]
Abosso	Ashanti Belt/ Kumasi Basin	Mesothermal vein type; Palaeoplacer type	~ 0.6 Moz Au	[18]
Anwia	Kumasi Basin	Mesothermal vein type	~ 0.15 Moz Au	[18]
Atewa Range	Kibi-Winneba Belt	Mesothermal vein type	NA	[18]
Ayanfuri	Central Ashanti Belt	Mesothermal vein type; Granitoid type	0.2 Moz Au	[18]
Bepkong	Wa-Lawra Belt	Mesothermal vein type	~ 0.1 Moz Au	[142]
Benso	Eastern limb of Ashanti Belt	Mesothermal vein type	NA	[143]
Besease	Northeastern margin of Ashanti Belt	Mesothermal vein type; Alluvial	~ 0.2 Moz Au	[18]
Bibiani	Sefwi Belt	Mesothermal vein type	~ 1.8 Moz Au	[18]
Bogosu	Southwestern margin of Ashanti Belt	Mesothermal vein type	> 2.5 Moz Au	[18, 112, 123, 144]
Brohani	Bui Belt	Palaeoplacer type	NA	[18]
Chirano	Central western margin of Sefwi Belt	Mesothermal vein-type; Granitoid type	~ 1.4 Moz Au	[18]
Dabokrom	West of Kibi-Winneba Belt, Cape Coast Basin	Mafic-hosted mineralization	~ 0.7 Moz Au	[18]
Damang	Southern Ashanti Belt	Mesothermal vein-type; Palaeoplacer type	~ 5 Moz Au	[18, 19, 145]
Dokrupe	Bole-Nangodi Belt	Mesothermal vein type	~ 0.18 Moz Au	[18]
Golden Ridge	Northeastern margin of Ashanti Belt	Mesothermal vein type	NA	[18]

(continued)

Table 2 (continued)

Name of deposit	Location	Type of mineralization	Total resource estimates	References
Iduapriem	Southwestern margin of Ashanti Belt	Palaeoplacer type	~ 2 Moz Au	[18]
Julie	Wa-Lawra Belt	Granitoid type	~ 0.8 Moz Au	[53]
Konongo	Northeastern end of Ashanti Belt	Mesothermal vein type	~ 0.9 Moz Au	[18, 144]
Kubi	Western margin of Ashanti Belt	Mesothermal vein type	~ 0.8 Moz Au	[18]
Mampon-Aboronye	Western flank of Ashanti Belt	Mesothermal vein type	~ 0.4 Moz Au	[18]
Mpeasem	Asankrangwa Belt	Mesothermal vein type	~ 0.8 Moz Au	[18]
Nkran Hill	Northwest of Ashanti Belt	Mesothermal vein-type; Granitoid type	> 2.5 Moz Au	[18]
Nkroful	West of Ashanti Belt, Kumasi Basin	Mesothermal vein type; Alluvial	NA	[18]
Ntotoroso	Sefwi Belt	Mesothermal vein-type; Granitoid type	~ 1.4 Moz Au	[18]
Obuasi	Southern flank of the Ashanti Belt	Mesothermal vein-type; Granitoid type	> 60 Moz Au	[131, 144, 146, 147]
Prestea	Southwest of Ashanti Belt	Mesothermal vein type	> 7.2 Moz Au	[123, 144]
Salman	Western boundary of Ashanti Belt	Mesothermal vein type	~ 1.2 Moz Au	[18]
Tarkwa	Southwestern Ashanti Belt	Palaeoplacer type	> 40 Moz Au	[18, 19]
Teberebie	Southern margin of Ashanti Belt	Palaeoplacer type	~ 2 Moz Au	[18]
Wassa	Southern margin of Ashanti Belt	Mesothermal vein type	~ 4 Moz Au (possibly up to 10 Moz Au)	[116, 143]

(continued)

Table 2 (continued)

Name of deposit	Location	Type of mineralization	Total resource estimates	References
Winneba-Mumford area	Southern extension of Kibi-Winneba Belt	Mesothermal vein-type; Palaeoplacer type	NA	[18]
Yamfo-Kenyase	Sefwi Belt	Mesothermal vein-type; Granitoid type	~ 6.2 Moz Au	[18]

found in the greenstone belts of the Birimian Supergroup and are spatially and genetically associated with fault and shear zones. This vein-type gold mineralization is found as free gold fractured zones of the veins or as invisible gold disseminated with sulfides within the host rocks. The typical mineral assemblage that features this mineralization style includes arsenopyrite, pyrite, chalcopyrite, pyrrhotite, marcasite, ankerite, calcite, chlorite, and tourmaline. Paleoplacer gold deposits, formed under supergene conditions, contain free gold hosted within conglomerates of the Banket Formation in the Tarkwaian Group. These deposits are typically accompanied by authigenic minerals such as ilmenite/rutile, magnetite, hematite, and sericite, with some evidence of hydrothermal modification. Alluvial gold deposits sourced from sediments along significant rivers in the country are also present.

With Ghana's long-standing history and expertise in gold production, this rich geological endowment positions the country as a critical player in the global gold market. Continued investment in exploration technologies and methods can lead to discovery of new mineral resources, ensuring the sustainability and longevity of Ghana's gold industry. This study recommends more research on gold mineralization supported by fieldwork and various high-precision analytical techniques to constrain the age and geochemical features of rocks in the geological provinces of Ghana.

Acknowledgements This study is part of the first author's PhD (Geological Engineering) seminar report submitted to the Graduate School of Natural and Applied Sciences of Niğde Ömer Halisdemir University in Türkiye. The authors thank everyone who contributed to improving the quality of this chapter. The Editors of the series are also duly acknowledged for their invaluable contributions that reshaped this work.

References

1. Opoku-Ntim I, Gyampo O, Andam AB (2019) Risk assessment of radon in some bottled water on the Ghanaian market. Environ Res Commun 1(10):105001. https://doi.org/10.1088/2515-7620/ab4568
2. Waters B, McKinley I (2010) Carbon capture and storage for a waste to energy plant: Accra, Ghana

3. Kpordzro RI (2018) Assessment of factors affecting indoor radon-222 concentration in Dome and its environs-Greater Accra Region of Ghana. Doctoral dissertation, University of Ghana
4. Wright JB, Wright JB (1985) The Pan African of West Africa—the eastern domain. In: Wright JB (ed) Geology and mineral resources of West Africa. Springer Netherlands, p 56–69. https://doi.org/10.1007/978-94-015-3932-6_6
5. Block S, Baratoux L, Zeh A, Laurent O, Bruguier O, Jessell M et al (2016) Paleoproterozoic juvenile crust formation and stabilisation in the south-eastern West African craton (Ghana); new insights from U-Pb-Hf zircon data and geochemistry. Precambrian Res 287:1–30. https://doi.org/10.1016/j.precamres.2016.10.011
6. Kuma J, Kim YU, Boamah D, Sakamoto I (2010) Gold potential of the Ashanti Belt of Ghana. J Sch Mar Sci Technol Tokai Univ 8:25–39
7. Sakyi PA, Manu J, Su BX, Kwayisi D, Nude PM, Dampare SB (2019) Geochemical and Sm–Nd isotopic evidence for the composition of the Palaeoproterozoic crust of the West African craton in Ghana. Geol J 54(6):3940–3957. https://doi.org/10.1002/gj.3387
8. Awunyo EK (2018) Petrography and geochemistry of Granitoid gneisses and Associated Rocks in parts of the Suhum Basin, Ghana: implications for petrogenesis and tectonic setting. Doctoral dissertation, University of Ghana
9. Nwaila GT, Durrheim RJ, Jolayemi OO, Maselela HK, Jakaitė L, Burnett M et al (2020) Significance of granite-greenstone terranes in the formation of Witwatersrand-type gold mineralisation–a case study of the neoarchaean black reef formation, South Africa. Ore Geol Rev 121:103572. https://doi.org/10.1016/j.oregeorev.2020.103572
10. Koffi YH, Wenmenga U, Djro SC (2016) Tarkwaian deposits of the Birimian belt of Houndé: petrological, structural and geochemical study (Burkina-Faso, West Africa). Int J Geosci 7(5):685–700. https://doi.org/10.4236/ijg.2016.75053
11. Tay CK, Hayford EK, Hodgson IO (2017) Application of multivariate statistical technique for hydrogeochemical assessment of groundwater within the lower Pra Basin, Ghana. Appl Water Sci 7(3):1131–1150. https://doi.org/10.1007/s13201-017-0540-6
12. Tay CK, Kortatsi BK, Hayford E, Hodgson IO (2014) Origin of major dissolved ions in groundwater within the lower Pra Basin using groundwater geochemistry, source-rock deduction and stable isotopes of 2H and 18O. Environ Earth Sci 71(12):5079–5097. https://doi.org/10.1007/s12665-013-2912-z
13. Amponsah I (2015) High-pressure Granulites and Eclogites of the Dahomeyide Suture Zone (Adaklu, Southeastern Ghana): petrogenetic evolution and tectonic implications. Doctoral dissertation, University of Ghana
14. Nuamah Daniel OB, Tandoh Kingsley K, Brako A (2019) Geochemistry of minor and trace elements in soils of Akuse area, Southeastern Ghana. Geosciences 9(1):8–17. https://doi.org/10.5923/j.geo.20190901.02
15. Jones WB (1990) The Buem volcanic and associated sedimentary rocks, Ghana: a field and geochemical investigation. J Afr Earth Sci (Middle East) 11(3–4):373–383. https://doi.org/10.1016/0899-5362(90)90016-8
16. Abu M, Sunkari ED, Şener M (2019) Untapped economic resource potential of the neoproterozoic to early Paleozoic Volta Basin, Ghana: a review. Nat Resour Res 28(4):1429–1445. https://doi.org/10.1007/s11053-019-09478-5
17. Le Mignot E, Reisberg L, André-Mayer AS, Bourassa Y, Fontaine A, Miller J (2017) Re-os geochronological evidence for multiple Paleoproterozoic gold events at the scale of the West African craton. Econ Geol 112(1):145–168. https://doi.org/10.2113/econgeo.112.1.145
18. Griffis RJ, Barning K, Agezo FL, Akosah FK (2002) Gold deposits of Ghana. Minerals Commission 432p
19. Pigois JP, Groves DI, Fletcher IR, McNaughton NJ, Snee LW (2003) Age constraints on Tarkwaian palaeoplacer and lode-gold formation in the Tarkwa-Damang District, SW Ghana. Mineral Deposita 38(6):695–714. https://doi.org/10.1007/s00126-003-0360-5
20. Kristinsdóttir BD (2013) U-Pb, O and Lu-Hf isotope ratios of detrital zircon from Ghana, west-African craton: formation of juvenile, Palaeoproterozoic crust. Dissertations in geology at Lund University

21. Petersson A, Scherstén A, Kristinsdóttir B, Kemp A, Whitehouse M (2018) Birimian crustal growth in the West African craton: U-Pb, O and Lu-Hf isotope constraints from detrital zircon in major rivers. Chem Geol 479:259–271. https://doi.org/10.1016/j.chemgeo.2018.01.021
22. Block S, Ganne J, Baratoux L, Zeh A, Parra-Avila LA, Jessell M et al (2015) Petrological and geochronological constraints on lower crust exhumation during Paleoproterozoic (eburnean) orogeny, NW Ghana, West African craton. J Metamorph Geol 33(5):463–494. https://doi.org/10.1111/jmg.12129
23. Arhin E (2013) Use of Regolith geochemistry to delineate gold mineralisation under cover: a case study in the Lawra Belt, NW Ghana. Doctoral dissertation, University of Leicester
24. Graham KM, Kwasi P, Wemegah DD, Boamah D (2013) Geological and structural interpretation of part of the Buem formation, Ghana, using aerogeophysical data. M.Sc. thesis. Kwame Nkrumah University of Science and Technology
25. Eisenlohr BN, Hirdes W (1992) The structural development of the early Proterozoic Birimian and Tarkwaian Rocks of Southwest Ghana, West Africa. J Afr Earth Sci (Middle East) 14(3):313–325. https://doi.org/10.1016/0899-5362(92)90035-B
26. Nyame FK (2013) Origins of Birimian (ca 2.2 ga) mafic magmatism and the Paleoproterozoic "Greenstone belt" metallogeny: a review. Isl Arc 22(4):538–48. https://doi.org/10.1111/iar.12047
27. Smith AJ, Henry G, Frost-Killian S (2016) A review of the Birimian supergroup-and Tarkwaian group-hosted gold deposits of Ghana. Episodes 39(2):177–197. https://doi.org/10.18814/epiiugs/2016/v39i2/95775
28. Kesse GO (1985) The mineral and rock resources of Ghana. A. A. Balkema Publishers, 610p
29. Dampare SB, Shibata T, Asiedu DK, Osae S, Banoeng-Yakubo B (2008) Geochemistry of Paleoproterozoic Metavolcanic Rocks from the Southern Ashanti volcanic belt, Ghana: petrogenetic and tectonic setting implications. Precambrian Res 162(3–4):403–423. https://doi.org/10.1016/j.precamres.2007.10.001
30. Feybesse JL, Billa M, Guerrot C, Duguey E, Lescuyer JL, Milesi JP et al (2006) The Paleoproterozoic Ghanaian Province: geodynamic model and ore controls, including regional stress modeling. Precambrian Res 149(3–4):149–196. https://doi.org/10.1016/j.precamres.2006.06.003
31. Leube A, Hirdes W, Mauer R, Kesse GO (1990) The early proterozoic Birimian supergroup of Ghana and some aspects of its associated gold mineralization. Precambrian Res 46(1–2):139–165. https://doi.org/10.1016/0301-9268(90)90070-7
32. Petersson A, Scherstén A, Gerdes A (2018) Extensive reworking of Archaean crust within the Birimian terrane in Ghana as revealed by combined zircon U-Pb and Lu-Hf isotopes. Geosci Front 9(1):173–189. https://doi.org/10.1016/j.gsf.2017.02.006
33. Karikari AY, Duah AA, Akurugu BA, Darko HF (2021) Assessing the impacts of artisanal mining on the quality Ofvf South-Western Rivers system in Ghana. Environ Monit Assess 193(11):715. https://doi.org/10.1007/s10661-021-09515-y
34. Pelig-Ba KB (2012) Geochemistry of crystalline rocks from the east of the Upper East region of Ghana. Res J Environ Earth Sci 4(5):534–545
35. Ahenkorah I, Awuah EM, Ewusi A, Affam M (2018) Geotechnical and petrographic characterisation of the Birimian granitoids in Southern Ghana as an aggregate for sustainable road construction. IJAERS 5(3):40–49. https://doi.org/10.22161/ijaers.5.3.7
36. Ganyaglo SY, Gibrilla A, Teye EM, Owusu-Ansah ED, Tettey S, Diabene PY et al (2019) Groundwater fluoride contamination and probabilistic health risk assessment in fluoride endemic areas of the Upper East region, Ghana. Chemosphere 233:862–872. https://doi.org/10.1016/j.chemosphere.2019.05.276
37. Hirdes W, Davis DW (2002) U-Pb geochronology of Paleoproterozoic Rocks in the southern part of the Kédougou-Kéniéba inlier. Precambrian Res 118(1–2):83–99. https://doi.org/10.1016/S0301-9268(02)00080-3
38. Abdul-Rashid A (2018) Modeling of KUNSU ORE body in the Kumasi Basin Northwest of Kumasi, Ghana. Doctoral dissertation, University of Ghana

39. Nunoo S, Hofmann A, Kramers J (2022) Geology, zircon U-Pb dating and εHf data for the Julie Greenstone Belt and associated rocks in NW Ghana: implications for Birimian-to-Tarkwaian correlation and crustal evolution. J Afr Earth Sci 186:104444. https://doi.org/10.1016/j.jafrearsci.2021.104444
40. Ansah SK (2018) Geostatistical estimation of a paleoplacer deposit with hard geological boundary: case study at Tarkwa Gold mine, Ghana. Doctoral dissertation, Memorial University of Newfoundland
41. Nyame FK (2008) Petrography and geochemistry of intraclastic manganese–carbonates from the~ 2.2 ga Nsuta deposit of Ghana: significance for manganese sedimentation in the Palaeoproterozoic of West Africa. J Afr Earth Sci 50(2–4):133–147. https://doi.org/10.1016/j.jafrearsci.2007.09.007
42. Kalsbeek F, Frei R (2010) Geochemistry of Precambrian sedimentary rocks used to solve stratigraphical problems: an example from the neoproterozoic Volta Basin, Ghana. Precambrian Res 176(1–4):65–76. https://doi.org/10.1016/j.precamres.2009.10.004
43. Deynoux M, Affaton P, Trompette R, Villeneuve M (2006) Pan-African tectonic evolution and glacial events registered in neoproterozoic to Cambrian cratonic and foreland basins of West Africa. J Afr Earth Sci 46(5):397–426. https://doi.org/10.1016/j.jafrearsci.2006.08.005
44. Boher M, Abouchami W, Michard A, Albarede F, Arndt NT (1992) Crustal growth in West Africa at 2.1 ga. J Geophys Res 97(B1):345–369. https://doi.org/10.1029/91JB01640
45. Affaton P (1973) Étude Géologique et Structurale du Nord-Ouest Dahomey, du Nord Togo et du Sud-Est de la Haute-Volta [thesis]. University of Aix-Marseille III, 223p. Available from: http://pascal-francis.inist.fr/vibad/index.php?action=getRecordDetail&idt=PASCALGEODEBRGM7620266331
46. Anani CY, Mahamuda A, Kwayisi D, Asiedu DK (2017) Provenance of sandstones from the neoproterozoic Bombouaka group of the Volta Basin, Northeastern Ghana. Arab J Geosci 10(21):1–15. https://doi.org/10.1007/s12517-017-3243-2
47. Affaton P, Sougy J, Trompette R (1980) The tectono-stratigraphic relationships between the upper Precambrian and lower Paleozoic Volta Basin and the Pan-African Dahomeyide orogenic belt (West Africa). Am J Sci 280(3):224–248. https://doi.org/10.2475/ajs.280.3.224
48. Carney JN, Jordan CJ, Thomas CW, Condon DJ, Kemp SJ, Duodo JA (2010) Lithostratigraphy, sedimentation and evolution of the Volta Basin in Ghana. Precambrian Res 183(4):701–724. https://doi.org/10.1016/j.precamres.2010.08.012
49. Anani CY, Bonsu S, Kwayisi D, Asiedu DK (2019) Geochemistry and provenance of neoproterozoic metasedimentary rocks from the Togo structural unit, Southeastern Ghana. J Afr Earth Sci 153:208–218. https://doi.org/10.1016/j.jafrearsci.2019.03.002
50. Attoh K, Nude PM (2008) Tectonic significance of carbonatite and ultrahigh-pressure rocks in the Pan-African Dahomeyide suture zone, Southeastern Ghana. SP 297(1):217–231. https://doi.org/10.1144/SP297.10
51. Darko JA, Ahenkorah I, Aakyiir MN (2019) Petrogenesis, geochemistry and structural evidence of the neoproterozoic Pan-African orogenic event in Ghana, West Africa. SN Appl Sci 1(4):1–14. https://doi.org/10.1007/s42452-019-0319-5
52. Yidana SM (2011) Hydrochemical characterization of aquifers using sequential multivariate analyses and geographic information systems in a tropical setting. J Environ Eng 137(4):258–272. https://doi.org/10.1061/(ASCE)EE.1943-7870.0000329
53. Amponsah PO, Salvi S, Béziat D, Siebenaller L, Baratoux L, Jessell MW (2015) Geology and geochemistry of the Shear-Hosted Julie Gold deposit, NW Ghana. J Afr Earth Sci 112:505–523. https://doi.org/10.1016/j.jafrearsci.2015.06.013
54. Mani R (1978) The geology of the Dahomeyan of Ghana. Geology of Ghana project. Ghana Geol Surv Bul 45
55. Antwi E (2016) Geochemical characterization and Petrogenesis of the scapolite–bearing gneisses in the Dahomeyide Suture Zone, South–Eastern Ghana. Doctoral dissertation, University of Ghana
56. Bonsu S (2015) Geochemistry of neoproterozoic phyllites from the Akuapim Range, Southeastern Ghana. Doctoral dissertation, University of Ghana

57. Leube A, Hirdes W, Mauer R (1986) The Birimian supergroup of Ghana: depositional environment, structural development and conceptual model of an Early Proterozoic suite. Tech Coop
58. Nyavor E (2016) Structural and petrologic evolution of the internal lithotectonic units of the Dahomeyide orogen in Southeastern Ghana. Doctoral dissertation, University of Ghana
59. Affaton P, Aguirre L, Ménot RP (1997) Thermal and geodynamic setting of the Buem volcanic rocks near Tiélé, Northwest Bénin, West Africa. Precambrian Res 82(3–4):191–209. https://doi.org/10.1016/S0301-9268(97)80686-9
60. Barry B, Obuobie E, Andreini M, Andah W, Pluquet M (2005) The Volta River basin. Comparative study of river basin development and management [rapport]. IWMI, CAWMA
61. Nti E (2005) Hydrochemical and isotopic characterization of groundwater in the Buem, Voltaian and Togo formations of the Volta Region, Ghana. Doctoral dissertation, University of Ghana
62. Kwayisi D, Lehmann J, Elburg M (2020) The architecture of the Buem structural unit: implications for the tectonic evolution of the Pan-African Dahomeyide orogeny, West Africa. Precambrian Res 338:105568. https://doi.org/10.1016/j.precamres.2019.105568
63. Asiedu DK, Atta-Peters D, Hegner E, Hegner A, Shibata T (2011) Palaeoclimatic control on the composition of Palaeozoic shales from Southern Ghana, West Africa. Ghana Min J 12(1). https://doi.org/10.4314/gm.v12i1.64364
64. Asare A, Appiah-Adjei EK, Ali B, Owusu-Nimo F (2021) Assessment of seawater intrusion using ionic ratios: the case of coastal communities along the Central Region of Ghana. Environ Earth Sci 80(8):1–14. https://doi.org/10.1007/s12665-021-09601-x
65. Asiedu DK, Hegner E, Rocholl A, Atta-Peters D (2005) Provenance of late Ordovician to early cretaceous sedimentary rocks from Southern Ghana, as inferred from Nd isotopes and trace elements. J Afr Earth Sci 41(4):316–328. https://doi.org/10.1016/j.jafrearsci.2005.05.003
66. Bär P, Riegel W (1980) Latest Ordovician to earliest Silurian microfloras from the lower Sekondi series of Ghana (W Africa) and their relation to those from the Itaim Formation of the Maranhao basin in NE Brazil. N Jahrb Geol Paläontol Abh 160(1):42–60 (in German with English abstract). https://doi.org/10.1127/njgpa/160/1980/42
67. Crow AT (1952) The Rocks of the Sekondi series of the gold coast (no 18). Published Under the Authority of the Government of the Gold Coast
68. Caputo MV, Crowell JC (1985) Migration of glacial centers across Gondwana during Paleozoic era. Geol Soc Am Bull 96(8):1020–1036. https://doi.org/10.1130/0016-7606(1985)96%3c1020:MOGCAG%3e2.0.CO;2
69. Anderson MM (1961) The late Palaeozoic glaciation of Gondwanaland in relation to West Africa and to northern South America during Carboniferous time. Geol Mag 98(6):449–457. https://doi.org/10.1017/S0016756800062087
70. King PL (1958) Basic palaeogeography of Gondwanaland during the Late Palaeozoic and Mesozoic eras. Q J Geol Soc 114(1–4):47–70. https://doi.org/10.1144/gsjgs.114.1.0047
71. Mensah MK, Chaloner WG (1971) Lower carboniferous lycopods from Ghana. Palaeontology 14(2):357–369
72. Adda GW (2013) The petroleum geology and prospectivity of the neo-Proterozoic, Paleozoic and Cretaceous sedimentary basins in Ghana. Search and Discov 10544
73. Dapaah-Siakwan S, Gyau-Boakye P (2000) Hydrogeologic framework and borehole yields in Ghana. Hydrogeol J 8(4):405–416. https://doi.org/10.1007/PL00010976
74. Machens E (1973) The geologic history of the marginal basins along the North Shore of the Gulf of Guinea. In: Nairn AE, Stehli FG (eds) The South Atlantic. Springer, pp 351–90. https://doi.org/10.1007/978-1-4684-3030-1_9
75. Annan W (2016) Palynological analysis of the St-5 and St-6 Wells, Offshore Tano Basin, Southwestern Ghana: implications for stratigraphy and paleoenvironment. Doctoral dissertation, University of Ghana
76. Kesse GO (1986) Oil and gas possibilities on-and offshore Ghana
77. Nazifi HM (2018) Investigation of subsurface geology in the Tano Basin of Ghana using seismic refraction method. Mühendislik Bilimleri Derg. Niğde Ömer Halisdemir Üniversitesi 7(3):1220–1224. https://doi.org/10.28948/ngumuh.502839

78. Arku B (2012) Hydrocarbon potential evaluation of shallow waters West of Accra Basin. Doctoral dissertation, University of Ghana
79. Anani CY, Anim RO, Armah BN, Atichogbe JF, Sakyi PA, Mahu E et al (2020) Petrography of detrital zircons from sandstones of the lower devonian accraian formation, SE Ghana: implications on provenance. Geol J 55(5):3716–3731. https://doi.org/10.1002/gj.3633
80. Saul JM, Boucot AJ, Finks RM (1963) Fauna of the Accraian series (Devonian of Ghana) including a revision of the gastropod Plectonotus. J Paleontol 1042–1053. Available from: https://www.jstor.org/stable/1301581
81. Anani CY, Kwayisi D, Agra NA, Asiedu DK (2018) Provenance of shales and sandstones from the Devonian Accraian group, Southern Ghana. Geosci J. 22(3):393–405. https://doi.org/10.1007/s12303-017-0066-9
82. Adeoye JA, Akande SO, Adekeye OA, Sonibare WA, Ondrak R, Dominik W et al (2020) Source rock maturity and petroleum generation in the Dahomey basin SW Nigeria: insights from geologic and geochemical modelling. J Petrol Sci Eng 195:107844. https://doi.org/10.1016/j.petrol.2020.107844
83. Kumapley NK (1989) The geology and geotechnology of the keta basin with particular reference to coastal protection. In: van der Linden WJ, Cloetingh SA, Kaasschieter JP, van de Graaff WJ, Vandenberghe J, van der Gun JA (eds) Coastal lowlands. Springer, pp 311–320. https://doi.org/10.1007/978-94-017-1064-0_24
84. Yidana SM (2010) Groundwater classification using multivariate statistical methods: Southern Ghana. J Afr Earth Sci 57(5):455–469. https://doi.org/10.1016/j.jafrearsci.2009.12.002
85. Jørgensen NO, Banoeng-Yakubo BK (2001) Environmental isotopes (18O, 2H, and 87Sr/86Sr) as a tool in groundwater investigations in the keta basin, Ghana. Hydrogeol J 9(2):190–201. https://doi.org/10.1007/s100400000122
86. Yidana SM, Chegbeleh LP (2013) The hydraulic conductivity field and groundwater flow in the unconfined aquifer system of the keta strip, Ghana. J Afr Earth Sci 86:45–52. https://doi.org/10.1016/j.jafrearsci.2013.06.009
87. Elbra A (2017) A history of gold mining in South Africa, Ghana and Tanzania. In: Governing African gold mining. Palgrave Macmillan, pp 67–103. https://doi.org/10.1057/978-1-137-56354-5_3
88. Hilson G (2002) Harvesting mineral riches: 1000 years of gold mining in Ghana. Resour Policy 28(1–2):13–26. https://doi.org/10.1016/S0301-4207(03)00002-3
89. Borquaye Y (2020) Co-management of Ghana's Gold: proposals for a policy and legislative framework to address the issue of galamsey, vol 28. NY Universidad Envtl LJ, p 361. Available from: https://heinonline.org/HOL/LandingPage?handle=hein.journals/nyuev28&div=21&id=&page=
90. Ofosu-Mensah EA (2017) Mining in Ghana and its connections with mining in the Brazilian diaspora. Extr Ind Soc 4(3):473–480. https://doi.org/10.1016/j.exis.2017.05.010
91. Ofosu-Mensah AE (2013) Traditional gold mining in the Akan States of Ghana. In: Proceedings of the XV balkan mineral processing congress. University of Mining and Geology; St. Rilski I. Publishing House, pp 1265–1269
92. Bosman W (1967) A new and accurate description of the Gold Coast of guinea. Frank Cass, pp 26–34. Available from: https://library.wur.nl/WebQuery/titel/368156
93. Arhin K (1978) Gold-mining and trading among the Ashanti of Ghana. J Afr 48(1):89–100. https://doi.org/10.3406/jafr.1978.1806
94. Kumah D, Nyarko EA (2018) Gold mining and its effects through the lens of an archaeologist: experiences from the Prestea area, southwestern Ghana. West Afr J Appl Ecol 26:133–148
95. Ferguson GE (1974) The papers of George Ekem Ferguson: a Fanti official of the Government of the Gold Coast, 1890–1897; edited with Introd. and notes by Kwame Arhin, vol 7. Afrika-Studiecentrum
96. Miller FP, Vandome AF, McBrewster J (2009) History of Ghana. Alphascript Publishing, Saarbrücken, Germany et al
97. Klein HS (2010) The Atlantic slave trade. Cambridge University Press. https://doi.org/10.1017/CBO9780511779473

98. Addy SNG (1998) Ghana: revival of the mineral sector. Resour Policy 24(4):229–239. https://doi.org/10.1016/S0301-4207(98)00035-X
99. Nugent P (2021) Nationalism and decolonization in the Gold Coast. In: Oxford research encyclopedia of African history. Oxford University Press. https://doi.org/10.1093/acrefore/9780190277734.013.616
100. Jedwab R, Moradi A (2012) Colonial investments and long-term development in Africa: evidence from Ghanaian railways [unpublished manuscript]. George Washington University and Sussex University
101. Ofosu-Mensah EA (2021) Mining (ASM) in Ghana: from precolonial. J Afr Pol Stud 27(1)
102. Akabzaa T, Darimani A (2001) Impact of mining sector investment in Ghana: a study of the Tarkwa mining region. Third World Netw 11(2):47–61
103. Jackson R (1992) New mines for old gold: Ghana's changing mining industry. Geography 77(2):175–178. https://doi.org/10.1080/20436564.1992.12220503
104. Hilson GM (2004) Structural adjustment in Ghana: assessing the impacts of mining-sector reform. Afr Today 53–77
105. Opoku-Dapaah E, Boko SH (2010) Liberalization of the mining sector in Ghana and its impact on the economy: a sociological approach. Back Track Sect-Led Growth Afr Implic Dev 229
106. Worlanyo AS, Alhassan SI, Jiangfeng L (2022) The impacts of gold mining on the welfare of local farmers in Asutifi-North District in Ghana: a quantitative and multi-dimensional approach. Resour Policy 75:102458. https://doi.org/10.1016/j.resourpol.2021.102458
107. Bloch R, Owusu G (2012) Linkages in Ghana's gold mining industry: challenging the enclave thesis [enclave thesis]. Resour Policy 37(4):434–442. https://doi.org/10.1016/j.resourpol.2012.06.004
108. Ghana chamber of mines. Report on the performance of the mining industry in Ghana (2014). Accra, Ghana 2015
109. Owusu O, Wireko I, Mensah AK (2016) The performance of the mining sector in Ghana: a decomposition analysis of the relative contribution of price and output to revenue growth. Resour Policy 50:214–223. https://doi.org/10.1016/j.resourpol.2016.10.006
110. Ntiamoah-Agyakwa Y (1979) Relationship between gold and manganese mineralizations in the Birimian of Ghana, West Africa. Geol Mag 116(5):345–352. https://doi.org/10.1017/S0016756800043983
111. Hirdes W, Senger R, Adjei J, Efa E, Loh G, Tettey A (1993) Explanatory notes for the geological map of Southwest Ghana 1: 100,000: sheets Wiawso. Asafo. 1993. Kukuom (0603B), Goaso (0603A), Sunyani (0703D) and Berekum (0703C); with 18 tables (p 139). Schweizerbart. Available from: https://library.wur.nl/WebQuery/titel/595812
112. Allibone A, Teasdale J, Cameron G, Etheridge M, Uttley P, Soboh A et al (2002) Timing and structural controls on gold mineralization at the Bogoso Gold Mine, Ghana, West Africa. Econ Geol 97(5):949–969. https://doi.org/10.2113/gsecongeo.97.5.949
113. Perrouty S, Aillères L, Jessell MW, Baratoux L, Bourassa Y, Crawford B (2012) Revised eburnean geodynamic evolution of the gold-rich Southern Ashanti belt, Ghana, with new field and geophysical evidence of pre-Tarkwaian deformations. Precambrian Res 204–205:12–39. https://doi.org/10.1016/j.precamres.2012.01.003
114. Perrouty S, Lindsay MD, Jessell MW, Aillères L, Martin R, Bourassa Y (2014) 3D modeling of the Ashanti Belt, Southwest Ghana: evidence for a litho-stratigraphic control on gold occurrences within the Birimian Sefwi group. Ore Geol Rev 63:252–264. https://doi.org/10.1016/j.oregeorev.2014.05.011
115. Agbenyezi TK, Adomako-Ansah K, Foli G, Sunkari ED, Bayari EE, Gawu SK (2024) A review of intrusion-hosted gold systems of the Palaeoproterozoic Birimian terrane of Ghana. Int J Earth Sci 113(7):1551–1570
116. Perrouty S, Jessell MW, Bourassa Y, Miller J, Apau D, Parra-Avila LA et al (2016) Geological setting of the Wassa gold deposit, SW Ghana. Ore Geol Rev 78:687–691. https://doi.org/10.1016/j.oregeorev.2015.09.024
117. Perrouty S, Jessell MW, Bourassa Y, Miller J, Apau D, Siebenaller L et al (2015) The Wassa deposit: a poly-deformed orogenic gold system in Southwest Ghana–implications for regional exploration. J Afr Earth Sci 112:536–547. https://doi.org/10.1016/j.jafrearsci.2015.03.003

118. Forson ED, Amponsah PO, Wemegah DD, Ahwireng MD (2024) Random forest-based mineral prospectivity modelling over the Southern Kibi-Winneba belt of Ghana using geophysical and remote sensing techniques. Appl Earth Sci 133(1):30–45
119. Sunkari ED, Oppong O, Agbenyezi TK (2024) Petrogenetic and geochemical characteristics of some auriferous granitoids in the Kumasi Basin, Ghana: implications for geodynamic settings and controls of orogenic gold mineralization in the Edikan Gold Mine. Deep Undergr Sci Eng
120. Appiah H, Norman DI, Boadi I (1991) The geology of the Prestea and Ashanti goldfields: a comparative study. In: Symposium Brazil Gold'91. pp 247–55
121. Oberthür T, Vetter U, Schmidt Mumm A, Weiser T, Amanor JA, Gyapong WA et al (1994) The Ashanti gold mine at Obuasi in Ghana: mineralogical, geochemical, stable isotope and fluid inclusion studies on the metallogenesis of the deposit. Geol Jahrb D 100:31–130
122. Tetteh AT, Issaka Y, Dodoo P, Atugba B, Tengnibuor A (2025) Assessing the impact of Ghana's geology on gold mining using geospatial techniques. Heliyon
123. Mumin AH, Fleet ME, Chryssoulis SL (1994) Gold mineralization in as-rich mesothermal gold ores of the Bogosu-Prestea mining district of the Ashanti Gold Belt, Ghana: remobilization of "invisible" Gold. Mineral Deposita 29(6):445–460. https://doi.org/10.1007/BF00193506
124. Graham V, Tetteh GM (2024) Petrology and geochemical relationship of Birimian Gabbro-Diorite-Granodiorite Dykes in Basaltic Rocks at Butre Area, SW Ghana. J Geosci Environ Prot 12(2):107–130
125. Adams SJ, Sunkari ED, Hanson JE, Tandoh KK, Tetteh SE, Nude PM, Asiedu DK, Dampare SB (2024) Geochemistry of the Winneba-Mankoadze pegmatites in southern Ghana: a clue to the petrogenesis of the pegmatites. Ore Energy Resour Geol 17:100051
126. Mumin AH, Fleet ME (1995) Evolution of gold mineralization in the Ashanti Gold Belt, Ghana: evidence from carbonate compositions and parageneses. Mineral Petrol 55(4):265–280. https://doi.org/10.1007/BF01165121
127. Klemd R, Hünken U, Olesch M (1996) Fluid composition and source of Early Proterozoic lode gold deposits of the Birimian volcanic belt, West Africa. Int Geol Rev 38(1):22–32. https://doi.org/10.1080/00206819709465321
128. Mumm AS, Oberthür T, Vetter U, Blenkinsop TG (1997) High CO_2 content of fluid inclusions in gold mineralisations in the Ashanti Belt, Ghana: a new category of ore forming fluids? Mineral Deposita 32(2):107–118. https://doi.org/10.1007/s001260050078
129. Junner NR (1935) Gold in the Gold Coast: Gold Coast Geol Surv Mem p 4
130. Yao Y, Robb LJ (2000) Gold mineralization in palaeoproterozoic granitoids at Obuasi, Ashanti region, Ghana: ore geology, geochemistry and fluid characteristics. S Afr J Geol 103(3–4):255–278. https://doi.org/10.2113/1030255
131. Shenberger DM, Barnes HL (1989) Solubility of gold in aqueous sulfide solutions from 150 to 350 C. Geochim Cosmochim Acta 53(2):269–278. https://doi.org/10.1016/0016-7037(89)90379-7
132. Hayashi KI, Ohmoto H (1991) Solubility of gold in NaCl-and H2S-bearing aqueous solutions at 250–350 °C. Geochim Cosmochim Acta 55(8):2111–2126. https://doi.org/10.1016/0016-7037(91)90091-I
133. Bowers TS (1991) The deposition of Gold and other metals: pressure-induced fluid immiscibility and associated stable isotope signatures. Geochim Cosmochim Acta 55(9):2417–2434. https://doi.org/10.1016/0016-7037(91)90363-A
134. Craw D, Teagle DA, Belocky R (1993) Fluid immiscibility in late-alpine gold-bearing veins, eastern and northwestern European alps. Mineral Deposita 28(1):28–36. https://doi.org/10.1007/BF00199007
135. John T, Klemd R, Hirdes W, Loh G (1999) The metamorphic evolution of the Paleoproterozoic (Birimian) volcanic Ashanti Belt (Ghana, West Africa). Precambrian Res 98(1–2):11–30. https://doi.org/10.1016/S0301-9268(99)00024-8
136. Sunkari ED, Gan AN, Fiadonu EB (2024) Geochemical vectors for targeting hydrothermal gold mineralization in the Damang area, southwestern Ghana: constraints from petrography,

multi-element geochemistry and multivariate statistical analysis. Ore Energy Resour Geol 100071. https://doi.org/10.1016/j.oreoa.2024.100071

137. Sestini G (1971) Sedimentology of a paleoplacer: the gold-bearing Tarkwaian of Ghana. In: Amstutz GC, Bernard AJ (eds) Ores in sediments: VIII, August 31, 1973. Heidelberg: International sedimentological congress. Springer, pp 275–305. https://doi.org/10.1007/978-3-642-65329-2_21
138. Strogen P (1991) The sedimentology, stratigraphy and structure of the Tarkwaian, western region, and its relevance to gold exploration and development. In: Proceedings of the international conference on the geology of Ghana, vol 3. Geological Society of Ghana
139. Hall AM, Thomas MF, Thorp MB (1985) Late quaternary alluvial placer development in the humid tropics: the case of the Birim diamond Placer, Ghana. J Geol Soc 142(5):777–787. https://doi.org/10.1144/gsjgs.142.5.0777
140. Macdonald G (1902) Gold in West Africa. Afr Aff 1:416–30. https://doi.org/10.1093/oxfordjournals.afraf.a093156
141. Dei LA (1972) The central coastal plains of Ghana: a morphological and sedimentological study. Geomorphology 16(4):415–431. https://doi.org/10.1127/zfg/16/1972/415
142. Amponsah PO, Salvi S, Béziat D, Baratoux L, Siebenaller L, Nude PM et al (2016) The Bepkong gold deposit, Northwestern Ghana. Ore Geol Rev 78:718–723. https://doi.org/10.1016/j.oregeorev.2015.06.022
143. Parra-Avila LA, Bourassa Y, Miller J, Perrouty S, Fiorentini ML, Campbell McCuaig TC (2015) Age constraints of the Wassa and Benso mesothermal gold deposits, Ashanti Belt, Ghana, West Africa. J Afr Earth Sci 112:524–535. https://doi.org/10.1016/j.jafrearsci.2015.05.017
144. Oberthür T, Weiser T, Amanor JA, Chryssoulis SL (1997) Mineralogical siting and distribution of Gold in quartz veins and sulfide ores of the Ashanti mine and other deposits in the Ashanti Belt of Ghana: genetic implications. Mineral Deposita 32(1):2–15. https://doi.org/10.1007/s001260050068
145. Tunks AJ, Selley D, Rogers JR, Brabham G (2004) Vein mineralization at the damang gold mine, Ghana: controls on mineralization. J Struct Geol 26(6–7):1257–1273. https://doi.org/10.1016/j.jsg.2003.11.005
146. Bowell RJ (1992) Supergene gold mineralogy at Ashanti, Ghana: implications for the supergene behaviour of gold. Mineral Mag 56(385):545–560. https://doi.org/10.1180/minmag.1992.056.385.10
147. Osae S, Kase K, Yamamoto M (1999) Ore mineralogy and mineral chemistry of the Ashanti gold deposit at Obuasi, Ghana. Resour Geol 49(1):1–11. https://doi.org/10.1111/j.1751-3928.1999.tb00027.x
148. Sunkari ED, Appiah-Twum M, Lermi A (2019) Spatial distribution and trace element geochemistry of laterites in Kunche area: implication for gold exploration targets in NW, Ghana. J Afr Earth Sci 158:103519. https://doi.org/10.1016/j.jafrearsci.2019.103519

Structural Controls on Gold Deposits in Ghana

Structural Controls on Orogenic Gold Mineralization: Evidence from Major Operating Mines in Ghana

Samuel Kwame Atta

1 Introduction

Orogenic gold deposits encompass all quartz-carbonate lode-gold deposits formed approximately 5–15 km deep indicating shallow to mid crust and usually within metamorphic terrains [1]. Forty percent of the world's gold production emanates from orogenic gold deposits [2]. These deposits are associated with metamorphic complexes of different ages in orogenic belts and most often have no relation to magmatism. The metamorphic model for orogenic gold ore formation postulates that gold-bearing aqueous-carbonic fluids which are integral products of devolatilization were deposited within seismogenic zones across greenschist-amphibolite boundaries [2]. These deposits usually have close geographical and time-related associations with collisional orogens or convergent margins. These collisions were preceded by synchronous oceanic arc or ridge volcanism and sedimentation [3]. Orogenic gold deposits are mostly found in areas not far from first-order transcrustal structures bordering tectonically adjacent metamorphic supracrustal sequences of all ages [4]. The observed relations between orogenic gold deposits and major structures demonstrate their importance in providing pathways for fluid deposition in the upper crust. McCuaig and Kerrich [4] have observed that the first-order geological structures with favourable prospects display the following features: (1) They are inosculating shear zones of brittle-ductile deformation. (2) They are segmented but several hundreds of kilometres long. (3) Double plunging folds and mega-boudinages are prevalent. (4) Both lineation and foliation textures are present, and (5) The country rocks have undergone metamorphism.

These first-order transcrustal faults are generally concordant having a deformational history that includes both shortening and high-angle reverse movements. The reverse movements exposed these faults to high fluid flux during compression as

S. K. Atta (✉)
Ghana Geological Survey Authority, Accra, Ghana
e-mail: sammy_atta@yahoo.com

E. D. Sunkari and S. Ali (eds.), *Geology and Gold Mineralization in Ghana*, Earth and Environmental Sciences Library, https://doi.org/10.1007/978-3-031-97163-1_2

their misorientation increased fluid pressure. Interestingly, these major first-order crustal-scale structures, despite having spatial relations with orogenic gold deposits, most often do not host the mineralization [4]. The mineralization is rather hosted in second or third-order structures, which are offshoots of the major first-order structure. According to McCuaig and Kerrich [4], this phenomenon is a result of the following: (1) Temperature gradient that occurred distal from the first-order structure. (2) The plasticity of first-order structures compared to the brittleness of second and third-order structures leading to greater fluid pressure fluctuations. (3) Low mean stress in rocks adjacent to first-order structures resulting in higher permeability and fluid flux. Typical quartz vein systems of orogenic gold deposits are predominantly composed of quartz with about 5% sulphide content and up to 15% carbonate minerals [5].

The primary aim of this study is to prove that orogenic gold mineralization in Ghana is primarily structurally controlled. These findings emanate exclusively from data and information obtained from gold deposits currently being exploited by major operating mines. Open pits and underground mine face provide excellent subsurface exposures and access to fresh rocks as well as in situ information which are often lacking at the surface in typical tropical rainforest climatic regions characterised by deep weathering and laterisation. This study further examined the defining structures influencing gold mineralization at the mines and their implication for enhancing exploration success.

2 The Paleoproterozoic Terrains of Ghana

The Birimian Supergroup and the coterminous Tarkwaian Group are Paleoproterozoic sequences in Ghana within which the important gold deposits are hosted [6]. The Birimian Supergroup encompasses metavolcanic and metasedimentary rocks on which the Tarkwaian Group sedimentary rocks overlie [7, 8]. Several generations of granitic plutons are present within the Birimian Supergroup and Tarkwaian Group [7, 9]. These rocks have undergone successions of deformations resulting in regional greenschist to amphibolite facies metamorphism [10–13]. The Birimian terrain of Ghana is distinguished by predominant northeast–southwest trending metavolcanic belts and intervening adjacent metasedimentary basins (Fig. 1). The Kibi-Winneba, Ashanti, Sefwi, Bui, Bole-Navrongo and Wa-Lawra are the notable volcanic belts while the basins are the Cape-Coast, Kumasi, Sunyani and Maluwe. The belts are underlain by basalts, andesites, gabbros and some ultramafic rocks [14–16]. Conversely, the basin sedimentary rocks consist of greywackes, siltstones, and shales interbedded with volcaniclastic rocks and locally graphite-rich phyllites [13]. Chemical sedimentary rocks such as chert, carbonates, manganese-rich sediments, and graphite-rich rocks occur at the belt-basin transition zones [7].

The coterminous Tarkwaian Group forms an unconformity with the Birimian Supergroup. The Group is made up of conglomerates, sandstones and phyllites [17]. Occupying the bottom of the Tarkwaian stratigraphy are ~ 700 m of conglomerates and sandstones of the Kawere unit (Table 1). Overlying the Kawere rocks are the

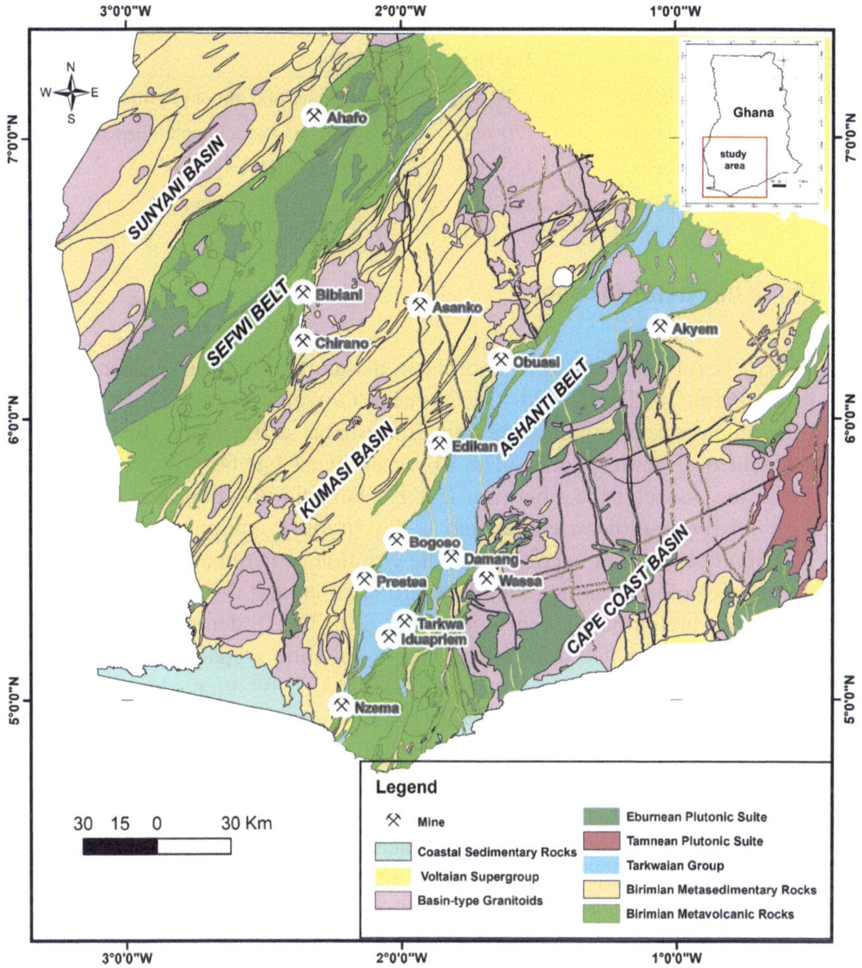

Fig. 1 Geology of south western Ghana indicating the location of all operating large-scale mines. *Source* Ghana Geological Survey Authority

approximately 600 m thick gold-bearing Banket conglomerates. The 400 m thick Tarkwa phyllites which occur above the Banket conglomerates also underlie the about 1300 m thick Huni sandstones. Hornblende-bearing Dixcove or belt type granitoids occur within the Birimian volcanic rocks while the two-mica Cape Coast and Winneba or basin-type granitoids are restricted to the Birimian sedimentary basins [18]. The basin-type granitoids locally grades into migmatites [7, 12]. The belt-type granitoid plutons consist of tonalite-trondhjemite-granodiorite rocks [TTGs]. Most of the granitic plutons in Ghana were emplaced during the Eburnean event at ~ 2100 Ma [7, 18, 19].

Table 1 Stratigraphy of the Tarkwaian Group (modified after Strogen [17])

Group	Series	Thickness (m)	Lithology
Tarkwaian	Huni sandstone	1370	Sandstones and minor phyllites
	Tarkwa phyllite	120–400	Chloritic and sericitic phyllites and schists
	Banket series	120–600	Quartzites and conglomerates
	Kawere conglomerate	250–700	Quartzites and conglomerates

3 The Eburnean Orogeny

Magmatism, deformation and metamorphism define the Eburnean orogenic events [12] in the Birimian. The 2.1–2.0 Ga Eburnean Orogeny resulted in the accretion of the Birimian volcanic belts, the formation of the sedimentary basins and the emplacement of granitic plutons. The Eburnean event has been divided into two distinct cycles, namely; the Eburnean I or Eoeburnean and Eburnean II [19, 20] based on existing radiometric ages and U–Pb zircon geochronology [7, 18, 21–24]. The Eburnean I event encompassed the eruption of Birimian mafic extrusives, emplacement of older granitic plutons, uplift, erosion, and deposition of Birimian sediments and a preliminary D_1-metamorphism [20]. The D_1 deformation caused N–S shortening which resulted in E–W oriented S_1-foliation subparallel to S_0-bedding observed in mica schists within the metavolcanics [13]. E–W oriented foliations have also been recognised in graphitic phyllites within the metasedimentary rocks [25]. The age of the Eburnean I event has been constrained to between 2195 and 2148 Ma. Nd-Sm isochron analysis of metamorphic garnet also produced an Eburnean-I metamorphism age of 2153 ± 13 Ma [23, 25]. Eburnean II, on the other hand, occurred between 2116 and 2088 Ma [20]. The Eburnean II orogeny was also characterised by deformation, metamorphism and granitic intrusions involving both Birimian and Tarkwaian rocks [20]. This event caused a NE–SW shortening producing subvertical crenulation cleavages and km-scale isoclinal folds with subvertical axial surfaces [13]. This Eburnean event included syn-kinematic emplacement of intrusives, D_2 and D_3 transcurrent deformation, metamorphism at ~ 2–3 kb [~ 6 km]/200–300 °C as well as spikes in gold mineralization [13]. The maximum observed regional metamorphism is generally up to upper greenschist facies. However, amphibolite facies metamorphism has been noticed at the contacts with late Eburnean granitoids and dolerite dykes where minerals such as kyanite and sillimanite in the Bonsa diamond field, garnet at the Nsuta manganese deposit, garnet-staurolite assemblage at the Damang gold mine serve as proofs [13, 26, 27]. Free gold mineralization in quartz veins and refractory gold encapsulated in arsenopyrite and/or pyrite formed between 2110 and 2095 Ma [20]. They mostly occur in bends and intersections within shear zones, minor faults, folds, etc. and are associated with carbonate, sericite and chlorite halos as well as albite alteration.

4 Local-Scale Structural Controls

Ahafo Gold Mine

The ~ 15-Moz Ahafo South gold deposits exploited at the Ahafo mine are located both close to and on the major tectonostratigraphic boundary where the Sefwi granite-greenstone belt of the Paleoproterozoic Birimian Supergroup borders the adjacent metamorphosed volcano-sedimentary rocks of the Sunyani-Comoé basin (Fig. 1). These deposits occur along a distance of almost 15 km on the Kenyase-Yamfo shear zone. All the deposits are confined by low-displacement, disjointed sinistral north-striking tear faults oblique to the Kenyase-Yamfo shear zone. The Ahafo South deposits encompass the Subika, Apensu, Awonsu, and Amoma deposits. While the Subika gold deposit is hosted entirely in dioritic plutons, granodiorite plutons act as hanging wall for the Awonsu, Apensu, and Amoma deposits with the volcano-sedimentary rocks as footwall (Fig. 2). The bulk of the gold mineralization at Subika is confined within the Subika magic fracture zone. The high-grade [> 3 g/t] ore lenses within the fracture zone are severally stacked laterally in quartz-carbonate veins separated by 2–5 m thick waste zones [28].

The most important geological structures to gold mineralization at the Ahafo mine are a number of N- to NNE-striking sinistral faults obliquely cross-cutting the belt-basin boundary. These discontinuities have segmented the boundary of the Sefwi belt into three sections and they all display significant offsets. These faults, which formed during the last stages of the Eburnean orogeny, formed oblique intersections because they behaved like tear faults. The Ahafo South gold deposits occur in a contractional jog setting with mineralization being hosted in smaller scale restraining jogs [29]. The local bends in the Kenyase-Yamfo fault system were caused by N- to NNE-striking tear faulting which also resulted in block rotation and competent rock mass failure [28]. Furthermore, NW–SE shortening caused sinistral displacement along the tear faults as well as reactivation and sinistral-reverse transpression along the belt-basin contact [12]. Boundaries between rigid intrusive bodies and mafic dykes in the hanging wall of the Kenyase-Yamfo shear zone represented sites of heterogeneous stress and inhomogeneous strain and hence depositional sites for fluids.

The deformational history at Ahafo, according to Masurel et al. [28], includes an initial D_{1Ah} deformational event which caused regional NNW–SSE bulk crustal shortening. This was followed by a D_{2Ah} WSW–ENE normal sinistral transtension during which high-grade rocks were focused into the relay zone between the Ketesso and Kenyase-Yamfo shear zones. Following afterwards was a transpressional tectonic event (D_{3Ah}) characterised by E–W shortening under regional greenschist facies metamorphic conditions. A late Eburnean transpressional deformational event D_{4Ah} at Ahafo resulted in sinistral-reverse displacement along the Kenyase-Yamfo shear zone which manifested as crosscutting N- to NNE-striking discontinuous sinistral faults. Finally, a D_{5Ah} deformational event developed ENE-striking dextral and NW-striking sinistral faults.

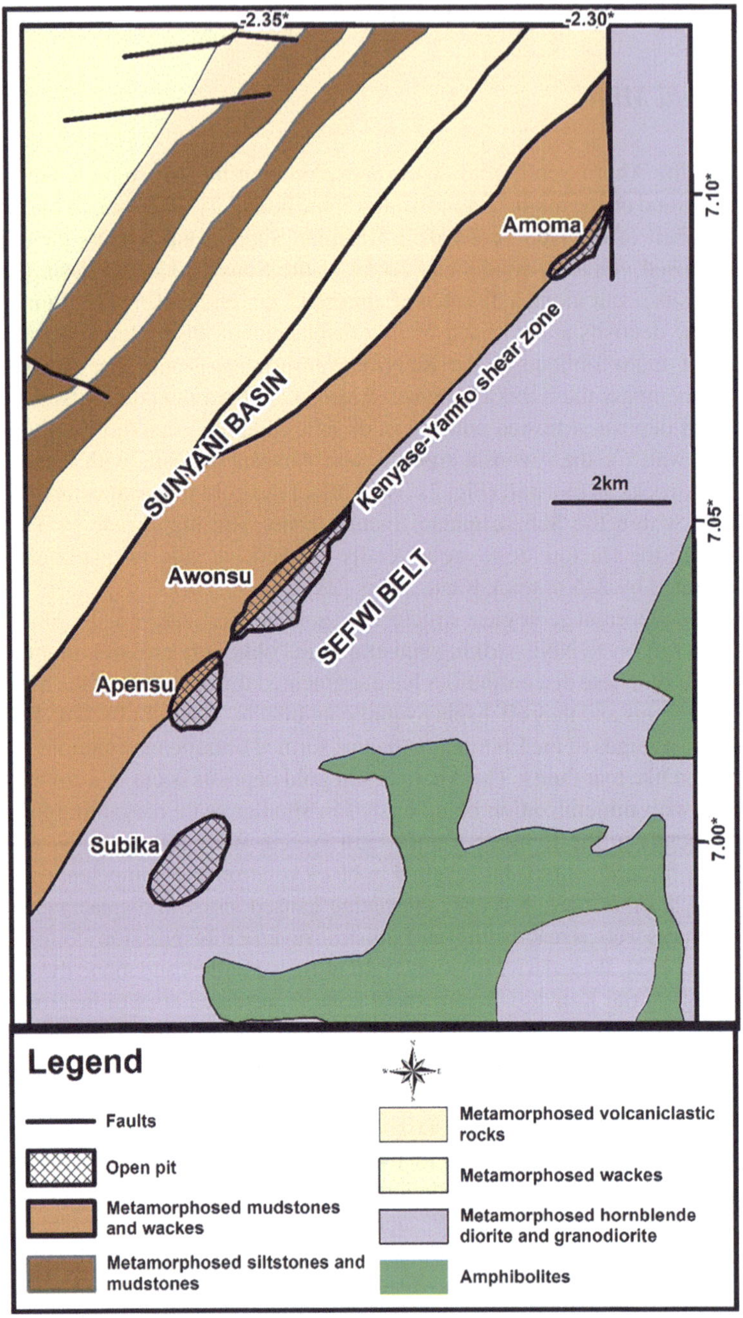

Fig. 2 Simplified geology of the Ahafo South gold deposits. Modified after Masurel et al. [28]

Wassa Gold Mine

The Wassa gold deposit which is exploited by the Wassa mine is located on the eastern side of the Ashanti volcanic belt (Fig. 1). The deposit is hosted in a sequence of metamorphosed basalts and andesites intercalated with metagreywackes, magnetite-rich argillites and rare black shale layers. The rocks are highly strained as a result of four observed folding events. Isoclinal F_{1W} folds and S_{1W} foliation manifested by chlorite, mica, carbonate and gold-bearing pyrite originated from the first D_{1W} deformational event. D_{2W} and D_{3W} deformational events also produced tight folds and crenulation cleavages [30]. The late D_{4w} deformational event resulted in the formation of crenulation cleavages and open to tight recumbent folds. The defining structure of the Wassa deposit is a kilometric-scale asymmetric synformal F_{3W} fold in which mineralization extends along both the NE–SW sub-vertical long limb and the E–W shallow S-dipping short limb (Fig. 3). Mineralization occurs within M_{1W}-related quartz veins with the highest grades seemingly occurring at the hinge of the F_{3W} fold [30]. S_{1W} stretched and aligned pyrites at the Wassa mine host tiny gold particles which formed during the primary gold mineralization event (M_{1W}) that characterised the first deformational event (D_{1W}). Secondary minerals present include ilmenite, brookite (polymorph of rutile) and rare tourmaline. The Wassa deposit was again locally remobilised with gold concentrating along late structural fabrics and veins such as the axial surfaces of late F_{4W} recumbent folds. The remobilised gold is associated with euhedral pyrite and minor chalcopyrite, arsenopyrite, pyrrhotite and sphalerite minerals.

Three generations of quartz veins occur in this mine. The first-generation veins are smoky to translucent and are associated with tourmaline, sericite, sulphides (primarily pyrite) and carbonate selvages [30]. These quartz veins have the highest gold grades found in the mine and were deformed by F_{3W} folding. The second-generation quartz veins are also smoky to translucent and contains anomalous gold values [30]. These veins are sub-parallel to the S_{3W} crenulation cleavage and slightly deformed by open F_{4W} folding. The third-generation quartz-carbonate veins are milky to pink and are associated with pyrite, carbonate and potassic alterations. These veins are either sub-parallel to S_{4W} crenulation cleavage or to S_{1W} foliation and cross-cut all four folds.

Edikan Gold Mine

The > 9 Moz Edikan gold mine is spatially associated with the Akropong fault zone in the metasedimentary successions of the Kumasi Basin (Fig. 1). These sedimentary rocks were intruded by several generations of granitoids, most of which were emplaced during the Eburnean orogeny between ca. 2115 and 2080 Ma [31]. The Esuajah North granitoid at Edikan, for example, crystallised at 2105 ± 3 Ma [22]. The

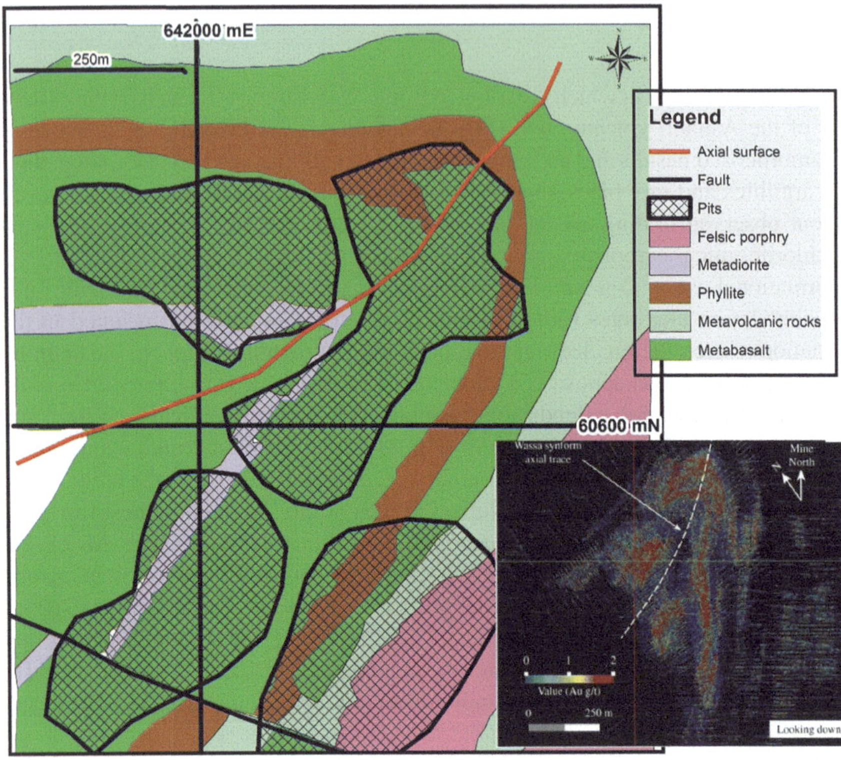

Fig. 3 Outlines of the open pits draped over the geology of the Wassa mine showing the presence of a northeast-striking kilometre-scale fold. (insert) Gold assay data visualised in Leapfrog 3D. Modified after Perrouty [30]

long axes of these granitic plutons are generally subparallel to the penetrative NE–SW striking structures observed in the metasedimentary country rocks. The Fetish, Bokitsi, and Chirawewa gold mineralization occur either within faults in the metasedimentary rocks or as veins within the granitoids. The Esuajah North and South gold mineralization occur in cylindrical granitoid plugs while the AG2-AG3-Fobinso gold deposits are hosted entirely in a granodiorite dyke (Fig. 4). Gold mineralization in the granitoids is characterised by disseminated sulphides and quartz-carbonate veins. While gold distribution is disseminated within the granitic host, nuggets are present in the fault-hosted mineralization.

According to Tourigny et al. [31], the preliminary D_{1Edk} convergence event produced bedding-parallel shearing and rarely preserved S_{1Edk} foliations in the hinges of west-striking recumbent folds F_{1Edk}. The granitoid-hosted gold-bearing quartz veins are presumed to have formed during the D_{3Edk} deformational event which followed the D_{2Edk} deformation. These quartz veins developed both within and adjacent to low-angle cross-cutting thrust faults. These N–S and NW–SE striking D_{3Edk} faults are perpendicular to lithological contacts and earlier formed steeply dipping

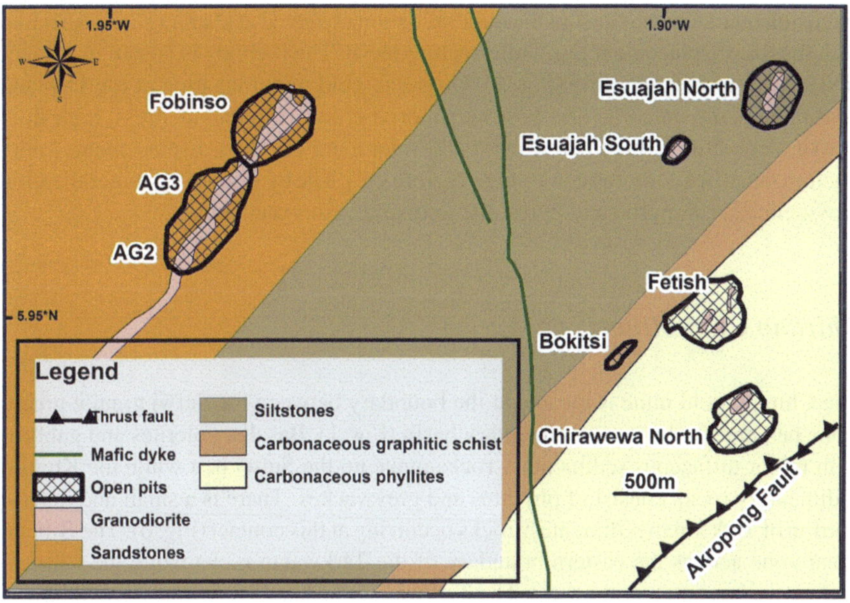

Fig. 4 Simplified geology of the Edikan mine. Modified after Tourigny et al. [31]

D_{2Edk} faults [31]. These mineralized D_{3Edk} low angle thrust fault system resulted from WNW–ESE compression that progressed into a strike-slip event which included both transpressional and transtensional events. Restraining jogs formed during the reactivation of reverse and sinistral strike-slip faults served as depositional sites for shallow and steeply dipping gold-quartz veins. Almost all the gold mineralization at Edikan occur either in quartz veins occupying low-angle thrust faults or shallow dipping oblique-extension veins within granitoids [31].

Damang Gold Mine

Unique to the Damang gold mine is the occurrence of two dissimilar styles of gold mineralization within the metasedimentary Tarkwaian Group (Fig. 1). Commonly associated with the quartz–lithic conglomerates of the Banket Series are stratabound palaeoplacer mineralization. The orogenic mineralization within these rocks resulted from a late-forming extensive low-displacement, fault–fracture mesh which formed during a compressional stress regime. Tunks et al. [32] showed that regional deformation in the Tarkwaian included an initial NW–SE shortening (D_{1D}) followed by N–S shortening (D_{2D}) resulting in the formation of E–W trending thrust faults with slight displacements. Steep D_{1D} faults were reactivated during D_{3D} shortening as well as the formation of new sets of low-angle thrust faults and associated flat-lying extension quartz veins. Most of the mineralization observed at Damang are associated with the low displacement D_{3D} fault fracture mesh. The orebody is hosted in a major NNE-trending F_{1D} anticline (Fig. 5). Orogenic gold mineralization at the Damang gold mine is the outcome of a post metamorphic, low differential stress, high fluid pressure condition which permitted the development of a low-displacement, fault–fracture mesh in a compressional stress regime [32]. The hydrothermal mineralization is associated with pyrite and pyrrhotite in quartz carbonate veins.

Chirano Gold Mine

The Chirano gold mine is located at the boundary between the Sefwi granite greenstone belt and the Kumasi sedimentary basin (Fig. 1). Basalts, dolerites and gabbros with minor tuffaceous sedimentary rocks make up the Sefwi belt while the Kumasi sedimentary rocks consist of phyllites and greywackes. There is a small and narrow portion of Tarkwaian sedimentary rocks occurring at this contact (Fig. 6). The Bibiani shear zone acts as the eastern boundary of the Tarkwaian rocks while the Chirano shear zone borders it at the west. The Tarkwaian rocks are composed of conglomerates, sandstones, and thin mudstone beds. Granitoid batholiths occur to the east and west of the Chirano shear zone. Occurring within this fault zone are smaller tonalite bodies and dolerite dykes [33]. The basalts, dolerites and tonalites have all been metamorphosed to greenschist facies assemblages [33].

The Chirano gold deposits are located mainly along the Chirano shear zone (CSZ) and another parallel structure 200 m to the west within mafic rocks. Open pit exposures indicate that mineralization is hosted mostly within strongly hydrothermally altered metabasalts and metaandesites. The Chirano shear zones resulted from folding and thrusting related to NW–SE shortening. This progressive deformation resulted in gentle folding of the Tarkwaian rocks into S- to SW-plunging folds and have also physically juxtaposed the Tarkwaian sedimentary rocks adjacent to Birimian mafic rocks (Fig. 6). Steeply plunging drag folds adjacent to the Chirano shear zone further orientated the folded Tarkwaian rocks. During the continuous shortening, pre-existing fault zones were reactivated in addition to newly developed faults zones. These structural discontinuities acted as conduits for granitic intrusions, associated quartz-albite alteration and later on for gold-bearing hydrothermal fluids [33].

According to Ackun-Wood et al. [34], the chronology of deformational events at Chirano began with a preliminary deformational event (D_{1C}) characterised by folding and thrusting followed by a secondary D_{2C} event which produced gentle N–S plunging open folds. The third deformational event (D_{3C}) resulted in the formation of steeply-dipping NNW-striking faults. These faults were intruded by tonalites and subsequently by hydrothermal fluids during which gold was precipitated. A fourth deformational event (D_{4C}) resulted in NE trending crosscut faults. The Chirano deposits lie adjacent to the Chirano shear zone except at the southernmost extent where the Akwaaba deposit lies directly on the shear zone (Fig. 6). Seven deposits (Mamnao, Sariehu, Obra, Tano, Paboase, Suraw and Akoti South) occur along a

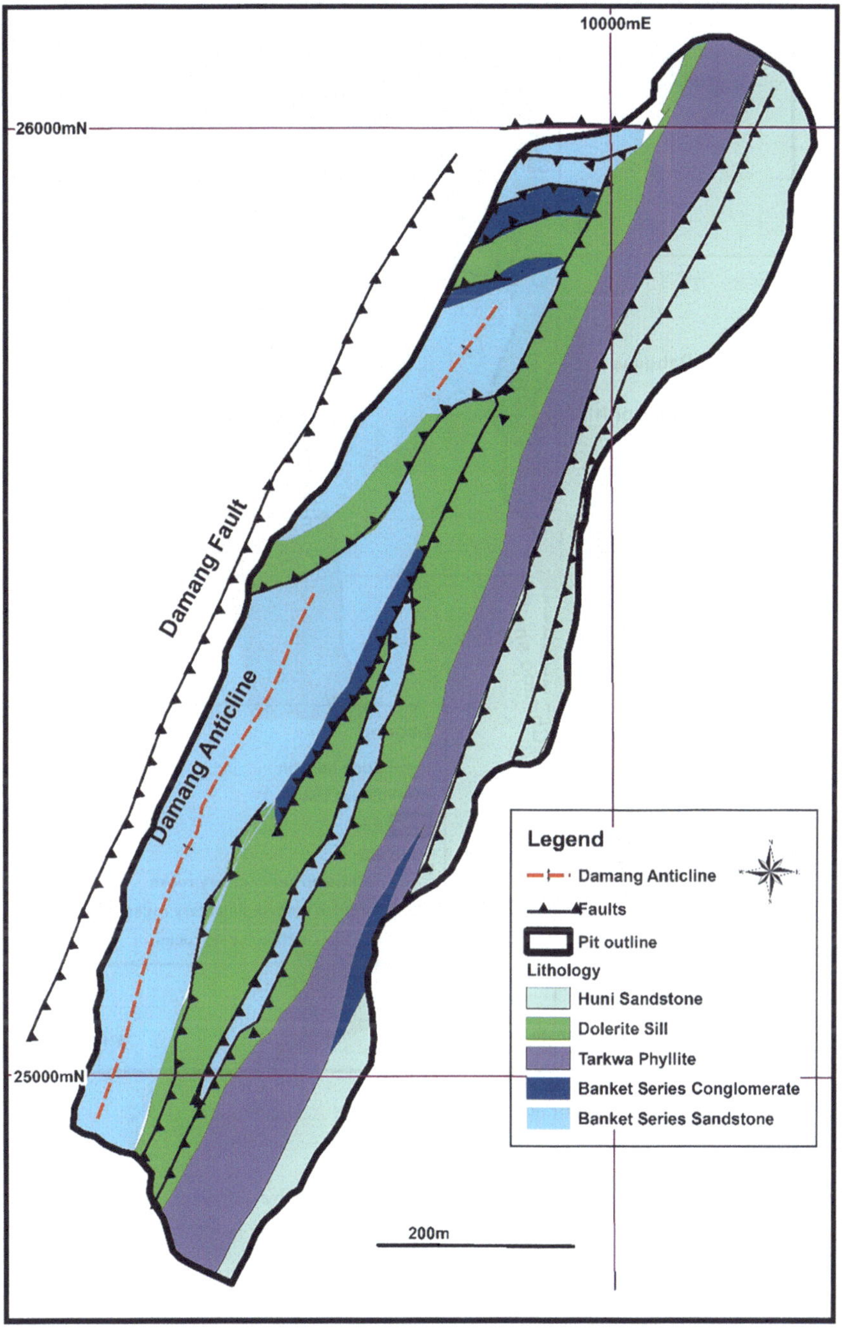

Fig. 5 Simplified geology of the main Damang pit. Modified after Tunks et al. [32]

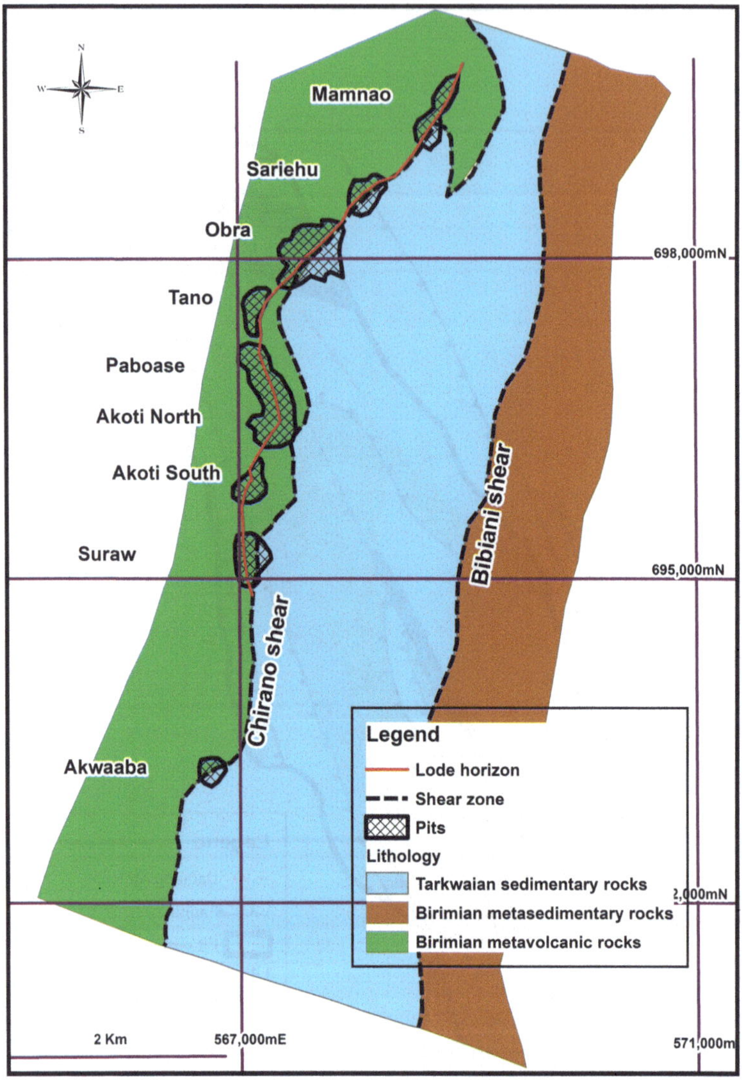

Fig. 6 Simplified geology of Chirano mine. Modified after Ackun-Wood et al. [32]

length of 10 km, approximately 200 m west of the CSZ, hosted in fractured and altered mafic volcanic rocks and referred to as the Chirano Lode Horizon [34].

Obuasi Gold Mine

The Paleoproterozoic Birimian Obuasi gold deposit exploited at the Obuasi mine is arguably the largest deposit in West Africa with approximately 62 Moz of gold. This deposit is hosted within the Kumasi basin metasedimentary rocks close to the boundary with the Ashanti volcanic belt (Figs. 1 and 7). The metasedimentary rocks, which have been intruded by several generations of felsic dykes and granitoids, consist of carbonaceous phyllites, volcaniclastic rocks, slates and psammites.

According to Fougerouse et al. [35], three deformational phases can be observed at the Obuasi mine. The initial deformational event (D_{1Ob}) manifested as a bedding-parallel fabric which is an indication of shearing concordant to bedding in the metasedimentary rocks. The ensuing D_{2Ob} NW–SE compressive deformation produced tight to isoclinal folding as well as intense sub-horizontal stretching [36]. This deformational event occurred concordant to the plane of a pervasive NE-striking

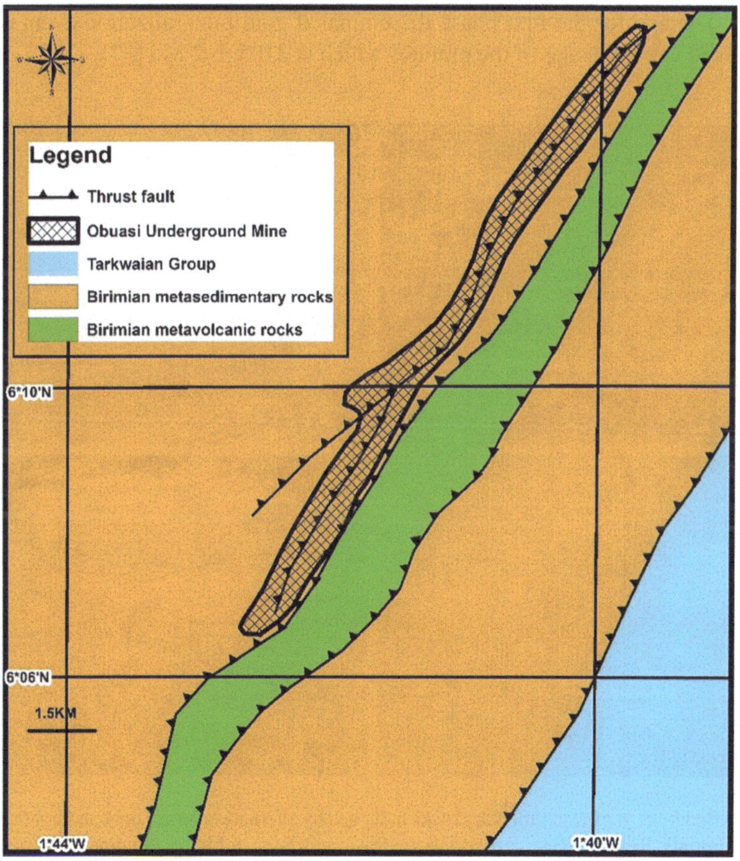

Fig. 7 Simplified geological map of the Obuasi mine. Modified after Fougerouse et al. [35]

subvertical foliation (S_{2Ob}). Finally, D_{3Ob} N–S shortening refolded previously formed structures and produced variably dipping S_{3Ob} cleavages with a peculiar ENE-strike, which is widespread throughout the deposit (Fig. 8). Fougerouse [37] found two kinds of mineralization at the Obuasi mine. The first kind are gold-bearing sulphides made up mainly of arsenopyrite disseminated within metasedimentary rocks. Disseminated gold-bearing sulphide mineralization at Obuasi is associated with D_{2Ob} deformation. S_{3Ob} folded strain shadows surrounding gold-bearing arsenopyrites provide evidence that the sulphides were formed during D_{2Ob} deformation. Up to 700 ppm of gold can be found within the epitaxial growth zones of arsenopyrite grains. The second type are free-milling gold hosted in quartz veins with width up to 25 m wide. The gold-bearing quartz veins are associated with D_{3Ob} NNW-SSE shortening which manifested as gently dipping E to NE foliations (S_{3Ob}). The gold in the veins are hosted in microcracks controlled by D_{3Ob} deformation where S_{3Ob} cleavages cross-cuts quartz veins. The three observed predictable ore shoot locations are contacts between volcaniclastic units and pre-D_{1Ob} felsic dykes, fault intersections, and F_{3Ob} fold hinges. Multiple stages of gold mineralization have been observed in the Obuasi deposit. The age for the first phase disseminated gold mineralization event can be constrained using the age of the granites which is 2105 ± 2 Ma [37].

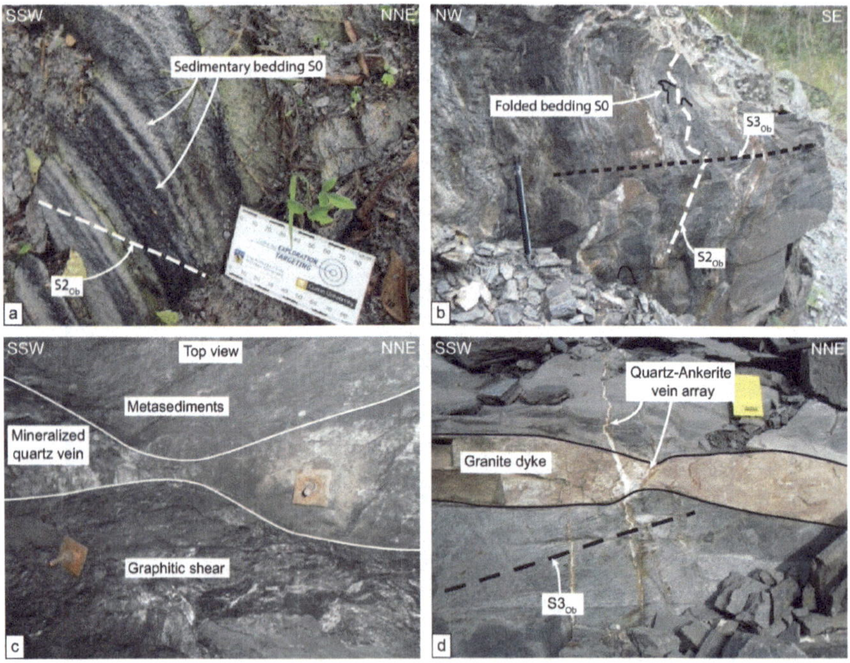

Fig. 8 **a** Beddings in metasedimentary rocks indicated by alternate layers of dark and light minerals S_{2Ob} overprints S_{0Ob}. **b** S_{3Ob} crosscut both S_{2Ob} and S_{0Ob} folds. **c** Boudinaged mineralized quartz vein adjacent to a graphitic shear at the Obuasi underground mine face. **d** Boudinaged and mineralized felsic dyke and subvertical quartz veins [37]

Bogoso Gold Mine

Gold mineralization at the Bogoso mine spans a length of 18 km along a major fault zone which separates Birimian sedimentary rocks of the Kumasi basin from adjacent Tarkwaian rocks (Fig. 1). The mine pits are positioned on approximately 200–400 m long ore shoots which occur in three dissimilar structural settings. The first are steeply plunging, comparatively high-grade orebodies which are largely restricted to left-hand dilational jogs which occur along strike with the controlling faults [20]. Minor cross-cut faults are observed to have intersected these jogs. The mineralization is largely restricted to fault-defining graphitic mylonite and imbricated slices of carbonate-altered rocks. The second type features mineralized quartz stockworks and disseminated auriferous pyrite and arsenopyrite formed in the vicinity of right-hand bends and splays in steeply dipping bounding faults. The third, which is characterised by smaller disseminated sulphide mineralized zones, occur in adjacent Birimian metasedimentary rocks where mineralized faults are discordant to beddings or foliations [20]. These mineralized faults also cross-cut the short limbs of tight, upright, east-trending folds. Majority of ore shoots in the Bogoso mine are contained in the most highly connected network of larger faults (Fig. 9). This fault network formed during an early phase of a deformational event consisting of a regional-scale shortening which included thrusting and isoclinal folding. The structural setting and geometry of the ore shoots suggest that mineralization occurred during the late stage of sinistral strike-slip fault movements. The occurrence of undeformed gold-bearing sulphides show that mineralization occurred during the last phase of ductile–brittle deformation. According to Allibone et al. [20], the deformational history at Bogoso includes an initial D_{1Bgs} deformational event which caused the formation of bedding-parallel foliation, isoclinal folds and thrust faults. This was followed by a D_{2Bgs} event characterised by upright, east-striking, SW-plunging, 100- to 300-m-scale folds. Lastly, a final D_{3Bgs} deformational event resulted in the sinistral strike-slip reactivation of previously formed thrust faults.

5 Implications for Gold Exploration

The defining structures at the major mines in Ghana have been summarised in Table 2. The regional-scale spatial distribution of gold deposits is immensely influenced by primary or first order transcrustal deformational structures at the Birimian belt-basin boundary. Second and third order discordant deformational structures, on the other hand, determine the location of mineralization at the deposit or mine-scale level.

The aspect of structural analysis during geological mapping must be taken seriously to provide meaning to interspersed observations and structural measurements taken during such campaigns due to paucity of outcrops. A focus on geological and structural characteristics similar to those already mentioned may increase the rate of discoveries of orogenic gold deposits. Comparable gold deposits may likely be

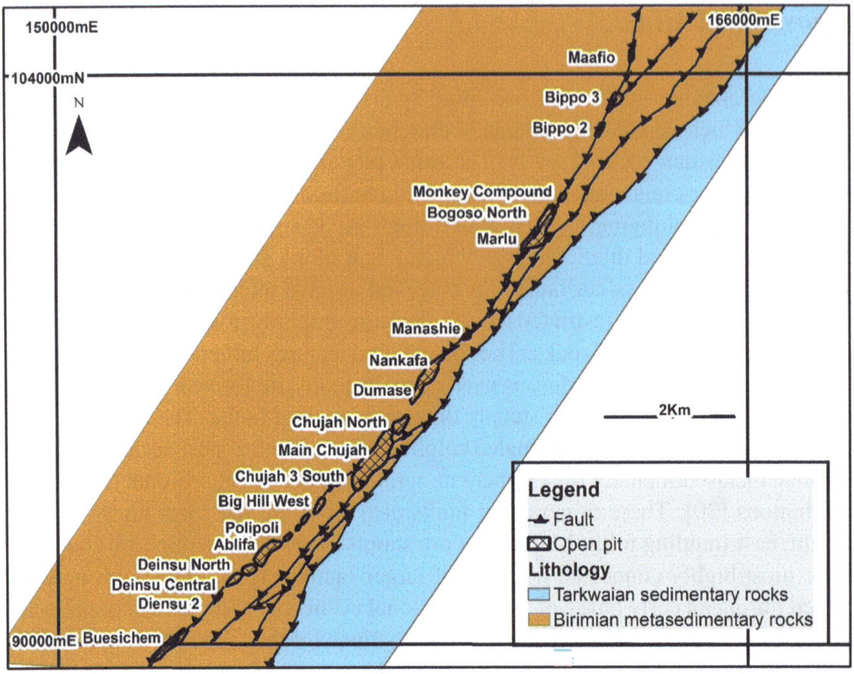

Fig. 9 Simplified geology of the Bogoso mine indicating the locations of the open pits. Modified after Allibone et al. [20]

discovered by analysing: (1) the chronology of deformational events, (2) temporal variations in the direction of the stress fields, and (3) variations in the strength of the various stresses accompanying each of these events.

It must, however, be noted that the prospective corridors occur along or close to the tectonic contacts between the Birimian metavolcanic belts and the adjacent metasedimentary basins [38]. Secondary and tertiary faults, folds and cleavages oblique or subparallel to the belt-basin contact must be the focus of geological investigations. A thorough chronological assessment of the deformational history of various terrains must also be conducted. All the aforementioned elements should influence ground selection in the search for potential brownfields.

Table 2 Defining structures of orogenic gold deposits in Ghana

Mine	Major fault or shear zone	Host rock	Defining or control structures	References
Ahafo	Kenyase-Yamfo	Diorites of the Sefwi granite greenstone belt	N- to NE oriented faults oblique to belt basin boundary	[28]
Wasssa	Wassa Akyempim-Akyem Ntronang	Metamorphosed basalts and andesites of the Ashanti belt	Mineralized hinges of kilometric-scale asymmetric synformal fold	[30]
Edikan	Akropong	Granitoids within the Kumasi basin	Low-angle thrust faults	[31]
Damang	Damang	Sandstones and conglomerates of the Banket series	Low displacement fault-mfracture mesh	[32]
Obuasi	Ashanti	Metasedimentary rocks of the Kumasi basin	Contacts between volcaniclastic units and pre-D_1 felsic dykes, fault intersections, and F_3 fold hinges	[35]
Chirano	Chirano	Basalts and basaltic andesites of the Sefwi granite greenstone belt	Fractures subparallel to primary shear zone	[34]
Bogoso	Ashanti	Metasedimentary rocks of the Kumasi basin	Concordant dilational jogs, crosscut or discordant faults, bends and splays	[20]

6 Conclusion

Most of the major mines in Ghana are distributed along the major tectonic boundaries between the Birimian volcanic belts and adjacent sedimentary basins which are marked by transcrustal faults. The regional-scale spatial distribution of orogenic gold deposits are strongly influenced and controlled by these primary or first order transcrustal deformational structures. These faults served as conduits for ore fluids which were deposited in second and third order discordant structures. These second and third order discordant deformational structures also determine the distribution of mineralization at deposit or mine-scale level.

7 Recommendations

Much attention must be given to structural analysis during geological investigations and mineral exploration to properly interrelate and interpret isolated observations and structural measurements taken during examination of outcrops. The focus should be on identifying areas with high intersection and density of geological structures along the belt-basin contact. The search for such structures may increase the rate of discoveries of similar orogenic gold mineralization.

Acknowledgements I truly appreciate the editors for their suggestions and recommendations which greatly improved this manuscript. I hereby also express my profound gratitude to Dr. Denis Fougerouse for permission to reuse aspects of his work. Credit to all researchers whose literary works form the basis of this chapter. I sincerely thank the Editors for their insightful comments and valuable contributions, which significantly improved the quality of this chapter.

References

1. Goldfarb RJ, Pitcairn I (2023) Orogenic gold: is a genetic association with magmatism realistic? Miner Deposita 58:5–35. https://doi.org/10.1007/s00126-022-01146-8
2. Chugaev AV (2024) Orogenic gold deposits of Northern Transbaikalia, Russia: geology, age, sources, and genesis. Geochem Int 62:909–978. https://doi.org/10.1134/S0016702924700484
3. Groves DI, Goldfarb RJ, Gebre-Mariam M, Hagemann SG, Robert F (1998) Orogenic gold deposits: a proposed classification in the context of their crustal distribution and relationship to other gold deposit types. Ore Geol Rev 13(1–5):7–27. https://doi.org/10.1016/S0169-1368[97]00012-7
4. McCuaig C, Kerrich R (1998) P-T-t-deformation-fluid characteristics of lode gold deposits: evidence from alteration systematics. Ore Geol Rev 12:381–453. https://doi.org/10.1016/S0169-1368[98]00010-9
5. Goldfarb RJ, Baker T, Dubé B, Groves DI, Hart CJ, Gosselin P (2005) Distribution, character, and genesis of gold deposits in metamorphic terrane. In: Economic geology 100th anniversary volume. https://doi.org/10.5382/AV100.14
6. Griffis RJ, Barning K, Agezo FL, Akosah FK (2002) Gold deposits of Ghana. Minerals Commission, Accra, Ghana
7. Leube A, Hirdes W (1986) The Birimian Supergroup of Ghana-depositional environment, structural development, and conceptual model of an early Proterozoic suite. Report No. 99529, Bundesanstalt für Geowissenschaften und Rohstoffe, Hannover
8. Sylvester PJ, Attoh K (1992) Lithostratigraphy and composition of 2.1 Ga greenstone belts of the West African craton and their bearing on crustal evolution and Archean-Proterozoic boundary. J Geol 100:377–393. https://doi.org/10.1086/629593
9. Hirdes W, Davis DW, Lüdtke G, Konan G (1996) Two generations of Birimian [Paleoproterozoic] volcanic belts in northeastern Côte d'Ivoire [West Africa]: consequences for the "Birimian controversy." Precambr Res 80:173–191. https://doi.org/10.1016/S0301-9268(96)00011-3
10. John T, Klemd R, Hirdes W, Loh G (1999) The metamorphic evolution of the Paleoproterozoic [Birimian] volcanic Ashanti belt [Ghana, West Africa]. Precambr Res 98:11–30. https://doi.org/10.1016/S0301-9268(99)00024-8
11. Milési J-P, Ledru P, Feybesse JL, Dommanget A, Marcoux E (1992) Early Proterozoic ore deposits and tectonics of the Birimian orogenic belt, West Africa. Precambr Res 58:305–344. https://doi.org/10.1016/0301-9268(92)90123-6

12. Feybesse J-L, Billa M, Guerrot C, Duguey E, Lescuyer J, Milési JP, Bouchot V (2006) The Paleoproterozoic Ghanaian province: geodynamic model and ore controls, including regional stress modeling. Precambr Res 149:149–196. https://doi.org/10.1016/j.precamres.2006.06.003
13. Perrouty S, Ailleres L, Jessell MW, Baratoux L, Bourassa Y, Crawford B (2012) Revised eburnean geodynamic evolution of the gold-rich Southern Ashanti Belt, Ghana, with new field and geophysical evidence of pre-Tarkwaian deformations. Precambr Res 204–205:12–39. https://doi.org/10.1016/j.precamres.2012.01.003
14. Attoh K (1982) Structure, gravity models, and stratigraphy of an Early Proterozoic volcanic sedimentary belt in northeastern Ghana. Precambr Res 18:275–290. https://doi.org/10.1016/0301-9268(82)90014-6
15. Dampare S, Shibata T, Asiedu D, Okano O, Manu J, Sakyi P (2009) Sr–Nd isotopic compositions of Paleoproterozoic metavolcanic rocks from the southern Ashanti volcanic belt, Ghana. Okayama Univ Earth Sci Rep 16:9–28
16. Asiedu DK, Agoe M, Amponsah PO, Nude PM, Anani CY (2019) Geochemical constraints on provenance and source area weathering of metasedimentary rocks from the Paleoproterozoic [~2.1 Ga] Wa-Lawra Belt, southeastern margin of the West African Craton. Geodinamica Acta 31(1):27–39. https://doi.org/10.1080/09853111.2019.1670414
17. Strogen P (1988) The structure and sedimentology of the Tarkwaian, and its relevance to gold mining exploration and development. In: International conference and workshop on the geology of Ghana with special emphasis on gold programme and abstracts, Accra, Ghana, 41
18. Taylor NP, Moorbath S, Leube A, Hirdes W (1992) Early Proterozoic crustal evolution in the Birimian of Ghana: constraints from geochronology and isotope geochemistry. Precambr Res 56:97–113. https://doi.org/10.1016/0301-9268(92)90086-4
19. Allibone A, McCuaig TC, Harris D, Etheridge M, Munroe S, Byrne D (2002) Structural controls on gold mineralization at the Ashanti gold deposit, Obuasi, Ghana. In: Goldfarb RJ, Nielsen RL (eds) Special publication 9: integrated methods for discovery: global exploration in the twenty-first century, vol 219. Society of Economic Geologists, pp 65–93. https://doi.org/10.5382/SP.09.04
20. Allibone A, Teasdale J, Cameron G, Etheridge M, Uttley P, Soboh A, Appiah-Kubi J, Adanu A, Arthur R, Mamphey J, Odoom B, Zuta J, Tsikata A, Pataye F, Famiyeh S, Lamb E (2002) Timing and structural controls on gold mineralization at the Bogoso gold mine, Ghana, West Africa. Econ Geol 97:949–969. https://doi.org/10.2113/gsecongeo.97.5.94
21. Davis DW, Hirdes W, Schaltegger U, Nunoo EA (1994) U-Pb constraints on deposition and provenance of Birimian and gold-bearing Tarkwaian sediments in Ghana, West Africa. Precambr Res 67:89–107. https://doi.org/10.1016/0301-9268(94)90006-X
22. Oberthür T, Vetter U, Davis DW, Amanor JA (1998) Age constraints on gold mineralisation and paleoproterozoic crustal evolution in the Ashanti belt of Southern Ghana. Precambr Res 89:129–143. https://doi.org/10.1016/S0301-9268(97)00075-2
23. de Kock GS, Armstrong RA, Siegfried HP, Thomas E (2011) Geochronology of the Birim Supergroup of the West African craton in the Wa-Bolé region of west-central Ghana: implications for the stratigraphic framework. J Afr Earth Sc 59:1–40. https://doi.org/10.1016/j.jafrearsci.2010.08.001
24. Block S, Baratoux L, Zeh A, Laurent O, Bruguier O, Jessell M, Ailleres L, Sagna R, Parra-Avila LA, Bosch D (2016) Paleoproterozoic juvenile crust formation and stabilisation in the South-Eastern West African craton (Ghana); new insights from U-Pb-Hf zircon data and geochemistry. Precambr Res 287:1–30. https://doi.org/10.1016/j.precamres.2016.10.011
25. Nunoo S, Manu P, Olarewaju VO, Asiedu DK, Nude PM (2016) Deciphering structures and deformation of the Obuom Gold Prospect, Central Ashanti Belt of Ghana: a lithostructural approach. J Afr Earth Sc 124:159–170. https://doi.org/10.1016/j.jafrearsci.2016.09.019
26. Nyame FK, Kase K, Yamamoto M (1998) Spessartine garnets in a manganiferous carbonate formation from Nsuta, Ghana. Resour Geol 48(1):13–22. https://doi.org/10.1111/j.1751-3928.1998.tb00003.x
27. Pigois J-P, Groves DI, Fletcher IR, McNaughton NJ, Snee LW (2003) Age constraints on Tarkwaian palaeoplacer and lode-gold formation in the Tarkwa-Damang district, SW Ghana. Miner Deposita 38:695–714. https://doi.org/10.1007/s00126-003-0360-5

28. Masurel Q, Morley P, Thébaud N, McFarlane H (2021) Gold deposits of the ~15-Moz Ahafo South Camp, Sefwi Granite-greenstone belt, Ghana: insights into the anatomy of an orogenic gold plumbing system. Econ Geol 116(6):1329–1353. https://doi.org/10.5382/econgeo.4829
29. Masurel Q, Eglinger A, Thébaud N, Allibone A, André-Mayer A, McFarlane H, Miller J, Jessell M, Ailleres L, Vanderhaeghe O, Salvi S, Baratoux L, Perrouty S, Begg G, Fougerouse D, Hayman PC, Wane O, Tshibubudze A, Parra-Avila L, Amponsah P (2022) Paleoproterozoic gold events in the southern West African craton: review and synopsis. Miner Deposita 57:1–25. https://doi.org/10.1007/s00126-021-01052-5
30. Perrouty S (2012) Structural evolution of the gold-rich Ashanti belt, SW Ghana. PhD thesis. Universite Paul Sabatier—Toulouse III
31. Tourigny G, Tranos MD, Masurel Q, Kreuzer O, Brammer S, Owusu-Ansah K, Yao D, Hayford T (2019) Structural controls on granitoid-hosted gold mineralization and paleostress history of the Edikan gold deposits, Kumasi Basin, southwestern Ghana. Miner Deposita 54:1033–1052. https://doi.org/10.1007/s00126-018-0858-5
32. Tunks AJ, Selley D, Rogers JR, Brabham G (2004) Vein mineralization at the Damang gold mine, Ghana: controls on mineralization. J Struct Geol 26:1257–1273. https://doi.org/10.1016/j.jsg.2003.11.005
33. Allibone A, Hayden P, Cameron G, Duku F (2004) Paleoproterozoic gold deposits hosted by Albite- and carbonate-altered Tonalite in the Chirano District, Ghana, West Africa. Econ Geol 99:479–497. https://doi.org/10.2113/gsecongeo.99.3.479
34. Ackun-Wood K, Kuma JSY, Yendaw JA (2016) The geometry and structural analysis of the gold deposits of Chirano mine. Ghana Mining J 16(2):10–19
35. Fougerouse D, Micklethwaite S, Miller J, Ulrich S, McCuaig TC (2015) The Obuasi gold deposit, Ghana: a West African giant. Ore Geol Rev https://doi.org/10.1016/j.oregeorev.2015.06.019
36. Fougerouse D, Micklethwaite S, Ulrich S, Miller J, Godel B, Adams DT, McCuaig TC (2017) Evidence for two stages of mineralization in West Africa largest gold deposit: Obuasi, Ghana. Econ Geol 112:3–22. https://doi.org/10.2113/econgeo.112.1.3
37. Fougerouse D (2015) Geometry and genesis of the giant Obuasi gold deposit, Ghana. PhD thesis, The University of Western Australia
38. Atta SK (2023) Mapping subsurface geological structures in the Birimian Supergroup, Ghana using airborne magnetic and radiometric data: implications for gold exploration. J Afr Earth Sci 2023(205):105003. https://doi.org/10.1016/j.jafrearsci.2023.105003

Petrogenesis of Auriferous Host Rocks in Major Gold Belts in Ghana

Petrographic and Geochemical Characteristics of Ayanfuri Gold-Bearing Granitoids (Abenabena Gap and Fetish Pits) in the Edikan Gold Mine, Kumasi Basin, Ghana: Implications for Petrogenesis

Etornam Bani Fiadonu, Endurance Korkor Buckle, and Emmanuel Daanoba Sunkari

1 Introduction

Granitoids are among the most abundant rock types in the continental crust, resulting from the partial melting of various source rocks and the segregation of any hypersthene-standard melt [1–3]. These rocks encompass a wide compositional range, from plagioclase-rich tonalites to alkali-rich syenites and monzonites to quartzolites [4–6]. Given this diversity, no classification scheme can comprehensively characterize granitoids in terms of their origin, compositional evolution, and geodynamic setting at the time of formation.

Classification approaches for granitoids often employ diverse parameters: (i) emplacement timing relative to tectonic events, distinguishing pre-, syn-, late-, and post-tectonic granitoids [7]; (ii) mineralogical proportions, such as alkali feldspar granites, granodiorites, and tonalites, based on the alkali feldspar to plagioclase

E. B. Fiadonu (✉) · E. K. Buckle · E. D. Sunkari
Department of Geological Engineering, Faculty of Geosciences, University of Mines and Technology, Tarkwa, Ghana
e-mail: ebfiadonu@umat.edu.gh

E. D. Sunkari
e-mail: edsunkari@umat.edu.gh

E. D. Sunkari
Department of Chemical Sciences, Faculty of Science, University of Johannesburg, Johannesburg, South Africa

Mining Engineering, Faculty of Integrated and Advanced Technology, Sir Padampat Singhania University, Udaipur, India

E. D. Sunkari and S. Ali (eds.), *Geology and Gold Mineralization in Ghana*, Earth and Environmental Sciences Library, https://doi.org/10.1007/978-3-031-97163-1_3

ratio [8]; (iii) chemical characteristics, identifying peraluminous, metaluminous, and peralkaline types based on alumina and alkali-saturation indices [5]; (iv) tectonic settings, including oceanic ridge, island arc, and post-orogenic granitoids [9]; and (v) thermal parameters, classifying granitoids as "cold" or "hot" based on radioelement content and radiogenic heat production [10]. The widely-used I-, M-, A-, and S-type classification (derived from igneous, mantle, and metasedimentary sources, respectively) remains popular [11].

Granitoids play an important role in gold mineralization, acting as both primary hosts and contributors to mineralising processes in numerous gold deposits worldwide [12, 13]. These rocks provide essential pathways for hydrothermal fluids, serve as sources of metals through magmatic differentiation, and create favourable structural and geochemical conditions for gold deposition. Globally, significant gold deposits associated with granitoids include the Kalgoorlie Super Pit in Australia, hosted by Archaean granitoid-greenstone terranes [14, 15]; the Pogo Gold Mine in Alaska, where mineralization is associated with reduced granitoids [16]; and the Grasberg Mine in Indonesia, where porphyry-style gold mineralization is linked to granitoid intrusions. These examples highlight the importance of granitoids in forming economic gold deposits and their contributions to global gold resources.

Granitoids significantly contribute to gold mineralization in Ghana, particularly within the Birimian Supergroup, which hosts some of the country's most productive gold belts [17–19]. These granitoids, typically classified as I-type and calc-alkaline, are associated with hydrothermal systems that transport and deposit gold along structural features such as shear zones and fractures. Examples of major gold deposits linked to granitoids in Ghana include the Edikan Gold Mine (formerly Ayanfuri Gold Project), where gold mineralization is closely associated with granitoid intrusions [20–23], and the Obuasi Gold Mine, where granitoids provide structural controls for gold emplacement in the Ashanti Belt. Additionally, the Bogoso-Prestea area shows mineralization linked to the interaction of granitoids with surrounding metavolcanic and metasedimentary rocks. These granitoids act as sources of heat and fluids, driving hydrothermal systems and creating geochemical traps for gold deposition, further emphasising their critical role in Ghana's gold endowment.

In Ghana, studies on basin-type granitoids have shown that granitoids in the Winneba area exhibit I-type characteristics [24], while S-type granitoids in the Kumasi basin also display features of I-type granitoids [25]. However, research on the Ayanfuri basin-type granitoids remains limited. The recent discovery of a significant shear zone, 10–20 km wide, cutting through the Kumasi basin suggests that structural influences may play a key role in shaping granitoid characteristics in this region [25]. Increased geological studies in these areas are vital for generating data to compare with the belt zones, enhancing understanding of the Birimian terrain's metallogenic history.

This study aims to address knowledge gaps by exploring the petrogenesis of basin-type granitoids and their relationship with I-type granitoids. I-type granitoids are especially significant due to their association with gold mineralization, more so than S-type granitoids. Previous studies have identified gold-bearing granitoids in the Edikan Gold Mine, yet the petrogenetic characteristics of these rocks did not cover the granitods in the Fetish and Abenabena Gap pits [21, 26]. Moreover, the petrographic and geochemical links critical to understanding granitoid formation processes are lacking. Therefore, geochemical and petrological analyses of samples from the Fetish and Abenabena Gap (AG) pits at the Edikan mine in Ayanfuri will be conducted to determine the granitoid types and reveal their petrogenesis.

2 Materials and Methods

Geology of the Study Area

The Edikan Gold Mine is located within the Ayanfuri and Nanankaw mining leases, spanning approximately 94 km^2 along the boundary between Ghana's Central and Western Regions (Fig. 1a). Situated in the Kumasi Basin of the Birimian Supergroup in southwestern Ghana, it lies about 16 km west of Dunkwa-Offin in the Central Region. Geographically, the mine is positioned between latitude 1° 51′ 00″ N and 2° 01′ 00″ N and longitude 5° 48′ 47″ W and 6° 10′ 00″ W (Fig. 1b).

It is accessible by a 106 km road from Kumasi to the north and a 185 km route from the port of Takoradi to the south [27]. The Kumasi Basin predominantly comprises Paleoproterozoic metasedimentary and metavolcanic rocks, formed in a back-arc tectonic setting bounded by active volcanic arcs to the west (Sefwi-Bibiani belt) and east (Ashanti belt) [28]. The Lower Birimian lithologies include schists, phyllites,

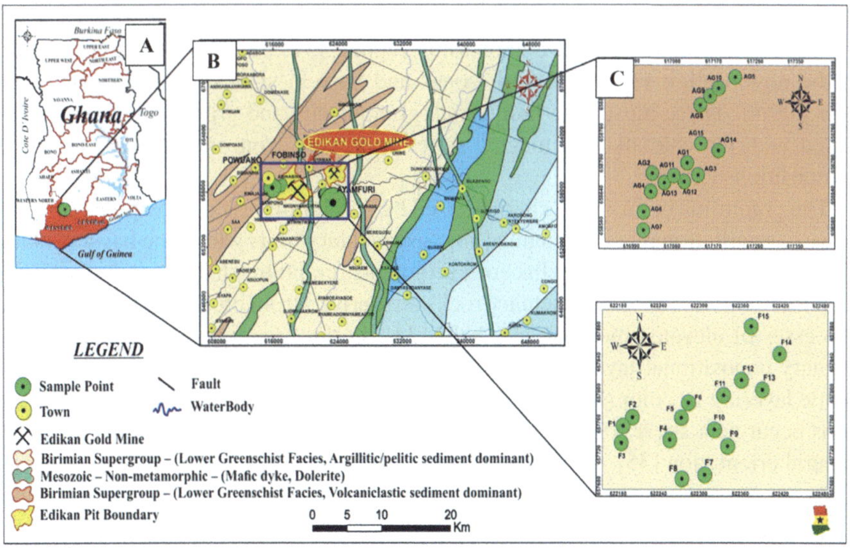

Fig. 1 Geological map of the study area showing the location of sample points

and greywackes, with feldspathic sandstones, Mn- and Si-rich sediments, and metamorphosed tuffs. Upper Birimian metavolcanics contain metamorphosed basaltic and andesitic lavas, hornblende-actinolite schists, and amphibolites alongside schists and greywackes, with minor Mn- and Si-rich chemical deposits [29]. Diabase and porphyry intrusions also penetrate the metavolcanic formations.

The basement of the Birimian Supergroup underlies much of the Kumasi Basin, marked by northeast-striking mafic metavolcanic belts separated from adjacent turbiditic units by major faults. Overlying these Birimian rocks are the younger Tarkwaian Group rocks, comprising conglomerates, sandstones, siltstones, and shales, which were deposited between 2150 and 2100 Ma in a foreland basin, sourced from adjacent greenstone belts [30]. The basin subsequently evolved into a back-arc basin, which was later subducted beneath active volcanic arcs [28]. As mantle plume activity intensified and the crust thinned, mafic magmas underplated the crust, resulting in asthenospheric upwelling [25]. Intermediate to felsic melts then intruded the Tarkwaian and Birimian lithologies as belt-type granitoids of the Eburnean Plutonic Suite. This phase marked the Kumasi Basin's second orogenic event, accompanied by the intrusion of basin-type granitoids and extensive deformation (D1–D5 phases) of the Birimian and Tarkwaian rocks [28].

The Edikan deposit is located within the sedimentary zones of the Kumasi Basin, near the western corridor of the Ashanti Greenstone Belt. The district hosts Paleoproterozoic Birimian metasediments and metavolcaniclastics metamorphosed to greenschist facies, along with various granitoids. These granitoids are auriferous and occur as sills, dykes, and ovoid plugs and include tonalitic, granitic, and leucogranitic plutons. The auriferous quartz veins and stockworks in the granitoids are hosted in fractures that were developed during the multi-stage deformations that accompanied the Eburnean Orogeny in the Birimian. The area is identified with network of brittle faults under hybrid compression regime [31, 32]. Basin-type granitoids in the region align with local fault zones and vary in size, sometimes extending over 200 m. These muscovite-biotite-rich granodiorite and granite suites are associated with the calc-alkaline magma series and range from intermediate to felsic compositions [31, 33].

The granitoids in this area, characterized by peraluminous compositions, suggest crustal assimilation into the magma [28]. Sedimentary units within the basin contain dacite, rhyolite, phyllites, argillaceous sediments, graphitic shale, greywackes, and graphitic schists. These sedimentary rocks are derived from erosion and lithification processes of elevated metavolcanic belts [34]. In the Ayanfuri metasediments, the primary depositional layers are often preserved, with bedding planes appearing as subtle layering or color shifts. Bedding planes typically strike ENE to NE, but rotations occur near shear zones or folded regions, leading to minor variations from the general orientation [35].

Sample Collection and Preparation

For this study, six rock samples were collected from Perseus Mining Ghana Limited (Edikan Mine) for petrographic and geochemical analysis. Thin sections of these samples were prepared at the Geological Engineering Laboratory of the University of Mines and Technology (UMaT) in July 2024 to facilitate microscopic mineral identification.

For thin section preparation, each rock sample was sliced into a 30 µm thick section, mounted on a glass microscope slide, and covered with a glass cover slip. The preparation steps were as follows:

(i) A slab was cut from each sample using a rock-cutting machine.
(ii) The resulting chip was ground by hand using abrasive papers of increasing grit (60, 80, 120, 240, 400, 600, and 1200) to achieve a smooth surface.
(iii) One side of the chip was labeled, and the other side was impregnated with epoxy resin. Excess epoxy was trimmed, and the sample was left to dry for 48 h.
(iv) After drying, the chip was further trimmed with a chip trim saw and polished to a thickness of 30 µm on a Hill Quist thin section machine.
(v) The prepared sections were washed, labeled, and examined under a petrographic microscope for mineral identification.

Thin and polished sections microscopy were conducted at the Petrology Laboratory in the Geological Engineering Department at the University of Mines and Technology, Tarkwa. Mineral identification in granitoid samples from the AG and Fetish pits focused on observing a range of mineralogical properties, including color, bireflectance, texture, isotropy or anisotropy, twinning, cleavage, extinction angles, and pleochroism. Transmitted and reflected lights microscopy were performed using LEICA DM750P and LEICA DM2700P microscopes, each equipped with Leica MC 120HD digital cameras connected to a computer for image capture and documentation. These combined petrographic and geochemical methods provide a comprehensive framework for understanding the origin, classification, and tectonic setting of the granitoids in the study area.

Geochemical Analysis

Approximately 5.0 g of each sample were ground and prepared for X-ray fluorescence (XRF) and Laser Ablation Inductively Coupled Plasma Mass Spectrometry (LA-ICP-MS) analysis, performed at ALS Geochemical Laboratory, an internationally accredited facility in Vancouver, Canada. These analyses targeted the major oxides, trace elements, and rare earth elements in each sample.

The volatile content in each sample was determined by Loss on Ignition (LOI) following the method described by Santisteban et al. [36]. LOI values, along with the concentrations of major and trace elements, were reported as weight percentages for oxides and in parts per million for trace elements.

3 Results

Petrography of the Abenabena Gap (AG) Pit

Petrographic analysis of granitoid samples from the AG pit revealed that they are predominantly composed of biotite, orthoclase, muscovite, plagioclase feldspar, and quartz (see Fig. 2a–f). Figure 2a–c are thin section photomicrographs corresponding to the field hand specimens shown in Fig. 2d–f, respectively. The thin section analysis of the sample in Fig. 2a confirmed the presence of quartz, muscovite, and plagioclase feldspar. The grains are medium grains averaging 1.5 to 4 mm and show interlocking texture of grains. This corresponds to the field sample shown in Fig. 2d, which exhibits a greyish color, coarse-grained texture, and visible microfractures, with feldspar and quartz as the main minerals. In Fig. 2b, the thin section of a sample in Fig. 2e, reveals biotite, plagioclase feldspar, quartz, orthoclase, muscovite, and opaque minerals. Its related field sample is greenish-grey, coarse-grained, and phaneritic in texture with interlocking grains, exhibiting microfractures and sericite alteration. The photomicrograph in Fig. 2c contains quartz, muscovite, sericite, and plagioclase feldspar under thin section. It has fine to medium grain sizes of the minerals and has micas to be dominant in the rock. Its corresponding hand specimen sample from the field as seen in Fig. 2f shows a dark greenish-grey, very fine-grained texture with vein-like microfractures and minor sericite alteration. These petrographic characteristics show that the Abenabena Gap granitoids are granodiorites, which can be complemented by their geochemical characteristics.

Petrography of the Fetish Pit

Petrographic analysis of granitoid samples from the Fetish Pit revealed that they are predominantly composed of biotite, orthoclase, muscovite, plagioclase feldspar, and quartz (see Fig. 3a–f). Figure 3a–c show the thin section photomicrographs corresponding to the respective hand specimen samples in Fig. 3d–f. The thin section in Fig. 3a, corresponding to the hand specimen in Fig. 3d, reveals the presence of plagioclase feldspar, muscovite, orthoclase, and opaque minerals. The sample is grey in color, coarse-grained (grain sizes ranging from 2 mm to 5 mm), and displays a phaneritic texture with weak chlorite and sericite alteration. Figure 3b, representing the thin section of the field sample in Fig. 3e, shows a similar grey color, coarse-grained phaneritic texture, with quartz veinlets and evidence of minor chlorite and sericite alteration. In Fig. 3c, the thin section photomicrograph of the sample shown

Fig. 2 Photomicrographs and hand-specimens of the granitoids from AG pit under crossed polarised light (opq = opaque minerals, bt = biotite, ms = muscovite, se = sericite, qtz = quartz, ort = orthoclase, plg = plagioclase)

in Fig. 3f, the rock appears darker in color, with medium- to coarse-grain, phaneritic texture, and also shows weak alteration by chlorite and sericite. The petrographic characteristics indicate that the Fetish Pit granitoids are granodioritic in composition, a conclusion further supported by their geochemical signatures.

Fig. 3 Photomicrographs and hand-specimens of the granitoids from Fetish pit under crossed polarised light (opq = opaque minerals, bt = biotite, ms = muscovite, ser = sericite, qtz = quartz, ort = orthoclase, plg = plagioclase, mc = microcline)

Ore Microscopy of Both AG and Fetish Pits

Gold mineralization in the AG and Fetish pits is predominantly associated with sulphide minerals such as pyrite and arsenopyrite (Fig. 4a–d), with occasional associations with chalcopyrite and galena. The gold and sulphide assemblages are typically found within fractures (Fig. 4b and c) and shear zones, indicative of structurally controlled deposition. Mineralization is closely related to quartz veins, stringers, and stockworks, as well as disseminations within the host granitoids.

Gold occurs primarily as very fine grains along sulphide grain boundaries and within fractures in sulphides, typically near vein margins. Coarse visible gold is occasionally observed within quartz veins. Higher-grade gold intercepts are often linked to zones containing coarse pyrite or arsenopyrite, and in some cases, chalcopyrite and galena.

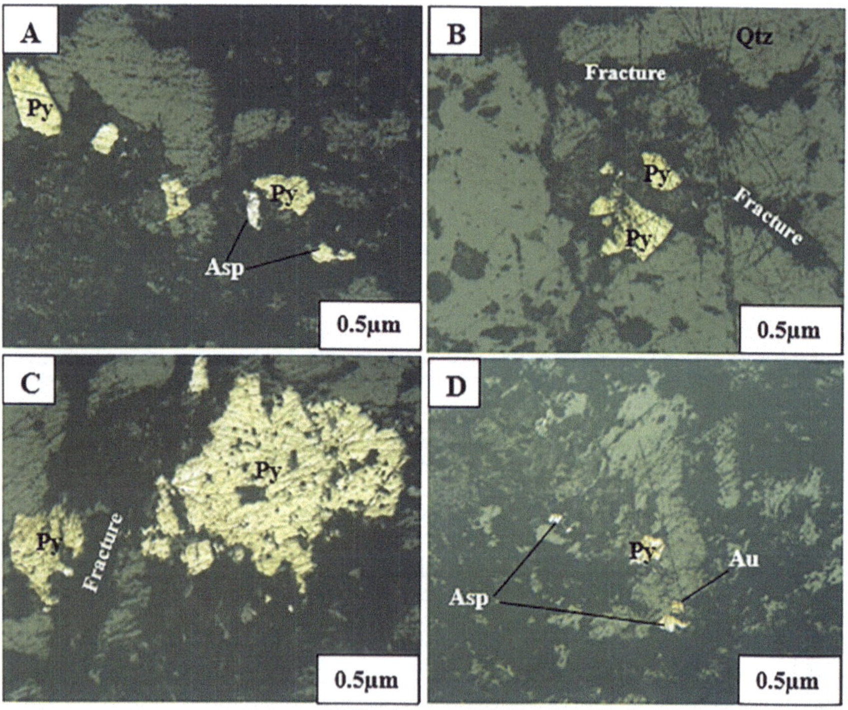

Fig. 4 Photomicrographs of the granitoids from the AG pit (**a** and **b**), and Fetish pit (**c** and **d**). Gold occurs in fractures in quartz veins (**b** and **c**), shear zones around the boundaries of the quartz veins, and it is associated mainly with pyrite and arsenopyrite (Asp = Arsenopyrite, Py = Pyrite, Au = Gold)

Geochemical Analysis

The chemical composition range of the granitoids (Table 1) and their statistical summaries (Table 2) with over 70 wt.% of SiO_2 places the granitoids in the felsic range and Na_2O (4.89–7.38 wt.%), Fe_2O_3 (1.84–2.53 wt.%) and K_2O (0.75–1.82 wt.%) suggests the granitoids are sodic granitoids. Moderate CaO content (1.35–2.54 and low TiO_2 (0.22–0.30 wt.%) are consistent with calc-alkaline magmatic series rocks. Their composition likely falls under granodiorite to tonalite granites. Predominance of Al_2O_3 (12.0–15.3 wt.%) relative to $Na_2O + K_2O$ (A/NK) and $CaO + Na_2O + K_2O$ (A/CNK) values indicates these are primarily metaluminous to peraluminous granites and typical characteristics of I-type granites (Tables 1 and 2).

Major element oxides versus silica were plotted in Harker diagrams (Fig. 5) to examine correlation patterns, such as negative and positive correlations between silica and other oxides. These trends help confirm a common origin for the granitoids. In the case of Al_2O_3, both AG and Fetish Pits samples show very weak correlations with SiO_2. Similarly, CaO also shows weak negative trends. Fe_2O_3 displays a more noticeable trend, especially in the Fetish Pit samples, where a moderate negative correlation suggests iron depletion with increasing silica. K_2O, on the other hand, shows a moderate positive correlation in AG samples. MgO trends are weakly negative in both sample sets, which is typical of fractional crystallization involving mafic minerals like olivine and pyroxene. However, the weak correlation suggests that other processes may be influencing the MgO content. Na_2O shows a weak positive trend in

Table 1 Geochemical data of major elements for AG and Fetish pits

Major oxides (wt.%)								
Sample ID	AG				Fetish			
	AG11	AG12	AG13	AG14	F11	F12	F13	F14
SiO_2	69.7	69.8	69.2	73.3	69.2	70	67.2	68.5
TiO_2	0.22	0.23	0.23	0.25	0.22	0.28	0.26	0.3
Al_2O_3	14.4	14.4	13.9	12	14.25	15.05	15.3	13.75
Fe_2O_3	2.02	2.41	2.26	2.35	2.25	1.84	2.46	2.53
MnO	0.02	0.02	0.02	0.02	0.02	0.02	0.03	0.02
MgO	0.61	0.64	0.63	0.48	0.71	0.53	0.82	0.95
CaO	1.97	2.03	1.91	1.52	2.24	1.35	2.28	2.54
Na_2O	5.4	4.89	6.49	5.45	5.16	7.38	6.22	5.58
K_2O	1.61	1.82	0.75	0.81	1.37	0.77	1.22	1.23
P_2O_5	0.06	0.06	0.09	0.08	0.07	0.12	0.1	0.07
LOI	3.05	3.27	3.08	2.42	3.75	1.89	3.84	4.26
Total	99.06	99.57	98.56	98.68	99.24	99.23	99.73	99.73
C	0.72	0.74	0.78	0.58	0.84	0.51	0.92	1.05
S	0.21	0.28	0.17	0.7	0.2	0.19	0.11	0.29

Table 2 Summary of concentration ranges of major elements in the granitoids for AG and Fetish pits

Oxide	AG pit granitoids (wt%)	Fetish pit granitoids (wt%)
SiO_2	69.2–73.3	67.2–69.2
TiO_2	0.22–0.25	0.22–0.30
Al_2O_3	12.0–14.4	13.75–15.3
Fe_2O_3	2.02–2.41	1.84–2.53
MgO	0.48–0.64	0.53–0.95
CaO	1.52–2.03	1.35–2.54
Na_2O	4.89–5.49	5.16–7.38
K_2O	0.75–1.82	0.77–1.37
P_2O_5	0.06–0.09	0.07–0.10

AG samples but a slightly negative one in Fetish samples. P_2O_5 and TiO_2 both exhibit very weak correlations with SiO_2 in both sample sets. Overall, the Harker diagrams (Fig. 5) suggest that while some oxides like K_2O and Fe_2O_3 show moderate trends indicative of magmatic differentiation, most oxides exhibit weak correlations with SiO_2. This points to a complex magmatic evolution involving multiple processes such as fractional crystallization, magma mixing, or source heterogeneity.

Granitoid Classification

Total Alkali-Silica (TAS) diagram (Fig. 6) by Cox et al. [37] was used to determine the type of plutonic rock for the granitoids at AG and Fetish granitoids. The samples plotted in the granite region with just two of the samples plotting in between the granodiorite and granite region.

The A/NK [$Al_2O_3/(Na_2O + K_2O)$] versus A/CNK[$Al_2O_3/(CaO + Na_2O + K_2O)$] plot by Shand [38] (Fig. 7) was used to distinguish between peraluminous, metaluminous, and peralkaline compositions of the AG and Fetish intrusives. Five samples plotted in the metaluminous field, while three samples plotted in the peraluminous field. Metaluminous compositions are defined by an A/NK value below 1.1 ($Al_2O_3 < Na_2O + K_2O$), whereas peraluminous compositions have an A/NK value above 1.1 ($Al_2O_3 > Na_2O + K_2O$). All samples were situated in the I-type granite field of the diagram. Notably, some samples were both peraluminous and I-type, indicating they belong to the category of peraluminous I-type granites.

The Alkali ($Na_2O + K_2O$), Iron Oxide (Fe_2O_3), and Magnesium Oxide (MgO) diagram, commonly referred to as the AFM diagram in Fig. 8, was used to classify the samples into magma series. The samples plot within the calc-alkaline magma series field, indicating a history of magma fractionation. This placement suggests early crystallization of Fe_2O_3 and MgO-rich minerals, such as olivine, which removed significant amounts of iron and magnesium from the melt, steering it toward a calc-alkaline composition. This process aligns with typical characteristics of calc-alkaline

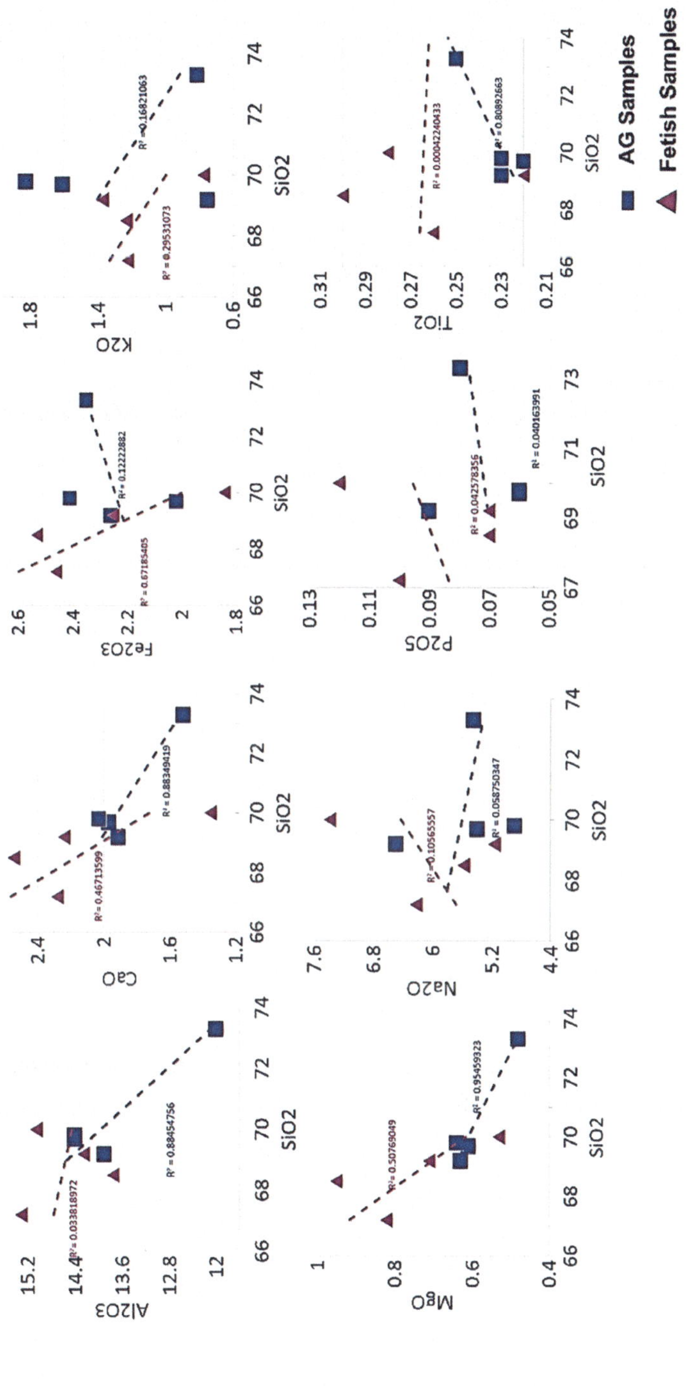

Fig. 5 Harker diagrams showing major oxides with silica variations

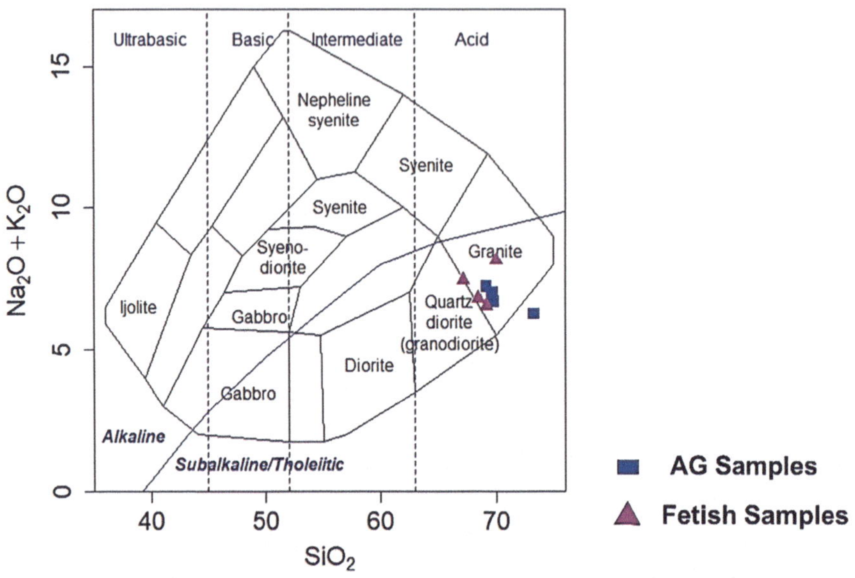

Fig. 6 Total Alkali-Silica (TAS, Na_2 + K_2O vs SiO_2) diagram for AG and Fetish granitoids [37]

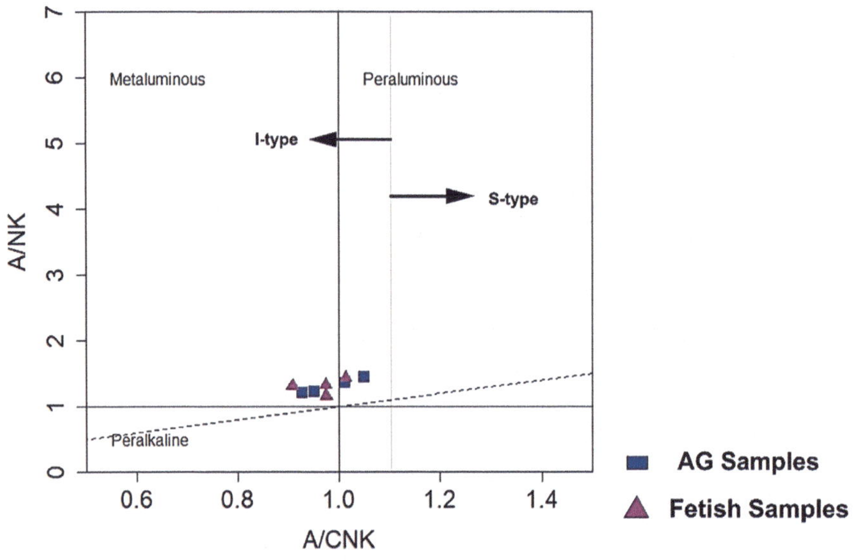

Fig. 7 A/CNK versus A/NK diagram [38] of the AG and Fetish granitoids

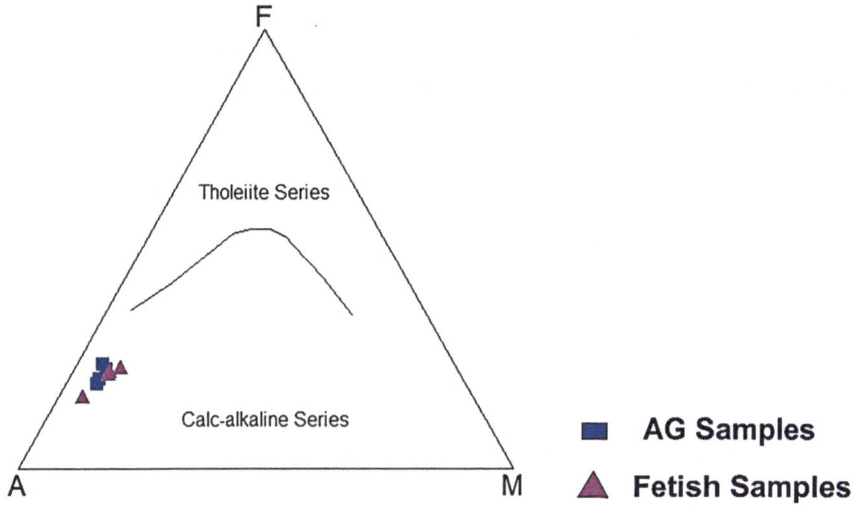

Fig. 8 Diagram of $Na_2O + K_2O$(A), Fe_2O_3* = total Fe expressed as FeO_3(F), MgO (M) of the AG and Fetish granitoids

magmas, where fractional crystallization and mineral differentiation play key roles in the magma's evolution.

Trace and Rare Earth Elements Geochemistry

The trace and rare earth element compositions are given in Tables 3 and 4, respectively. The geochemical data in Fig. 9, revealed enrichment in Rubidium (Rb), Potassium (K), and Lanthanum (La), accompanied by negative anomalies in Barium (Ba), Strontium (Sr), and Niobium (Nb), which are typically associated with S-type granites. These anomalies suggest the assimilation of crustal materials during magma evolution. The AG and Fetish intrusive samples display distinct characteristics, including low potassium (K) content and positive Ba and Sr anomalies. This geochemical signature indicates minimal to no assimilation of crustal materials, preserving the I-type granite characteristics [39]. Furthermore, the samples exhibit low Rare Earth Element (REE) enrichment, further supporting the interpretation of limited crustal interaction in the magmatic processes.

Table 3 Geochemical data of trace elements for AG and Fetish pits

Trace elements (ppm)								
Sample ID	AG				Fetish			
	AG11	AG12	AG13	AG14	F11	F12	F13	F14
Ag	0.029	0.116	0.05	0.28	0.066	0.046	0.014	0.077
As	472	341	19.25	7270	20.1	3270	77	143.5
Ba	440	480	338	262	351	197	429	367
Be	0.88	0.8	1.72	0.98	1.01	0.78	1.16	1.21
Bi	0.038	0.038	0.049	0.328	0.253	0.055	0.025	0.033
Cd	0.032	0.03	0.028	0.153	0.033	0.487	0.036	0.055
Ce	18.95	18.25	20.2	21.4	13.25	14.8	16.45	14.6
Co	3.99	4.88	4.02	5.4	5.48	6.3	5.94	5.17
Cr	19.8	20.9	23.7	21.4	283	26.1	27.8	218
Cs	2.37	2.81	1.48	1.69	2.73	1.1	2.24	2.12
Cu	10.55	9.15	36.8	26.1	28.8	4.68	11.3	22
Ga	20.6	20.6	19.4	19.55	19.8	17.9	20.1	18.2
Ge	< 0.05	0.05	0.06	0.05	< 0.05	< 0.05	< 0.05	< 0.05
Hf	2.12	1.96	2.13	1.765	1.8	2.1	1.935	1.7
In	0.013	0.02	0.013	0.011	0.016	0.017	0.016	0.02
La	10.1	9.67	10.45	11.3	6.68	7.48	8.33	7.42
Li	4.8	4.3	8.6	4.7	18.3	3.8	11	8
Mn	153	177.5	186.5	144	155	158	190	172.5
Mo	1.58	2.14	1.9	1.74	2.47	2.64	2.19	1.19
Nb	1.9	1.87	2.46	1.94	1.575	1.765	2.38	1.8
Ni	9.86	12.25	11.55	12.15	12.05	19.05	13.2	10.05
Pb	3.64	3.63	6.62	18.5	5.9	2.4	3.45	5.17
Rb	45.1	50.1	24.2	29.7	33.8	23.3	35.7	34.7
Re	< 0.002	< 0.002	< 0.002	< 0.002	< 0.002	< 0.002	< 0.002	< 0.002
Sb	0.22	0.17	0.47	5.67	0.43	0.55	0.29	1.19
Sc	3.21	3.37	3.71	2.84	3.37	2.5	4.34	4.1
Se	< 0.2	< 0.2	0.2	2.7	0.3	2.2	0.3	0.3
Sn	0.61	0.63	1.17	0.91	0.85	1.23	1.11	1.06
Sr	412	382	462	302	510	487	546	461
Ta	0.13	0.11	0.14	0.09	0.09	0.06	0.13	0.11
Te	0.05	0.04	0.04	0.5	0.09	1.1	0.04	< 0.04
Th	1.865	1.505	1.755	1.855	1.15	1.04	1.405	1.33
Ti	960	910	1100	900	1070	1320	1340	1290
Tl	0.247	0.264	0.133	0.135	0.195	0.118	0.193	0.203
U	0.8	0.57	0.69	0.62	0.5	0.37	0.54	0.39

(continued)

Table 3 (continued)

Trace elements (ppm)								
Sample ID	AG				Fetish			
	AG11	AG12	AG13	AG14	F11	F12	F13	F14
V	26.5	26.7	26.6	25.1	26.9	22.2	35.9	38.6
W	1.185	1.32	2.27	4.34	0.664	1.975	5.14	2.3
Y	2.53	2.27	3.24	2.53	2.73	2.94	3.2	2.32
Zn	37.8	43.3	37.1	38.8	44.2	71.1	44.1	47.3
Zr	71.3	70.1	72.5	60.9	63.7	84.4	71.9	63.4
Ti/Y	379.446	400.881	339.506	355.731	391.941	448.979	418.75	556.034
Au ppm	0.206	0.335	0.025	1.11	0.054	0.266	0.032	0.731
ppb	206	335	25	1110	54	266	32	731

Table 4 Geochemical data of rare earth elements for AG and Fetish pits

Rare earth elements (ppm)								
Sample ID	AG PIT				Fetish PIT			
	AG11	AG12	AG13	AG14	F11	F12	F13	F14
Ce	21.2	20.8	21.1	18.1	15.7	12.7	18.2	17
Dy	0.68	0.83	0.88	0.63	0.62	0.76	0.81	0.74
Er	0.34	0.35	0.44	0.33	0.33	0.38	0.41	0.39
Eu	0.47	0.43	0.41	0.41	0.45	0.4	0.5	0.47
Gd	1.38	1.42	1.38	0.96	1.24	1.04	1.26	1.06
Ho	0.14	0.13	0.16	0.13	0.12	0.12	0.15	0.15
La	11.2	11.1	10.7	9.3	7.9	6.3	9.6	8.6
Lu	0.03	0.02	0.05	0.03	0.02	0.03	0.03	0.03
Nd	8.7	8.5	9.1	7.7	6.9	6.1	7.8	7.3
Pr	2.41	2.32	2.39	2.05	1.77	1.5	2.1	1.92
Sm	1.8	1.84	1.75	1.54	1.49	1.24	1.52	1.28
Tb	0.14	0.17	0.18	0.13	0.15	0.13	0.16	0.14
Tm	0.03	0.03	0.03	0.02	0.02	0.03	0.04	0.03
Y	3.4	3.5	4	3.1	3	3.4	3.8	3.2
Yb	0.32	0.28	0.31	0.27	0.33	0.29	0.33	0.31

Tectonic Implications and Petrogenesis

The tectonic setting of the granitoids is inferred from the tectonic discrimination plots in Fig. 10. In the Rb/Y + Nb diagram, all samples plot within the Volcanic Arc Granite (VAG) field, while the Nb/Y plot positions them within the Syn-Collisional Granite (Syn-COLG) and VAG fields [43]. These results indicate that the granites

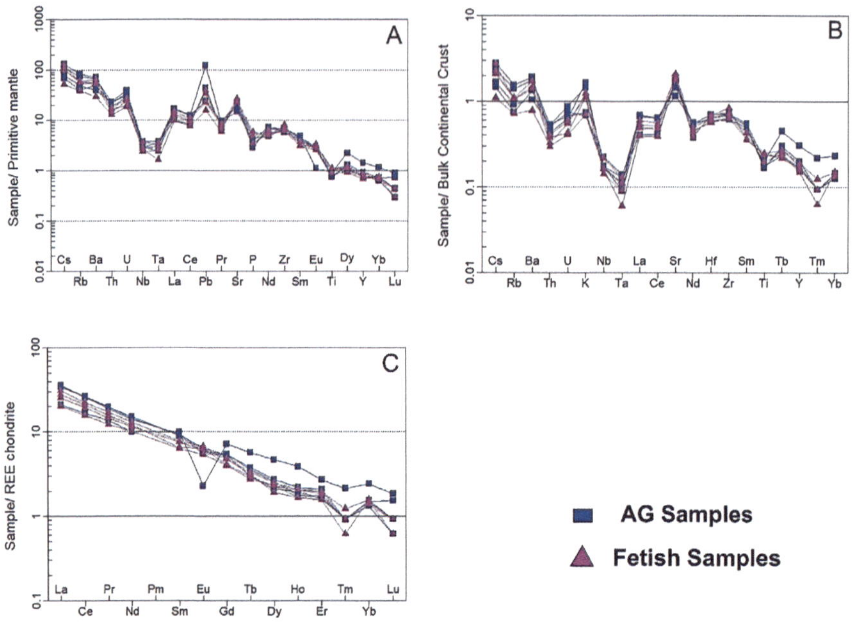

Fig. 9 Multi-element diagrams of Magma source and differentiation. **a** Primitive mantle spider plot normalised [40], **b** bulk continental spider plot [41] and **c** REE chondrite spider plot [42]

formed in a subduction-related tectonic environment, suggesting magmatic activity associated with an active convergent margin or orogenic processes.

The R1-R2 tectonic discrimination diagram reveals that the granites were emplaced in a syn-collisional tectonic setting. The positioning of the granites within the syn-collisional field indicates that the magmas were generated contemporaneously with the orogenic processes occurring at the time. This suggests a direct link between magma genesis and the tectonic events driving crustal deformation and collision.

4 Discussion

The petrographic and geochemical analyses of the granitoids from the AG and Fetish pits provide insights into their mineralogical compositions and petrogenesis. The AG granitoids predominantly consist of muscovite, biotite, plagioclase, quartz, and orthoclase (Fig. 2), the Fetish granitoids share a similar mineralogical assemblage, containing plagioclase, muscovite, quartz, biotite, and orthoclase (Fig. 3). Biotite is the only mafic mineral present in both sets of granitoids, consistent with the characteristics of typical I-type granites. These mineral assemblages, coupled with

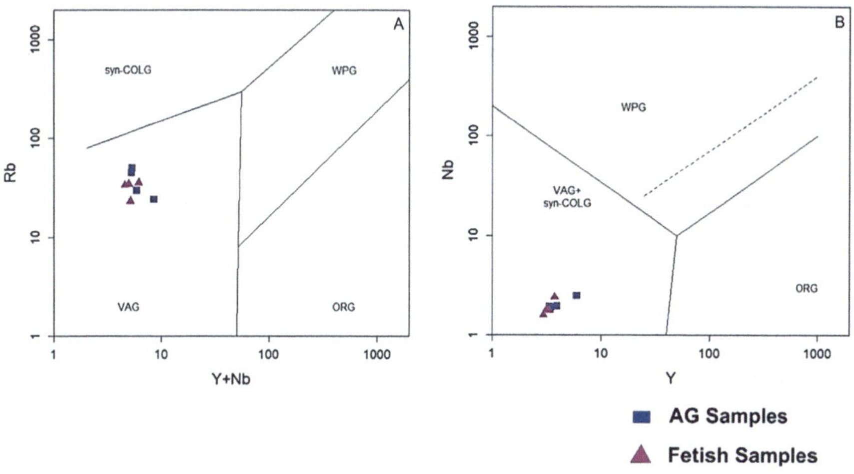

Fig. 10 Tectonic discrimination diagram for the granitoids Rb versus Y + Nb and Nb versus Y [43]

field observations and textural analyses, confirm the intermediate to felsic nature of the granitoids.

Magma Evolution and Geochemical Trends

The Harker diagrams (Fig. 5) reveal significant geochemical trends, such as very weak correlations of Al_2O_3 with SiO_2, suggesting progressive crystallization of minerals like pyroxene, feldspars, and micas during magma evolution. The decreasing alkali content (Na_2O and K_2O) reflects the removal of these elements through feldspar and mica crystallization, while accessory minerals like TiO_2 and P_2O_5 exhibit an increasing trend, characteristic of late-stage crystallization. High correlation coefficients among oxides, whether positive or negative, further imply a shared magmatic history and fractionation processes [32]. The samples plot within the calc-alkaline magma series on the AFM diagram (Fig. 8), reinforce the evidence for magma fractionation and early crystallization of Fe_2O_3- and MgO-rich minerals, such as olivine. The A/NK versus A/CNK diagram (Fig. 7) indicates that most samples fall within the I-type granite field. A subset of the granitoids, however, also exhibits peraluminous characteristics. Peraluminous granites are typically associated with elevated aluminum concentrations relative to Na, Ca, and K, often forming in regions with limited crustal interaction [45]. Conversely, the metaluminous samples suggest formation in settings with low metasedimentary influence, preserving their I-type signature [39]. Trace and rare earth element (REE) analyses highlight the geochemical distinction between the AG and Fetish intrusives. The granitoids display low K_2O contents and positive Ba and Sr anomalies (Fig. 9b), indicative of minimal crustal assimilation and the retention of their I-type characteristics. The low REE enrichments (Fig. 9c) further suggest limited crustal interaction, with minimal partial melting or contamination by crustal materials.

Tectonic Implications

The tectonic discrimination diagrams (Fig. 10) position the granitoids within the volcanic arc granites (VAG) and syn-collisional granite (COLG) fields, consistent with a subduction-related setting. The R1-R2 diagram (Fig. 11) corroborates this interpretation, showing that the granitoids formed in a syn-collisional environment. This tectonic setting implies that the granitoids were emplaced contemporaneously with regional orogeny. The lack of significant interaction with metasedimentary rocks during magma evolution further supports their preservation as I-type granites.

Overall, the AG and Fetish granitoids represent calc-alkaline, I-type granites formed in a subduction-related, syn-collisional tectonic setting. Their geochemical and petrographic characteristics suggest limited crustal assimilation and a magmatic history dominated by fractional crystallization. These findings provide crucial evidence for the tectono-magmatic evolution of the region and its association with subduction and orogenic processes.

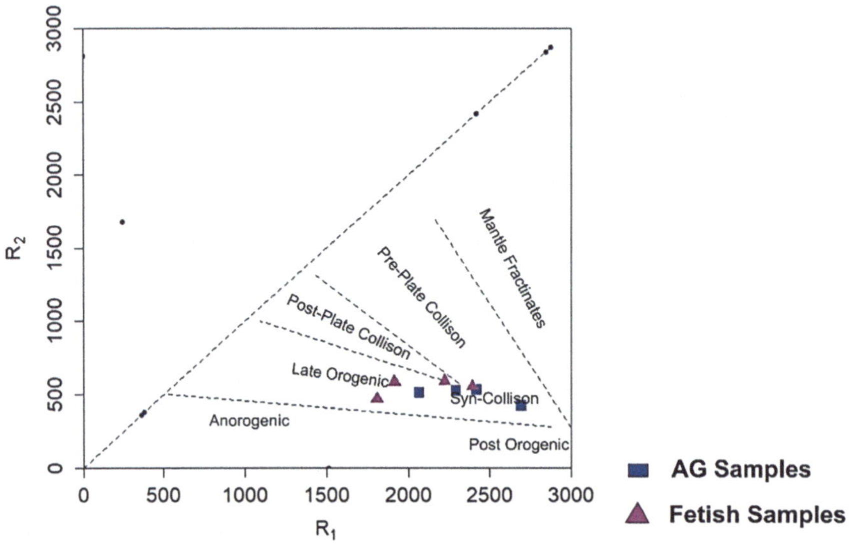

Fig. 11 R1[4Si-11(Na + K)-2(Fe + Ti)]-R2(6Ca + 2Mg + Al) discrimination diagram of AG, Fetish [44]

Petrogenetic Implications Related to Gold Mineralization

The gold mineralization observed in the granitoids from the AG and Fetish pits demonstrates a strong relationship between their petrogenesis and the associated sulfide assemblages. The granitoids, classified as calc-alkaline I-type granites, exhibit metaluminous to peraluminous characteristics, indicative of their formation in a subduction-related tectonic setting. Such settings are known for their potential to generate gold-bearing magmatic and hydrothermal systems, as evidenced by the mineralogical and geochemical features of the studied granitoids. The petrogenetic processes of these granitoids suggest a dual role in gold mineralization: (1) as a primary source of gold and related metals and (2) as a structural and lithological host for hydrothermal fluids. During the late stages of magmatic differentiation, volatile-rich fluids exsolved from the crystallizing granitoids become enriched in metals, including gold, as well as sulfur, arsenic, and other trace elements. These magmatic-hydrothermal fluids migrate through fractures, shear zones, and other structural pathways, precipitating gold and sulfide minerals as they cool or interact with the surrounding rocks.

Gold mineralization in these granitoids is primarily associated with pyrite, arsenopyrite, and occasionally chalcopyrite and galena, reflecting a sulfur-rich environment. The presence of quartz veins, stringers, and stockworks within the granitoids, as well as disseminated mineralization, indicates a strong hydrothermal influence. Gold deposition occurs along fractures, at sulfide grain boundaries, and within altered zones near vein margins, suggesting a close genetic link between hydrothermal fluid evolution and structural controls.

The low K_2O content and positive Ba and Sr anomalies in the geochemical data imply minimal crustal assimilation, preserving the I-type signature of the granitoids. This characteristic indicates that the magmatic source was relatively juvenile, derived from a mantle wedge enriched by subduction processes. The association of gold with sulfides and the structural alignment along shear zones and faults support a syn- to post-orogenic gold mineralization model, where regional tectonics created pathways for fluid migration and deposition.

Globally, granitoid-hosted gold deposits, such as the Kalgoorlie Super Pit in Australia and the Obuasi deposit in Ghana, share similar petrogenetic characteristics. In the case of the AG and Fetish pits, the subduction-related calc-alkaline magma series provided not only the necessary metal sources but also the thermal and structural framework for gold deposition.

In summary, the petrogenetic implications of the studied granitoids highlight their dual role as both a source and a host for gold mineralization. Their tectonic setting, magmatic differentiation processes, and subsequent hydrothermal activity collectively contributed to the formation of gold-bearing sulfide mineral assemblages, with significant implications for exploration and resource evaluation.

5 Conclusion

This study investigated the petrographic and geochemical characteristics of granitoids from the Abenabena Gap (AG) and Fetish Pits in the Edikan Gold Mine, with a focus on their petrogenesis and implications for gold mineralization. The granitoids were found to be predominantly I-type granites with minor occurrences of peraluminous I-type varieties, formed from a calc-alkaline magma series in a subduction-related tectonic setting. Petrographic studies revealed a mineral assemblage dominated by muscovite, biotite, quartz, plagioclase, and orthoclase, with biotite as the primary mafic mineral, showing that the granitoids are predominantly granodiorites. Geochemical analysis confirmed the granitoids' metaluminous nature, with high correlations among silica and major oxides indicating a shared magmatic history. The low potassium and total rare earth element (REE) concentrations suggest minimal crustal assimilation during magma evolution, preserving the granitoids' I-type characteristics. Tectonic discrimination diagrams positioned the granitoids in volcanic arc granite (VAG) and syn-collisional regions, confirming their formation during orogenic activity.

Gold mineralization in the granitoids was observed to be closely associated with pyrite, arsenopyrite, chalcopyrite, and galena, occurring within fractures, shear zones, and quartz veins. The spatial association of gold with sulfide minerals and hydrothermal alteration zones, particularly along vein margins and fractures, indicates a hydrothermal origin for the mineralization. The findings suggest that the granitoids played a significant role as host rocks and potential sources for gold-bearing fluids, with mineralization processes controlled by fluid flow through structural pathways.

This research contributes to the understanding of the petrogenesis and metallogenesis of the granitoids in the Kumasi Basin, providing insights into the processes governing gold mineralization in similar geological settings globally. The results have important implications for mineral exploration, highlighting the role of structural controls and hydrothermal processes in localizing gold deposits within granitoid terrains.

6 Recommendations

To deepen the understanding of the mineralized fluids and their physicochemical characteristics in the granitoids of the AG and Fetish pits, further studies are recommended. These studies should focus on fluid inclusion analysis, which can reveal the temperature, pressure, salinity, and composition of the hydrothermal fluids responsible for gold mineralization. Isotopic studies, including sulfur (S), oxygen (O), hydrogen (H), and lead (Pb) isotopes, are essential for tracing the sources of the fluids and their interaction with the host granitoids and surrounding rocks. Additionally, detailed petrographic and microstructural analysis of alteration minerals associated with gold and sulfide assemblages can provide insights into the conditions and timing of mineralization.

Geochemical modeling of trace elements and rare earth elements (REEs) within the granitoids and the mineralized zones should also be conducted to determine their partitioning behavior during fluid-rock interactions. Advanced spectroscopic techniques, such as Raman spectroscopy or laser ablation-inductively coupled plasma-mass spectrometry (LA-ICP-MS), can be employed to analyze the composition of fluid inclusions and sulfide minerals at micro scales. These studies will provide critical data on the thermodynamic and chemical environment of gold deposition, aiding in the reconstruction of the ore-forming processes and guiding further exploration efforts.

Acknowledgements This study is part of the BSc project work of the second author at the University of Mines and Technology, Ghana. The authors are grateful to Perseus Mining Ghana Limited for the permission to do this research on their deposit. We are also thankful to the Editors for their fruitful and feasible suggestions on this chapter.

References

1. Elms HC, Barker SJ, Morgan DJ, Wilson CJN, Charlier BLA (2023) Processes and timescales of magmatic rejuvenation and residence prior to post-caldera rhyolitic eruptions: Ōkataina volcanic centre (Aotearoa, New Zealand). J Volcanol Geotherm Res 437:107790. https://doi.org/10.1016/j.jvolgeores.2023.107790
2. Brown M (2013) Granite: from genesis to emplacement. Geol Soc Am Bull 125(7–8):1079–1113. https://doi.org/10.1130/B30877.1
3. Moyen JF, Janoušek V, Laurent O, Bachmann O, Jacob JB, Farina F, Fiannacca P, Villaros A (2021) Crustal melting versus fractionation of basaltic magmas: part 1, granites and paradigms. Lithos 402–403:106291. https://doi.org/10.1016/j.lithos.2021.106291
4. Streckeisen A (1974) Classification and nomenclature of plutonic rocks recommendations of the IUGS subcommission on the systematics of igneous rocks. Geol Rundsch 63(2):773–786. https://doi.org/10.1007/BF01820841
5. Bonin B, Janoušek V, Moyen JF(2020) Chemical variation, modal composition and classification of granitoids. SP Geol Soc 491(1):9–51. https://doi.org/10.1144/SP491-2019-138
6. Condie KC, Puetz SJ, Spencer CJ, Roberts NMW (2024) Four billion years of secular compositional change in granitoids. Chem Geol 644:121868. https://doi.org/ https://doi.org/10.1016/j.chemgeo.2023.121868
7. Hussein AAA, Ali MM, El Ramly MFA (1982) Proposed new classification of the granites of Egypt. J Volcanol Geotherm Res 14(1–2):187–198. https://doi.org/10.1016/0377-0273(82)90048-8
8. Bahariya GA (2021) An overview on the classification and tectonic setting of Neoproterozoic granites of the Nubian shield (Eastern Desert, Egypt). Geochemistry https://doi.org/10.5772/intechopen.95904
9. Adepoju MO, Asiwaju-Bello YA (2021) Geochemical classification and geotectonic setting of granitic gneisses from southeastern margin of Western Nigeria basement. J Geol Res 3(2):1–6. https://doi.org/10.30564/jgr.v3i2.2831
10. Pleitavino M, Carro Pérez ME, García Aráoz E, Cioccale MA (2021) Radiogenic heat production in granitoids from the Sierras De Córdoba, Argentina. Geotherm Energy 9(1):16. https://doi.org/10.1186/s40517-021-00198-9

11. Dumanlilar Ö, Aydal D, Dumanlilar H, Şen PA (2006) Geological and geochemical characteristics of granitoids in the eastern Tauride Belt, southeastern Turkey. İnt Geol Rev 48(5):443–465. https://doi.org/10.2747/0020-6814.48.5.443
12. Nguimatsia FWD, Bolarinwa AT, Yongue RF, de Dieu NJ, Olajide-Kayode JO, Olisa OG, Abdu-Salam MO, Kamga MA, Djou ES (2017) Diversity of gold deposits, geodynamics and formation conditions: a perspective view. Open J Geol 07(11):1690–1709. https://doi.org/10.4236/ojg.2017.711113
13. Phillips N (2022) Formation of gold deposits, vol 10. Springer
14. Nichols SJ, Hagemann SG (2014) Structural and hydrothermal alteration evidence for two gold mineralisation events at the new celebration gold deposits in Western Australia. Aust J Earth Sci 61(1):113–141. https://doi.org/10.1080/08120099.2014.879072
15. Moyen JF (2020) Archean granitoids: classification, petrology, geochemistry and origin. Geol Soc Lond Spec Publ 489(1):15–49. https://doi.org/10.1144/SP489-2018-34
16. Taylor RD, Graham GE, Lowers HA (2024) Hydrothermal monazite and xenotime chemistry as genetic discriminators for intrusion-related and orogenic gold deposits: implications for an orogenic origin of the Pogo gold deposit, Alaska. Miner Deposita 59(5):949–967. https://doi.org/10.1007/s00126-023-01240-5
17. Smith AJB, Henry G, Frost-Killian SA (2016) Review of the Birimian Supergroup-and Tarkwaian group-hosted gold deposits of Ghana. Episodes 39(2):177–197. https://doi.org/10.18814/epiiugs/2016/v39i2/95775
18. Atta SK (2023) Mapping subsurface geological structures in the Birimian Supergroup, Ghana using airborne magnetic and radiometric data: implications for gold exploration. J Afr Earth Sci 205:105003. https://doi.org/10.1016/j.jafrearsci.2023.105003
19. Goldfarb RJ, André-Mayer AS, Jowitt SM, Mudd GM(2017) West Africa: the world's premier Paleoproterozoic gold province. Econ Geol 112(1):123–143. https://doi.org/10.2113/econgeo.112.1.123
20. Agbenyezi TK, Adomako-Ansah K, Foli G, Sunkari ED, Bayari EE, Gawu SKY (2024) A review of intrusion-hosted gold systems of the Palaeoproterozoic Birimian Terrane of Ghana. Int J Earth Sci (Geol Rundsch) 113(7):1551–1570. https://doi.org/10.1007/s00531-024-02455-1
21. Sunkari ED, Oppong O, Agbenyezi TK (2024) Petrogenetic and geochemical characteristics of some auriferous granitoids in the Kumasi Basin, Ghana: implications for geodynamic settings and controls of orogenic gold mineralization in the Edikan gold mine. Deep Undergr Sci Eng. https://doi.org/10.1002/dug2.12128
22. Masurel Q, Morley P, Thébaud N, McFarlane H (2021) Gold deposits of the 15-Moz Ahafo South Camp, Sefwi granite-greenstone belt, Ghana: insights into the anatomy of an orogenic gold plumbing system. Econ Geol 116(6):1329–1353. https://doi.org/10.5382/econgeo.4829
23. Wilson MC, Amedjoe CG, Gawu SK (2022) Structures of Birimian and Tarkwaian rocks at North-West New Drobo–implication on deformation. Malays J Geosci (MJG) 6(1):36–44. https://doi.org/10.26480/mjg.01.2022.36.44
24. Nyarko ES, Aseidu DK, Osea S, Dampare S, Zakaria N, Hanson J, Osei J, Enti-Brown S, Tulasi D (2012) Geochemical characteristics of the Basin-type granitoids in the Winneba area of Ghana. Proc Int Acad Ecol Environ Sci 2(3):177
25. Brako BA, Foli G, Adomako-Ansah K, Aikins D, Dery S, Gawu SK (2020) Petrography and geochemistry of some Paleoproterozoic granitoids at the North-Eastern margin of the Kumasi Basin in Ghana. Earth Sci Malays (ESMY) 4(2):118–126. https://doi.org/10.26480/esmy.02.2020.118.126
26. Tourigny G, Tranos MD, Masurel Q, Kreuzer O, Brammer S, Owusu-Ansah K, Yao D, Hayford T (2019) Structural controls on granitoid-hosted gold mineralization and paleostress history of the Edikan gold deposits, Kumasi Basin, Southwestern Ghana. Miner Deposita 54(7):1033–1052. https://doi.org/10.1007/s00126-018-0858-5
27. Tiile RN, Darfour N (2016) Mine water management–a study at Edikan gold mine (EGM). J Energy Nat Resour Manage 3(2):81–84. https://doi.org/10.26796/jenrm.v3i2.91

28. Chudasama B, Porwal A, Kreuzer OP, Butera K (2016) Geology, geodynamics and orogenic gold prospectivity modelling of the Paleoproterozoic Kumasi Basin, Ghana, West Africa. Ore Geol Rev 78:692–711. https://doi.org/10.1016/j.oregeorev.2015.08.012
29. Hayford E, Anani C Kutu JM Armah T (2013) Aspects of the chemical composition of the Birimian gold fluid. J Earth Sci Geotech Eng 3(4):87–106
30. Jessell MW, Amponsah PO, Baratoux L, Asiedu DK, Loh GK, Ganne J (2012) Crustal-scale transcurrent shearing in the paleoproterozoic Sefwi-Sunyani-Comoe region, West Africa. Precambr Res 212–213:155–168. https://doi.org/10.1016/j.precamres.2012.04.015
31. Tourigny G, Tranos MD, Masurel Q, Kreuzer O, Brammer S, Owusu-Ansah K, Yao D, Hayford T (2018) Structural controls on granitoid-hosted gold mineralisation and paleostress history of the Edikan gold deposits, Kumasi Basin, southwestern Ghana. 1033–1052.https://doi.org/10.1007/s00126-018-0858-5
32. Agbenyezi TK, Foli G, Gawu SK (2020) Geochemical characteristics of gold-bearing granitoids at Ayanfuri in the Kumasi Basin, Southwestern Ghana: implications for the orogenic related gold systems. Earth Sci Malays 4(2):127–134. https://doi.org/10.26480/esmy.02.2020.127.134
33. AkalC (2019) Mechanism of emplacement and origin of the Ildır lava dome in the Karaburun Peninsula, Western Anatolia (Turkey). J Asian Earth Sci 179. https://doi.org/10.1016/j.jseaes.2019.04.012
34. Leube A, Hirdes W, Mauer R, Kesse GO (1990) The early proterozoic Birimian Supergroup of Ghana and some aspects of its associated gold mineralization. Precambr Res 46(1–2):139–165. https://doi.org/10.1016/0301-9268(90)90070-7
35. Agbenyezi TK, Foli G, Brako BA, Okon EE, Gawu SKY (2022) Paleoweathering, provenance and tectonic setting of metasedimentary rocks at Ayanfuri area in the Paleoproterozoic Kumasi Basin in Ghana: evidence from petrography and geochemistry. J Sediment Environ 7(3):519–538. https://doi.org/10.1007/s43217-022-00107-5
36. Santisteban JI, Mediavilla R, López-Pamo E, Dabrio CJ, Blanca Ruiz Zapata M, José Gil García M, Castaño S, Martínez-Alfaro PE (2004) Loss on ignition: a qualitative or quantitative method for organic matter and carbonate mineral content in sediments? J Paleolimnol 32(3):287–299. https://doi.org/10.1023/B:JOPL.0000042999.30131.5b
37. Cox KG, Ball JD, Pankhurst RJ (1979) The interpretation of igneous rocks. Allen & Unwin, p 450
38. Shand SJ (1943) Eruptive rocks: their genesis, composition, classification, and their relation to ore-deposits with a chapter on meteorite. Wiley
39. Collins WJ (1998) Evaluation of petrogenetic models for Lachlan fold belt granitoids: implications for crustal architecture and tectonic models. Aust J Earth Sci 45(4):483–500. https://doi.org/10.1080/08120099808728406
40. McDonough WF, Sun SS (1995) The composition of the earth. Chem Geol 120(3–4):223–253. https://doi.org/10.1016/0009-2541(94)00140-4
41. Taylor SR, McLennan SM (1995) The geochemical evolution of the continental crust. Rev Geophys 33(2):241–265.https://doi.org/10.1029/95RG00262
42. Boynton WV (1984) Cosmochemistry of the rare earth elements: meteorite studies. In: Developments in geochemistry, vol 2. Elsevier, pp 63–114. https://doi.org/10.1016/B978-0-444-42148-7.50008-3
43. Pearce JA, Harris NBW, Tindle AG (1984) Trace element discrimination diagrams for the tectonic interpretation of granitic rocks. J Petrol 25(4):956–983. https://doi.org/10.1093/petrology/25.4.956
44. Batchelor RA, Bowden P (1985) Petrogenetic interpretation of granitoid rock series using multicationic parameters. Chem Geol 48(1–4):43–55. https://doi.org/10.1016/0009-2541(85)90034-8
45. Chappell BW, White AJR (2001) Two contrasting granite types: 25 years later. Aust J Earth Sci 48(4):489–499. https://doi.org/10.1046/j.1440-0952.2001.00882.x

Petrographical Studies of the Subika Deposit, Ahafo, Ghana: Insights into the Litho-Characterization, Ore Textures and Role of Host Rocks During Hydrothermal Gold Mineralization

Kofi Adomako-Ansah, Raphael Boakye, and Adelaide Addo-Kuffuor

1 Introduction

Mafic rocks, are important host to gold mineralization in many parts of the world (e.g., [1–3]). Gabbro is mafic plutonic igneous rock with more than 30% mafic minerals [4, 5]. The mafic minerals are essentially pyroxene and hornblende. Diorite, on the other hand, is an intermediate plutonic igneous rock with less than 30% mafic minerals constituted by hornblende and biotite and with a phaneritic texture [5]. Sometimes gabbroic rocks containing only pyroxene and hornblende can have less than 30% mafic mineral composition. In such cases, it becomes difficult to distinguish them from diorite if the mineralogical constituents are unclear. Mineralogical studies by Zoheir et al. [1] identified gold in the gabbro-diorite complex in Southeast Egypt, using laser-ablation inductively coupled plasma-mass spectrometry (LA-ICP-MS) and electron probe micro-analysis (EPMA). Similarly, Ghose [2] identified native gold in olivine gabbro in the Naga Hills of India using electron probe micro-analysis (EPMA). Saad et al. [3] found that gold was deposited in hydrothermally-altered gabbroic rocks through the interaction between the ore fluid and the wall rock in the Tirek area of western Algeria. Hydrothermal alteration is a complex process that involves changes in the mineralogy, texture and chemistry of a rock as a result of interaction with hot aqueous fluids under physico-chemical conditions [6]. Consequently, the original minerals in the rock are replaced by new minerals (alteration minerals), which are stable at the conditions of alteration [7, 8]. The nature of the

K. Adomako-Ansah (✉) · R. Boakye
Department of Geological Engineering, University of Mines and Technology, Tarkwa, Ghana
e-mail: kadomakoansah@umat.edu.gh

R. Boakye · A. Addo-Kuffuor
Ahafo Mine, Newmont Ghana Limited, Kenyase, Ghana
e-mail: Adelaide.AddoKuffour@newmont.com

E. D. Sunkari and S. Ali (eds.), *Geology and Gold Mineralization in Ghana*, Earth and Environmental Sciences Library, https://doi.org/10.1007/978-3-031-97163-1_4

alteration is controlled by factors such as the lithological characteristics of the wall rocks and the composition, temperature, pressure and chemistry of the fluids [7, 9]. When these dark-appearing (i.e. mafic to intermediate) rocks are hydrothermally altered, it becomes increasingly difficult to decipher which rock unit was involved in the alteration and subsequent mineralization process. Fortunately, once the unaltered host rock is identified, it becomes easier to track the altered version within proximity of the same rock unit that has been crosscut by the hydrothermal quartz-carbonate veins.

Newmont Ghana Gold Limited Ahafo South Mine is situated within the Sefwi greenstone belt (Fig. 1) of Ghana, one of the five NE-SW trending volcanic belts of the Birimian Supergroup that form elongate bodies parallel to the regional strike [10, 11]. The mine operates both surface and underground mining. There are four deposits, namely Apensu, Amoma, Awonsu and Subika, of which the last two are still operational. Subika is the most southern among the four and approximately 2 km southeast of the Apensu deposit (Fig. 2) [10, 11]. The Subika deposit is the only deposit within the Ahafo area thought to contain gabbroic rocks. Although it was common knowledge that the altered dioritic rocks are mineralized, the gabbroic rocks in Subika are considered unmineralized (Company geologists, pers. Comm.). However, some altered and mineralized rocks that have not been well characterized were logged as gabbro. Therefore, it was essential to petrographically characterize these rocks in order to confirm whether they are the hydrothermally altered and mineralized versions of the host gabbroic rocks or not. Furthermore, gold mineralization in the Subika deposit is thought to be associated with fine pyrites but not with the coarse pyrite (Company geologists, Pers. Comm.). The differences between the two forms of pyrite and gold mineralization at Subika should be ascertained with scientific evidence, particularly from ore microscopic studies. This is the fastest and cheapest technique that can reveal the individual grain information in comparison to whole-rock geochemistry (pXRF) that was used at the mine. Thus, this study employed petrography based on microscopic observations in order to (a) characterize the host rocks and establish their lithological constitution, (b) identify any role(s) that these host rocks played in the gold mineralization process at Subika, and (c) clarify the relationship among coarse pyrites, fine pyrites and gold mineralization in the study area. Achieving these objectives will improve our understanding of the ore mineralization mechanism(s) at Subika and its implications on ore genesis and exploration targeting in the region. Then, applications from this study could be made for other hydrothermal deposit systems across Ghana and the world.

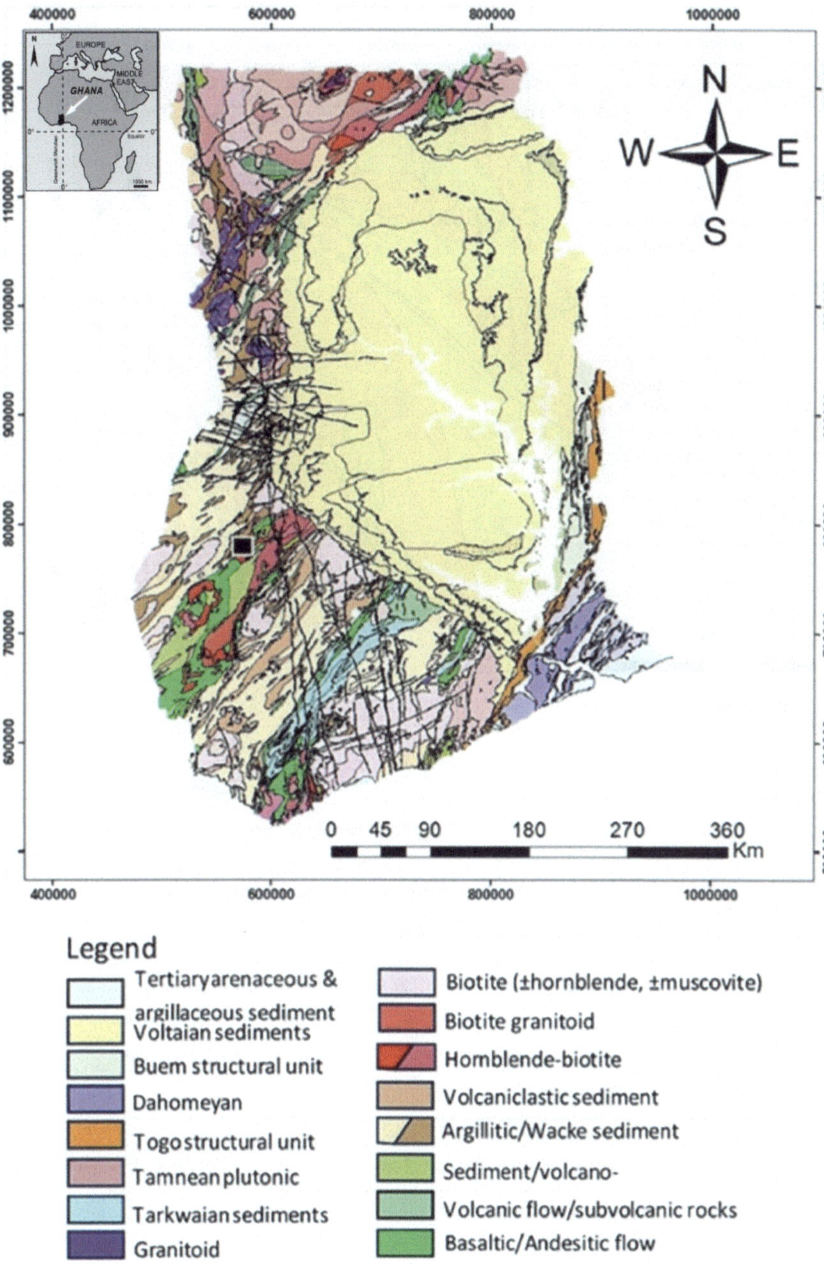

Fig. 1 Geology Map of Ghana showing the location of the Ahafo Mine area (black box). *Insert* Map of Africa showing the location of Ghana

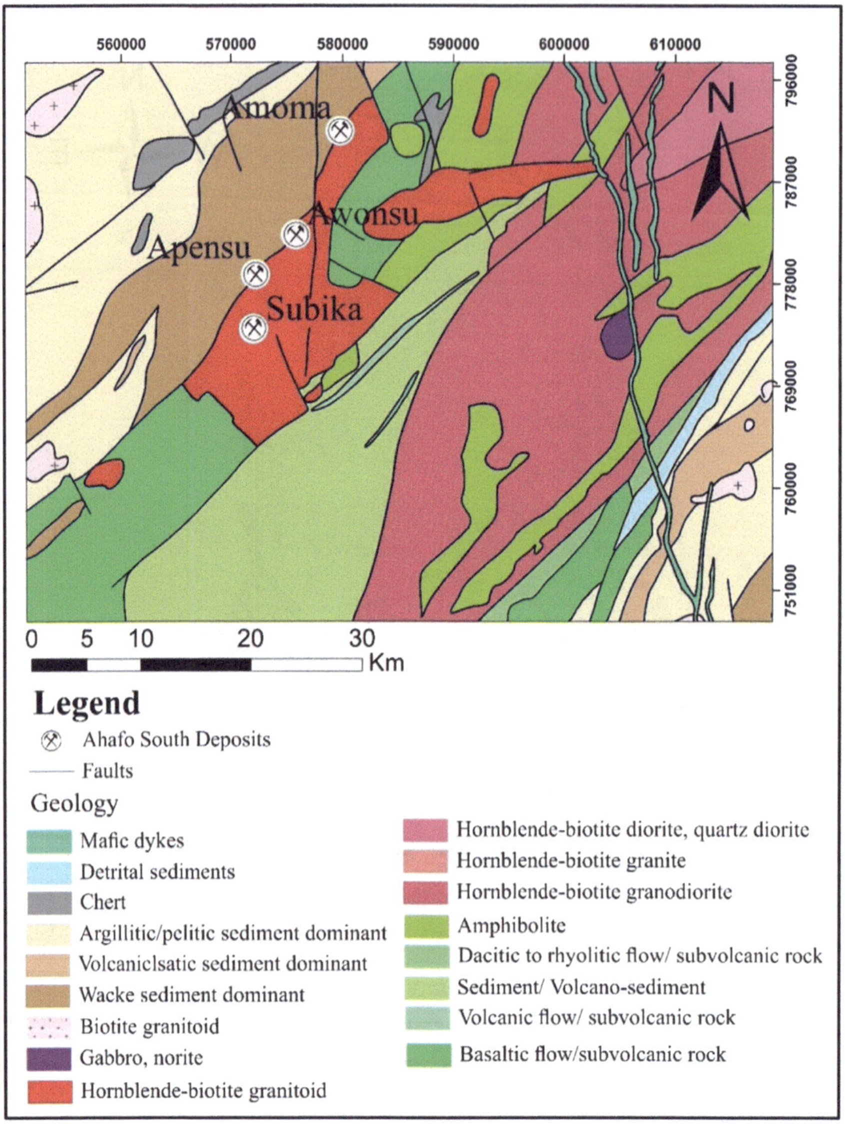

Fig. 2 Regional geological map of the Ahafo area showing the four deposits of the Ahafo South Mine (after [11]). Kenyase Thrust is parallel to the NE–SW trending structure that is along the contact between the Sunyani Basin and the Sefwi belt

2 Materials and Methods

Geological Background and Gold Mineralization

Ghana is situated within the eastern domain of the Leo-Man Shield which is located in the southernmost part of the West African Craton [12, 13]. The West African Craton was stabilized in the early Proterozoic during the Eburnean Orogeny. Allibone et al. [14, 15] proposed two orogenic events, which have been identified in Ghana. The first event is known as Eburnean I and is thought to be associated with the eruption of the Birimian (meta)volcanic belts and the intrusion of granites, followed by regional metamorphism between 2200 and 2150 Ma [14, 16, 17]. The second orogenic event is called Eburnean II which is characterized by regional NW–SE extension and the formation of the Birimian metasedimentary basins between 2150 and 2116 Ma [15]. Thus, the architecture of the Birimian Supergroup became characterized by a series of sub-parallel, NE–SW volcanic belts with intervening basins, both of which are intruded by extensive granitoids of Proterozoic age [12, 16, 18–20]. The rocks in this terraine are isoclinally folded and form an overthrust fold with an axis trending northeast and near vertical axial planes dipping from 65° to 90° [21].

The Sefwi belt is one of the five NE-SW trending volcanic belts of the Birimian supra-crustal domain of Paleoproterozoic age, which has been metamorphosed, folded and intruded by granitoids [12, 19, 20, 22]. Masurel et al. [11] indicated that regional peak metamorphic conditions of upper greenschist to lower amphibolite facies occurred in the central and northwest parts of the belt as compared to low to mid greenschist facies in the southeastern part. Contacts between the volcanic belts and the sedimentary basins are all major shear zones characterized by ductile fabrics and superimposed by brittle shear zones. These brittle shear zones host the major gold deposits [12, 16, 22]. The Sefwi and Ashanti belts are the most gold-productive greenstone belts in Ghana, and this is evident by the operations of the major large-scale gold mines on them.

The Ahafo South Mine, for which the Subika deposit forms a part, is located in the northern part of the Sefwi belt (Fig. 2). The Subika deposit is hosted predominantly by diorite and minor gabbro. Due to their gradational contacts, these igneous bodies are believed to belong to a composite pluton that probably underwent magmatic differentiation. On the other hand, crosscutting the host Subika pluton are a series of abundant fine- to medium-grained greenish mafic dykes. According to Masurel et al. [11], these dykes vary in thickness from 10 cm to 1.5 m and occur in two sets, which consist of ENE-striking, steeply S-dipping dykes and ENE- to NE-striking, shallow S-dipping dikes with rarer N-dipping conjugates. The earliest identifiable structural feature at Subika, which can be correlated with a subhorizontal magmatic foliation marked by flattened mafic enclaves (e.g., microgabbro and amphibolite xenoliths) and hornblende in the host diorite, is also crosscut by the mafic dykes [11].

Detailed presentation of the structural architecture for the Subika gold deposit is beyond the scope of this study. Readers are kindly referred to the article by Masurel et al. [11] and unpublished thesis of Baah-Danso [23] for comprehensive discussions on the structural elements governing the Subika gold deposit. Only the major highlights from these documents on the geological controls for the gold mineralization are presented here. The regional shear zone controlling mineralization in the area is the Kenyasi Thrust [10, 11, 23], which lies along the western contact between the Sefwi metavolcanic belt and the Sunyani metasedimentary basin (Fig. 2; [11, 23, 24]). The first three deposits (Apensu, Amoma, and Awonsu) are located on the Kenyasi Thrust (Fig. 2) whilst the Subika deposit sits on a parallel but separate, steeply dipping N050/60SE Magic Fracture Zone (MFZ) (Fig. 3), within the hanging wall of the Kenyasi Thrust Zone [11, 23, 24]. Generally, there is a clear spatial association among the NE-striking major fractures (e.g. the Kenyase Thrust and Magic Fracture Zone), NE-ENE-striking shallow-dipping mafic dykes and shear zones, and economic gold mineralization at Apensu, Awonsu, and Subika (Figs. 3 and 4) [11, 23]. A simplified conceptual model for the Subika gold deposit, modified from Masurel et al. [11] and Baah-Danso [23], is illustrated in Fig. 4 and discussed below. At Subika, the NE- to ENE-striking shallow-dipping shear zones (e.g., Victor, Kalbaas) can be traced over 1 km across benches in the open pit (Fig. 3). These low-angle shear zones vary in thickness between 1 and 3 m and obliquely cut the MFZ. Coincidentally, many of the mafic dykes share the same orientations as the shallow-dipping shear zones, and partly cut the MFZ as well (Fig. 4). The MFZ, which ranges from 1 to 60 m in width, is the localization and concentration of the high grade (2–5 g/t) gold mineralization [10, 11, 23]. Thus, it is believed that the MFZ is the conduit through which the ore-forming fluid accessed the Subika area (Fig. 4; [11, 23]). As observed in the field, coincidence of low-angle shear zones with mafic dykes at Subika reflects strain refraction induced by the presence of these less competent dykes [11, 25]. The brittle behavior of diorite wall rocks led to the formation of fracture permeability, promoting the tapping of ore-forming fluids [11, 25]. Thus, the sections of the crosscutting NE-ENE shallow-dipping shear zones and mafic dykes served as impediments that caused a temporal "damming" of the ore-forming fluid and consequently, buildup of localized pressure within the fluid (Fig. 4). During such temporal periods, the ore-forming fluids interacted with the mafic dykes or in part, flowed along the low angled structures that are occupied by these mafic dykes. Either way, the interaction by the ore-forming fluid likely caused the hydrothermal alteration and gold mineralization in the dykes within the intersections between the NE-ENE- and NE-trending structural corridors (Fig. 4). Following the over pressure-build up at the "dam" sites, fault/fracture reactivation, created the opening for the fluid to continue its upward propagation along the MFZ conduit until it is impeded by another set of shallow-dipping mafic dykes (Fig. 4; [11]). Then, the process of alteration and gold mineralization is repeated. In this case, sudden pressure drop in the fluid after the fracturing and/or shearing has occurred is often cited as the dominant cause of gold deposition in the quartz-carbonate veins [26–29]. At Subika, this is evidenced by shallow dipping gold-bearing veins along the shallow- to moderately dipping shear zones and in extensional to stockwork vein arrays adjacent to these shear zones [11]. In addition, the gold mineralization in the wall rocks, as shown from the present study, suggests that alteration of the host rocks also played an important role in the gold mineralization process at Subika.

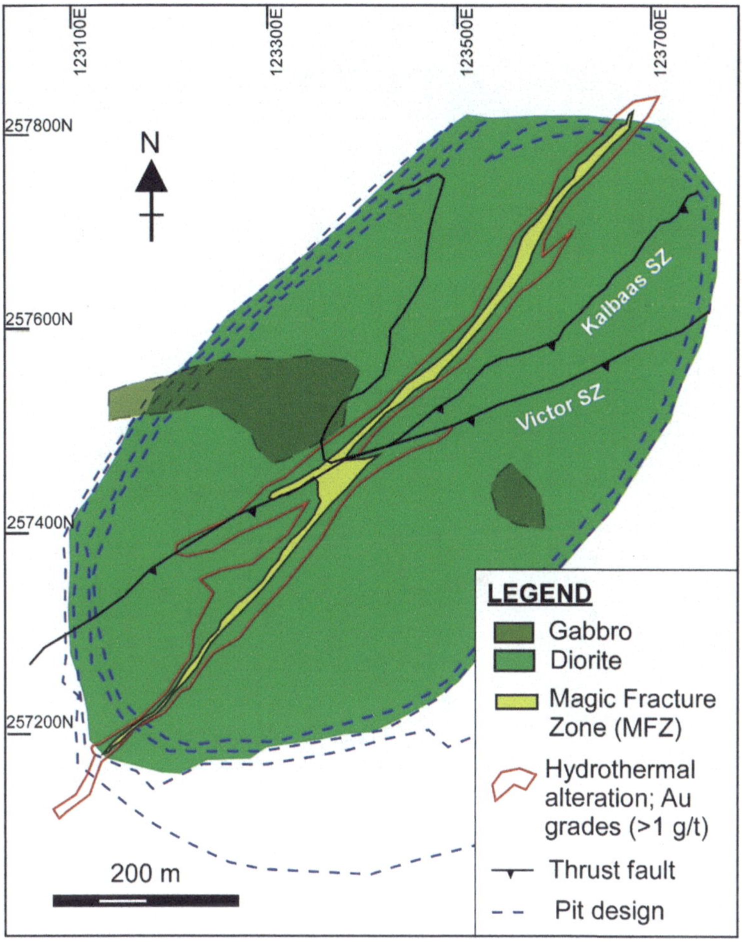

Fig. 3 Plan view showing the geologic setting of the Subika open pit (modified after [23] and [11])

The alteration minerals enveloping the mineralization in the area are silica, albite, sericite, chlorite, massive euhedral pyrites, and finely disseminated pyrites [11]. Baah-Danso [23] and Masurel et al. [11] identified a distal chlorite alteration and proximal sericite-carbonate-sulfide alteration to be related to the crosscutting quartz-carbonate veins and gold mineralization at Subika. Hydrothermal alteration mineral intensity is directly linked to brittle shears and vein density [11, 23]. The association with the quartz veins is plausible because host rocks devoid of quartz-carbonate veins generally exhibit negligible gold concentrations, while those in proximity to the quartz-carbonate veins demonstrate elevated gold grades.

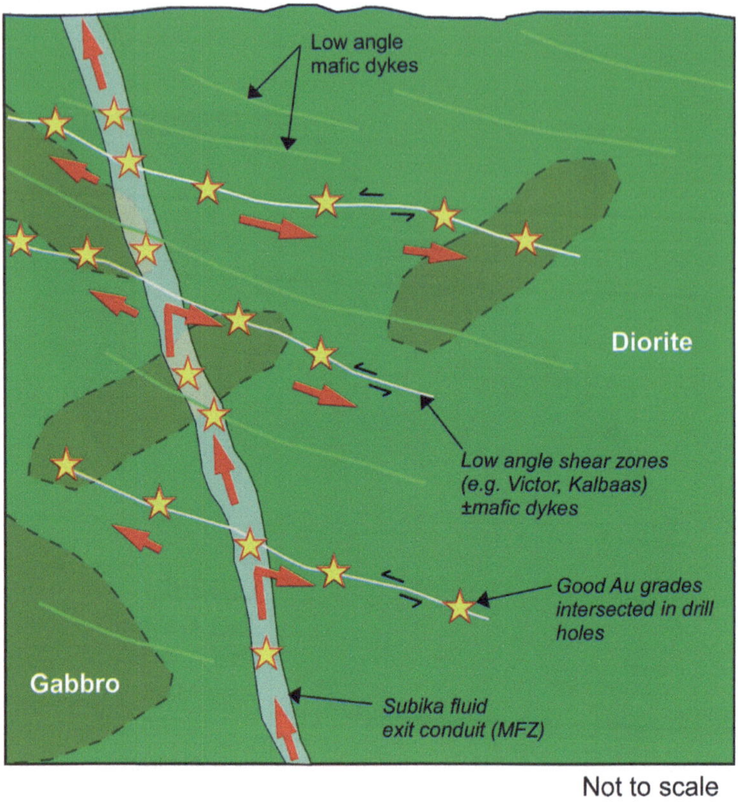

Fig. 4 Idealized ore genetic model, illustrating structural controls on ore-shoot location and geometry at Subika (modified after [11] and [23]). *Note* Red arrow = flow direction of ore-forming fluid

Field Survey and Sample Collection

An eight-week internship period was used to investigate the geological characteristics of the deposit and obtained samples for petrographic analysis. The samples were chosen based on their degree of alteration to accurately represent the altered and least altered zones in the Subika open pit. A total of 122 samples of grab and drill core samples were obtained from the mine pit (Fig. 3) and core shed of the exploration department and studied. For further laboratory analysis, fourteen dioritic grab samples (from B1 to B14) were selected. Out of these, four least altered samples (B1 to B4) were picked from the unmineralized hanging wall, three least altered (B5, B6 and B8) from the unmineralized footwall and seven altered samples (B9 to B14) from the ore zone of the pit. Eight core gabbroic rocks (from R1to R8) were also obtained from the core shed of the exploration department, of which three (R2, R5 and R6) were altered and five (R1, R3, R4, R7 and R8) were least altered. Thus, 22 samples were sent to the University of Mines and Technology, Tarkwa for laboratory analysis.

Petrographic Studies

Six altered (B9, B10, B12, B15, R2 and R6) and five least altered (B4, B5, B6, R4 and R7) samples were selected for the preparation of thin sections. Nine altered samples (B9, B10, B11, B12, B14, B15, R2, R5 and R6) were selected for polished sections. To identify the constituent minerals, eighteen polished sections of altered samples as well as twenty-six thin sections (altered and least altered) of gabbro and diorite were studied under the transmitted and reflected-light microscopy at the petrological laboratory of the Geological Engineering Department of the University of Mines and Technology (UMaT), Tarkwa. LEICA DM750P and LEICA DM 2700P microscopes with Leica MC 120HD digital cameras attached and linked to a computer were used for studying the thin sections under transmitted light. LEICA DM2700P with Leica MC 120HD digital camera attached to the microscope was used to study the polished sections under reflected light to identify the opaque minerals within the gabbroic rocks.

Data Analysis

The data obtained from the petrographical studies of samples were evaluated using Microsoft Excel spreadsheet and GCDKit software [24]. The steps below were used to analyze the data.

(a) The data was grouped into rows and columns in Microsoft Excel spreadsheet, with each columns representing a particular element and each row representing a sample.
(b) The data was labeled and organized with sample IDs in the first column and element names in successive columns.
(c) The Excel spreadsheet was opened and the data for the modal percentage of constituent minerals was imported by copying and pasting the file containing the data.
(d) After finishing all the necessary processing and visualization of the data in Excel, it was exported in a GCDK compatible format and then GCDK was launched to import the produced data file.
(e) The data file was loaded in the select data file. A notice was then displayed indicating the number of samples loaded. Relevant symbols and colors from the Plot Setting Menu were used before the data was plotted. The excel sheet was edited with numbers, symbols and colors chosen, which was then saved. The newly saved data was chosen and re-entered. The data was then plotted. The

plot menu, which contained the required plots (QAP and classification) were used. The QAP plot was selected and afterwards a window appeared which involved entering the number of samples to be plotted. Plots which were selected appeared, saved as JPEG files and edited with CorelDRAW software.

3 Results

Petrography

Least Altered Gabbro

In hand specimen, the least altered gabbroic rocks showed dark-green color with medium-to-coarse grained (phaneritic) texture (Fig. 5a). They were composed of abundant mafic minerals and plagioclase. No foliations were seen in the rocks. Some of these rocks had minor veins/veinlets running through them. Microscopically, the rock is composed of pyroxene, hornblende, plagioclase, biotite and quartz (Table 1 and Fig. 6a). Secondary minerals found were chlorite, carbonate, quartz and opaque minerals (Fig. 6a–c). Pyroxene showed second order violet-blue-green–red interference colors. Hornblende displayed second to third order brown-green interference colors with an inclined extinction angle of 15°. They exhibited hypidiomorphic texture for subhedral to anhedral grains. The plagioclase occurred as subhedral elongate crystals with polysynthetic twining. They showed first order white to gray colors with an inclined extinction angle of 45°. Biotite showed second to third-order brown-green interference colors with parallel extinction. In Fig. 6a–c, quartz-carbonate veinlets cut through the rock and chlorite was observed to replace pyroxene and hornblende in the selvages of the quartz-carbonate vein. There were opaque minerals in the quartz-carbonate vein (Fig. 6a–c). Largely, rare to no opaque minerals were found outside the veins (Fig. 6a–c).

Altered gabbroic samples had light grey color with many disseminated sulfides. They exhibited moderate alteration in the presence of large quartz-carbonate veins (Fig. 5b). Altered samples were light grey to greenish grey in color with quartz-carbonate veins and associated sulfide disseminations. They were moderately-to-strongly altered (Fig. 5b). In thin sections, the rocks were strongly altered by carbonates with dissemination of many opaque minerals (Fig. 6d). The primary texture in the rock has been destroyed. The pervasive carbonate alteration observed in Fig. 6d had the same features as the one associated with quartz-carbonate veinlets in the least altered gabbroic rocks (Fig. 6a–c) described.

Least Altered Diorite

Unaltered dioritic samples in hand specimen were coarse grained with speckled black and white color (Fig. 7a); and composed of plagioclase and mafic minerals.

Fig. 5 Photographs of samples showing **a** least altered Gabbro and **b** altered Gabbro

Table 1 Average modal percentages of the minerals in least altered gabbro

Mineral	Percentage (%)
Pyroxene	35
Hornblende	30
Plagioclase	20
Biotite	5
Quartz	10

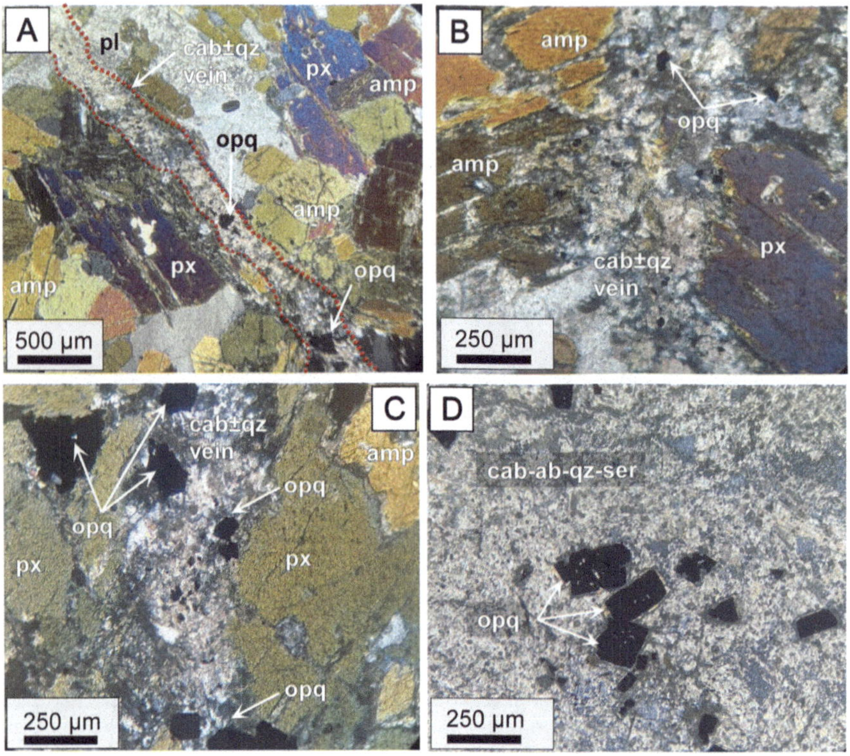

Fig. 6 Photomicrographs of gabbroic samples showing **a**–**c** pyroxene, hornblende, biotite, plagioclase and quartz in least altered gabbro cut by thin quartz-carbonate veinlets. Opaque minerals form within the interaction between the veinlets and the host rock minerals. Some of the host rock minerals appear brecciated in (**b**). **d** Strongly altered gabbro showing features of opaque minerals and alteration that is similar to the alteration caused by veinlets in (**a**)–(**c**). *Abbreviations* px—pyroxene, amp—amphibole, bio—biotite, pl—plagioclase, ab—albite, qz—quartz, ser—sericite, cab—carbonates, opq—opaque minerals (pyrite and gold)

Microscopic studies revealed plagioclase (~50%), biotite (~30%), hornblende (~10–15%) as the major mineralogical constituents of the least altered diorite (Table 2 and Fig. 8a, b). The texture of the rock is xenomorphic. Plagioclase, the dominant mineral, has been replaced in some sections of the diorite by sericite and epidote (Fig. 7a), even in the absence of quartz-carbonate veins (Figs.7a and 8a, b). Sericite was colorless under plane polarized light, and in cross polars, it showed low-to-high interference colors. Epidote on the other hand, showed a pale-yellow color with high relief in plane polarized light and high interference colors (i.e. pink or bright yellow) in cross polarized light.

Altered samples were light grey in color with quartz veining and sulfide distributions (Fig. 7b). They were moderately altered with some portions replaced by silica (Fig. 7b). In thin section observations, the primary textures of the altered samples were also obliterated (Fig. 8c, d). The plagioclase and mafic minerals are totally

Fig. 7 Photographs of samples showing **a** least altered diorite and **b** altered diorite

Table 2 Average modal percentages of minerals in least altered diorite

Mineral	Percentage (%)
Pyroxene	5
Hornblende	15
Plagioclase	50
Biotite	25
Quartz	5

replaced by carbonate minerals, sericite, and rare chlorite. The sulfides are found in the carbonate-sericite alteration (Fig. 8c, d), which is to the same as in the altered portions of the gabbro (Fig. 6a–d).

Ore Mineralogy and Textures

Microscopic studies of polished sections of altered gabbroic and dioritic samples under reflected light revealed four main ore minerals: pyrite, chalcopyrite, rutile and gold (Fig. 9). At least, two types of pyrites were distinguishable based on distinct morphology, texture and mineral associations. These pyrites are coarse-grained pyrites (denoted as py 1; Figs. 9a–d, 10a, b) and fine-grained pyrites (denoted as py 2; Fig. 10c, d).

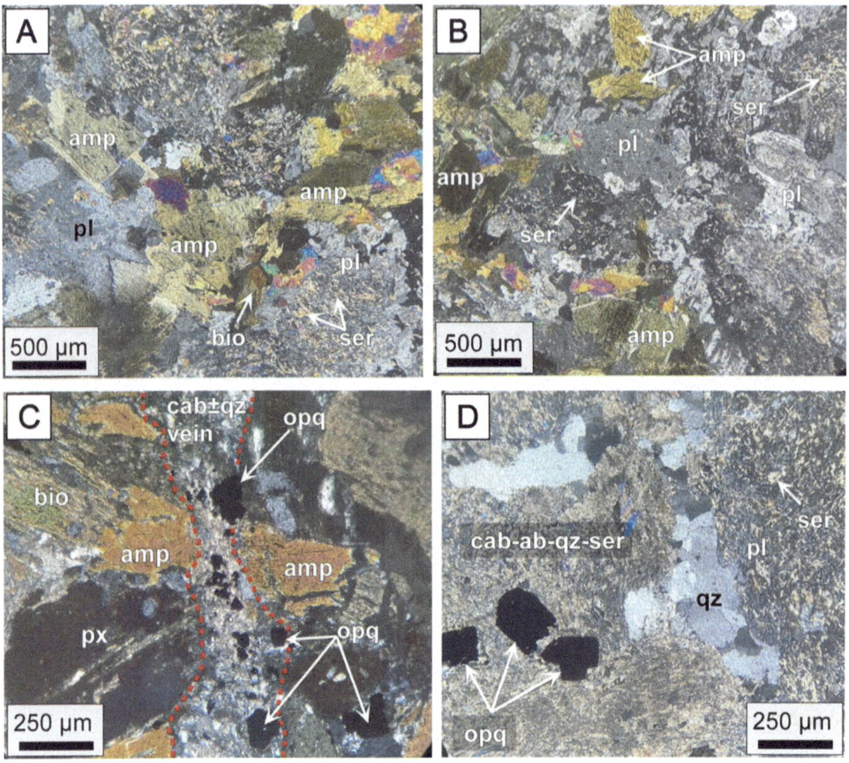

Fig. 8 Photomicrographs of dioritic samples showing **a** & **b** amphibole (amp), biotite (bio), plagioclase (pl) and sericite (ser) in least altered diorite. **c** Least altered diorite cut by quartz-carbonate veinlet. Opaque minerals also form within the interaction between the veinlet and the host diorite minerals. **d** Strongly altered diorite showing features of opaque minerals and alteration that is similar to the alteration caused by veinlet in (**c**). *Abbreviations* px—pyroxene, amp—amphibole, bio—biotite, pl—plagioclase, ab—albite, qz—quartz, ser—sericite, cab—carbonates, opq—opaque minerals (i.e. pyrite and gold)

4 Discussion

Mineralogy and Role of Host Rocks in the Gold Mineralization Process at Subika

Studies of least altered gabbroic and dioritic rocks under thin sections revealed the modal composition of the rocks. To confirm the lithological constituents, the mineral modal abundances for these least altered samples were plotted on a QAP diagram [30] using the GCD toolkit software [30]. In the QAP diagram (Fig. 11), least altered gabbroic and dioritic samples mainly plot in the field of gabbro and diorite, respectively. Epidotization and sericitization alteration assemblages in some of the diorite

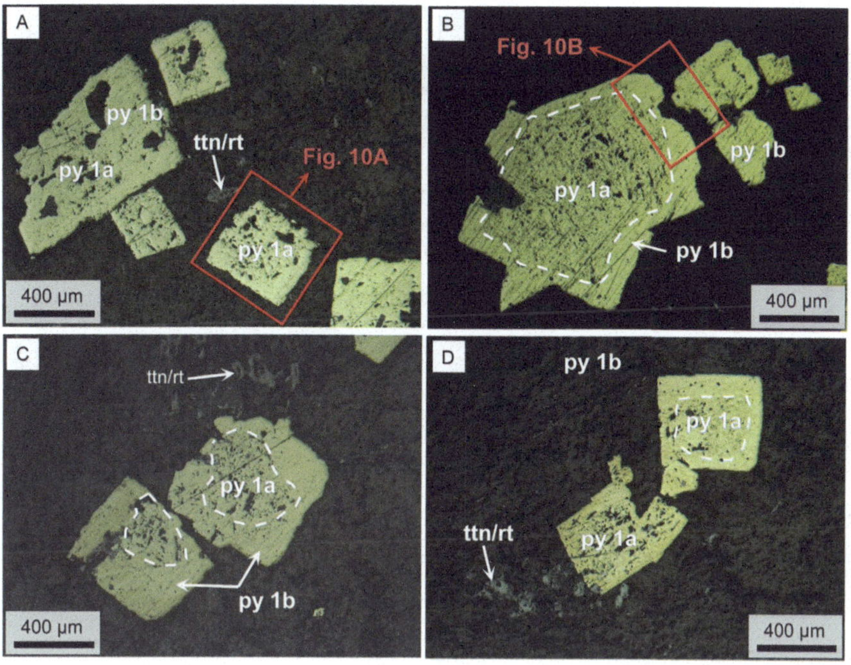

Fig. 9 Photomicrographs of Altered Samples Showing Coarse-Grained pyrite (py 1) with Porous Core (py 1a) and Massive Rim (py 1b) in **a** & **b** altered diorite, **c** & **d** altered gabbro. *Abbreviation* ttn—titanite, rt—rutile

samples suggest that the earlier alteration assemblage of epidote and fine sericite was likely due to metamorphism, which occurred before the hydrothermal gold mineralization event(s) in the area. Figure 12 illustrates the paragenetic sequence of minerals at Subika [11, 23]. The sequence presented here does not include the tectonic and/or structural fabrics reported by Masurel et al. [11] or Baah-Danso [23]. The first column shows the magmatic substage, which is identified by the least altered minerals of the Subika diorite-gabbro composite. The second column represents the metamorphic substage during which greenschist facies metamorphic minerals, dominated by epidote, sericite and chlorite, were formed. These two substages define the Pre-Ore Stage at Subika. The Syn-Ore Stage is characterized by the distal and proximal hydrothermal assemblages and the mineralization process associated with this Stage is expatiated below.

At Subika, the gabbro contains more pyroxene but less hornblende and biotite than diorite (Tables 1 and 2, respectively). Hornblende and biotite are enriched in Fe in comparison to pyroxene. This means that empirically, gabbro is less enriched in Fe than diorite for interaction with hydrothermal ore-forming fluid to form the pyrite-gold assemblages observed from ore microscopy. It was observed from the petrographic analysis that the altered gabbroic and dioritic rocks have the same

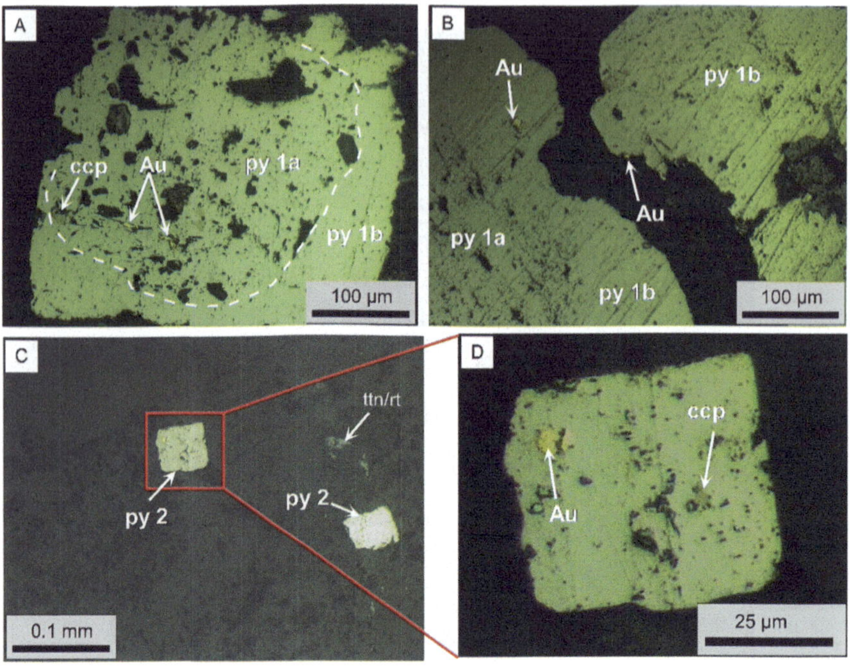

Fig. 10 Gold occurrences in pyrite **a** & **b** as hole-fill inclusions in core of coarse pyrite (py 1a) as well as open space fills in gangue minerals adjacent to pyrite, **c** & **d** as tight inclusions in fine pyrite (py 2). *Abbreviation* ttn—titanite, rt—rutile

alteration features and pyrite-gold association in comparison with the thin quartz-carbonate veinlets in the least altered gabbro (Figs. 6d and 8d). This implies that hydrothermal fluids of same composition passed through both lithologies. The quartz-carbonate veins imply that the mineralizing hydrothermal fluid was rich in CO_2 and SiO_4^{4-} [27]. The association of Au and pyrite means that Au was most likely transported in the fluid as a sulfide complex of gold, e.g. $Au(HS)_2^-$ (Eq. 1). The observance of opaque minerals (i.e. pyrite ± chalcopyrite ± gold) within the quartz-carbonate vein in the least altered gabbro and the subsequent sulfide dissemination associated with the strongly altered gabbroic and dioritic rocks (Figs. 6d and 8d) suggest that the quartz-carbonate veins are directly related to the formation of the opaque minerals (pyrite ± chalcopyrite ± gold) during hydrothermal alteration of the host rocks.

Thus, the gold mineralization process in the area began with fracturing of host rock(s) (Fig. 13) along the MFZ and accessed by the hydrothermal fluid through MFZ fracture (Figs. 4 and 13). As the fluid propagates upwards, it interacts with the host dioritic and gabbroic rocks, and reacts with the primary minerals in the wall rock as illustrated in Fig. 13. During fluid-rock interaction, CO_2 in the fluid reacts with Ca and Mg from the mafic minerals (e.g. pyroxene, hornblende and biotite) to form the carbonates in the altered rock (Eq. 1). Iron from hornblende, biotite

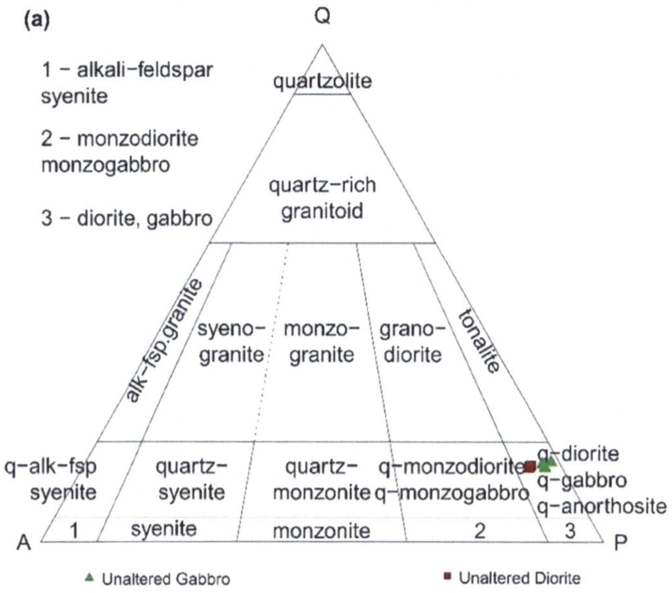

Fig. 11 QAP diagram (after [30]) showing plots of least altered gabbro and least altered diorite from the Subika Deposit

and minor amounts from pyroxene) reacts with sulfur from the gold-sulfur complex to form pyrite (FeS_2) and the associated Au (Figs. 9 and 10), in accordance to Eq. 1. Equation 1 is one of the possible reactions deduced from the textural relations observed in Figs. 6, 8, 9, and 10, in which the opaque minerals (i.e. pyrite and gold) are formed proximal to or within the quartz-carbonate veins that crosscut the rock-forming mineral assemblages constituting the host rocks. The $H^+_{(aq)}$ and $SiO_4{}^{4-}{}_{(aq)}$ in solution combines with oxygen to form water and the precipitation of SiO_2 as quartz in the veins (Eq. 1). These fluid-rock reactions result in the common observation of carbonatization, sericitization and sulfidation zones surrounding quartz-carbonate veins. The original rock texture is destroyed due to the intense heat and chemical reactions involved during the hydrothermal alteration process, which resulted in the replacement of primary minerals by alteration minerals (Figs. 6 and, 8c, d). The reaction (Eq. 1) is common for most hydrothermal alteration. As per the nature of the wall rock constituents, hornblende and biotite can also be used in Eq. 1 in lieu of pyroxene.

$$\underset{\text{(pyroxene)}}{CaFeSi_2O_6} + CO_2 + \underset{\text{(gold-sulfide complex)}}{Au(HS)_2^-} + H^+ \rightarrow \underset{\text{(carbonate)}}{CaCO_3} + \underset{\text{(pyrite)}}{FeS_2} + 2SiO_2 + H_2O + Au \tag{1}$$

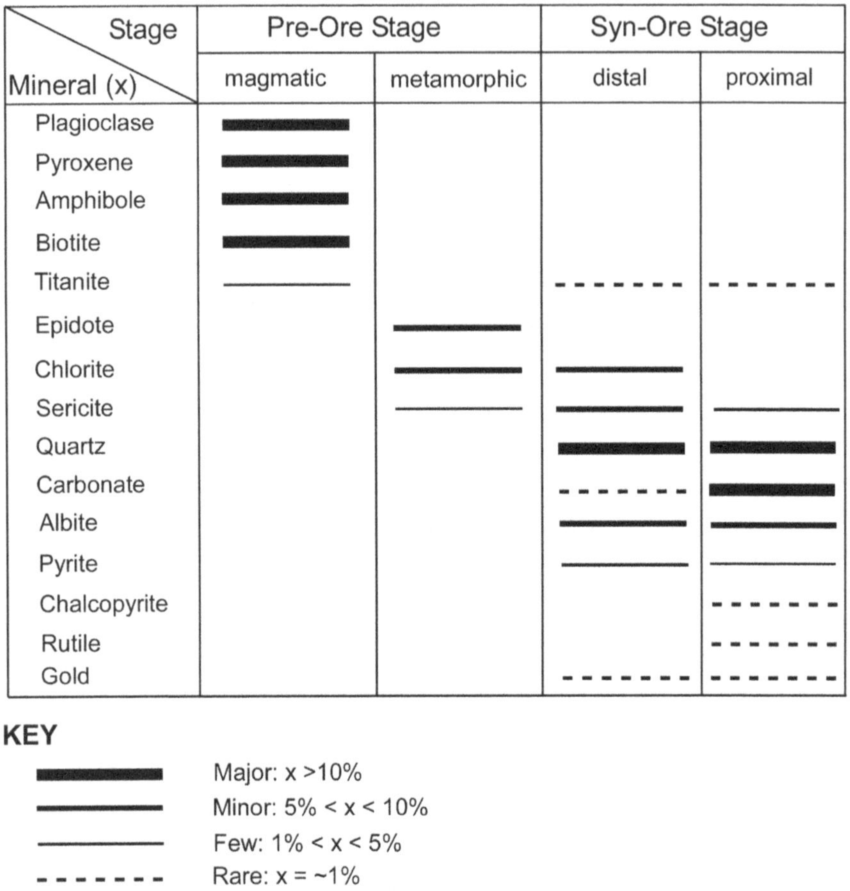

Fig. 12 Paragenetic sequence of minerals at Subika

Association of Pyrite-Types with Gold: The Implications

Pyrite is the dominant sulfide mineral in the Subika deposit. Petrographic observation reveals fine-grained and coarse-grained pyrite based on distinct morphology, texture and mineral associations. The key textural and morphological features as well as associated minerals of the individual types of pyrite are discussed below.

Coarse-grained pyrites occur as euhedral to subhedral grains with grain size greater than 500 μm. They are characterized by a porous core (py 1a) and a massive rim (py 1b) (Figs. 9 and, 10a, b). They are associated with chalcopyrite and rutile. Some of the porous core of the coarse pyrite contain Au (Fig. 10a). Therefore, Au occurs as hole-fills (Fig. 10a) and in the open-space adjacent to the pyrite (Fig. 10b), which indicates that Au mineralization was later than the formation of the coarse pyrite py 1.

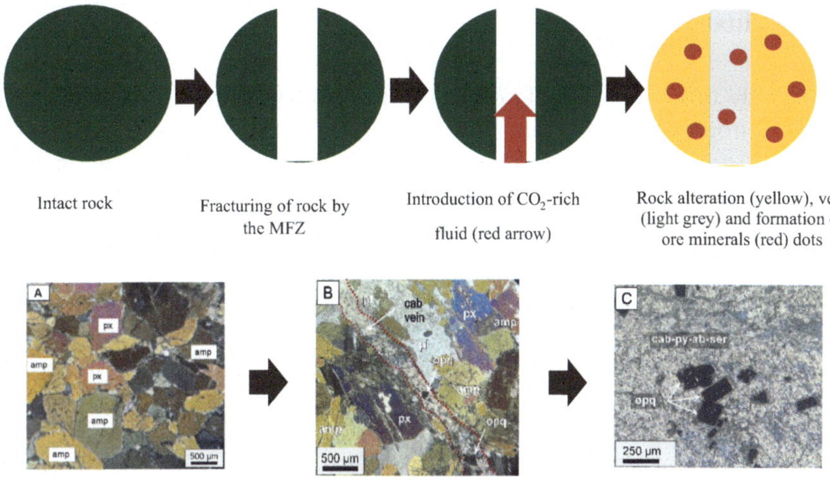

Fig. 13 Simplified ore-forming scenario in mafic rocks of the Subika deposit, signifying the litho-chemical control on the formation of ore minerals during fluid-rock interaction

Fine-grained pyrites are characterized by euhedral grains with grain size less than 500 µm (Fig. 10c, d). They have massive texture with no clear porosity or cavities i.e. they are non-porous. However, Au was observed as inclusions in this fine-grained pyrite, with no inclusion of gangue mineralogy as described for py 1. Therefore, the nature of inclusion of Au in the py 2 (Fig. 10d) suggests that Au was incorporated into py 2 during the precipitation of py 2, and thus, formed at the same time as the fine-grained pyrite.

The contrasting pyrite textures suggest a change in the physico-chemical conditions during ore formation. The porous core may indicate an unstable environment that allowed for rapid fluid movement and gas escape, leading to the formation of voids [31–33]. In contrast, the massive texture at the rim could indicate a more stable environment with less fluid activity, leading to a denser packing of the pyrite crystals [32–34]. The textural variations could also reflect changes in pH, temperature, activity of sulfur/oxygen, rate of cooling, and/or the availability of sulfur or iron [32–34]. Further analytical studies would be required to establish whether one or a combination of factors controlled the observed ore textures at Subika. Nevertheless, it is very clear that both pyrite types, coarse- and fine-grained, are host to gold grains albeit differences may be in the nature of occurrence of gold with the sulfides. This observation is in contrast with the idea by company geologists in Ghana that only fine-grained pyrite is closely associated with gold in the deposits. Therefore, care must be taken in handling rock samples from drill core, mine/pit exposures or outcrops that contain coarse pyrite grains.

5 Conclusion

In this study of the Subika gold deposit in the northern part of the Sefwi Belt, (Ahafo), petrographic analysis has aided in the (a) characterization of the host rocks of the Subika deposit, (b) elucidation of the role of these host rocks during gold mineralization and, (c) determination of the relationship between gold occurrences and at least two pyrite-types contained in these host rocks. The conclusions from the findings are summarized below:

(i) Transmitted light microscopy revealed that the least altered samples of the host rocks were constituted by varying amounts of the following dominant minerals: pyroxene, hornblende, biotite, and plagioclase. The plots of the modal abundance data on QAP diagram indicated that the host rocks are gabbro and diorite.
(ii) The dominant alteration minerals associated with the gold mineralization process included sericite, carbonates, chlorite, albite, minor pyrite and rare chalcopyrite. Epidotization in diorite is considered a metamorphic alteration, prior to the hydrothermal gold mineralization event.
(iii) Reflected light (ore) microscopic studies of altered gabbro and diorite samples revealed two types of pyrite: coarse pyrite (>500 μm; py 1) and fine pyrite (<500 μm; py 2). The coarse-grained pyrites were porous in the core (py 1a) and massive (py 1b) at the rim and inclusions of gold occurred as cavity- and fracture-fills in them. On the other hand, the py 2 are massive and non-porous in nature but with inclusions of gold.
(iv) Even though it is difficult to establish the paragenetic sequence of pyrite formation between py 1 and py 2, the nature of occurrence of gold in the various pyrite grains reflects the timing of gold mineralization with respect to these sulfides. Gold precipitation was synchronous with the formation of py 2 but post-dated py 1. Therefore, both fine pyrite and coarse pyrite can host gold, which is in contrast to the general idea that coarse pyrite is not associated with gold mineralization in Ghana.
(v) This study reveals that the characterization of the hydrothermal ore system at Ahafo and surrounding deposits must be carefully reviewed to enhance a better understanding of the ore formation mechanism in the deposits.

6 Recommendation

Care must be taken in handling rock samples from drill core, mine/pit exposures or outcrops that contain coarse pyrite grains so that these rocks are not necessarily misclassified as waste. It will be important to carry out mineralogical studies (using petrographical and QEMSCAN techniques from time to time to establish the nature of gold occurrences in such rocks that bear the coarse-grained pyrites.

Acknowledgements We are thankful to the Management of Newmont Ghana Ltd for the access to the mine to collect samples and publish these results. Part of this work was presented at the 8th Biennial Mining and Minerals International Conference in Tarkwa on August 2024 and the Ghana Institution of Geoscientists Conference (GhIGCON) in Sunyani on October 2024. We thank the Editors for their fruitful comments and suggestions on this chapter.

References

1. Zoheir BA (2012) Microchemistry and stable isotope systematics of gold mineralization in a gabbro-diorite complex, SE Egypt. Microchem J 103:148–157. https://doi.org/10.1016/j.microc.2012.02.003
2. Ghose NC (2014) Occurrence of native gold and gold-silver alloy in the olivine gabbro of layered cumulate sequence of Naga Hills ophiolite, India. Curr Sci Assoc 106(8):1125–1130
3. Saad W, Aissa DE, Watanabe K, Taguchi S (2018) Gold deposits associated with the gabbroic rocks at Tirek area, Western Hoggar, Algeria: fluid inclusion study. Arab J Geosci 11(26):1–10. https://doi.org/10.1007/s12517-017-3366-5
4. Popal A, Kakar IM, Khan M (2019) Geology and petrography of gabbroic rocks from Khanozai Ophiolite, Northwestern Pakistan. Int Res J Earth Sci 7(3):10–22
5. Haldar SK, Tišljar J (2014) Introduction to mineralogy and petrology. Elsevier, Netherlands, pp 93–120
6. Lagat J (2008) Hydrothermal alteration mineralogy in geothermal fields with case examples from Olkaria Domes Geothermal Field, Kenya, Short Course III on exploration for geothermal resources, organized by UNU-GTP and KenGen, Lake Naivasha, Kenya, p 26
7. Pirajno F (2009) Hydrothermal processes and mineral systems. Spinger, Dordrecht, pp 73–164
8. Adomako-Ansah K, Ishiyama D Allen R (2018) Origin of Zn–Pb–Sb–Au mineralization adjacent to the Paleoproterozoic Boliden Au-rich VMS deposit, Sweden: evidence from petrographic and oxygen isotope characteristics. Mineralium Deposita 53:683–700. https://doi.org/10.1007/s00126-017-0765-1
9. Adomako-Ansah K, Mizuta T, Ishiyama D, Hammond NQ (2017) Nature of ore-forming fluid and formation conditions of banded iron formation-hosted gold mineralization in the Amalia Greenstone Belt, South Africa. Ore Geol Rev 89:609–626. https://doi.org/10.1016/j.oregeorev.2017.06.021
10. Newmont Ghana Gold Limited Ahafo South Mine (2006) Ahafo Geology, Unpublished Report, p 57
11. Masurel Q, Morley P, Thébaud N, McFarlane H (2021) Gold deposits of the ~15-Moz Ahafo South Camp, sefwi granite-greenstone belt, Ghana: insights into the anatomy of an orogenic gold plumbing system. Econ Geol 116(6):26. https://doi.org/10.5382/econgeo.4829
12. Leube A, Hirdes W, Mauer R, Kesse GO (1990) The early proterozoic Birimian supergroup of Ghana and some aspects of its associated gold mineralization. Precambr Res 46:139–165. https://doi.org/10.1016/0301-9268(90)90070-7
13. Feybesse, JL Billa M, Guerrot C, Duguey E, Lescuyer JL, Milési JP, Bouchot V (2006) The paleoproterozoic Ghanaian province: geodynamic model and ore controls, including regional stress modeling. Precambrian Res 149:149–196. https://doi.org/10.1016/j.precamres.2006.06.003
14. Allibone A, Teasdale J, Cameron G, Etheridge M, Uttley P, Soboh A, Appiah-Kubi L, Adanu A, Arthur R, Mamphey J, Odoom B, Zuta J, Tsikata A, Pataye F, Famiyeh S, Lamb E (2002) Timing and structural controls on gold mineralization at the Bogoso gold mine", Ghana, West Africa. Econ Geol 97:949–969. https://doi.org/10.2113/gsecongeo.97.5.949
15. Allibone A, Hayden P, Cameron G, Duku F (2004) Paleoproterozoic gold deposits hosted by albite- and carbonate-altered tonalite in the Chirano district, Ghana, West Africa. Econonomic Geology 99:479–497. https://doi.org/10.2113/gsecongeo.99.3.479

16. Perrouty S, Aillères L, Jessell MW, Baratoux L, Bourassa Y, Crawford B (2012) Revised Eburnean geodynamic evolution of the gold-rich southern Ashanti Belt, Ghana, with new field and geophysical evidence of pre-Tarkwaian deformations. Precambr Res 204–205:12–39. https://doi.org/10.1016/j.precamres.2012.01.003
17. Dzikunoo EA, Kazapoe RW, Agbetsoamedo JE (2021) An integrated structural and geophysical approach to defining the structures of part of the nangodi greenstone belt, Northeastern Ghana. J African Earth Sci 180:104–238. https://doi.org/10.1016/j.jafrearsci.2021.104238
18. Chudasama B, Porwal A, Kreuzer OP, Butera K (2016) Geology, geodynamics and orogenic gold prospectivity modelling of the Paleoproterozoic Kumasi Basin, Ghana, West Africa. Ore Geol Rev 78:692–711. https://doi.org/10.1016/j.oregeorev.2015.08.012
19. McFarlane H, Thébaud N, Parra-Avila L, Armit R, Spencer C, Ganne J, Aillères L, Baratoux L, Betts P, Jessell M (2019) Onset of the supercontinent cycle: evidence for multiple oceanic arc accretion events in the Paleoproterozoic Sefwi greenstone belt of the West African craton. Precambr Res 335:105450. https://doi.org/10.1016/j.precamres.2019.105450
20. McFarlane H, Aillères L, Betts P, Ganne J, Baratoux L, Jessell M, Block S (2019) Episodic collisional orogenesis and lower crust exhumation during the Paleoproterozoic Eburnean orogeny—evidence from the Sefwi greenstone belt, West African craton. Precambr Res 325:88–110. https://doi.org/10.1016/j.precamres.2019.02.012
21. Dzigbodi-Adjimah K, Nana Asamoah D (2009) The geology of the gold deposits of prestea gold belt of Ghana. Ghana Mining J 11:7–18. https://doi.org/10.4314/gm.v11i1.53268
22. Jessell MW, Amponsah PO, Baratoux L, Asiedu DK, Loh GK, Ganne J (2012) Crustal-Scale transcurrent shearing in the paleoproterozoic Sefwi-Sunyani-Comoe Region, West Africa. Precambr Res 212:155–168. https://doi.org/10.1016/j.precamres.2012.04.015
23. Baah-Danso E (2011) The structural evolution of the Subika deposit, Ahafo, Sefwi belt, Ghana. M.Sc. thesis, Perth, The University of Western Australia, p 82
24. Takyi-Kyeremeh K, Wemegah DD, Preko K, Menyeh A (2019) Integrated geophysical study of the subika gold deposit in the sefwi belt, Ghana. Cogent Geoscience 5(1):1–17. https://doi.org/10.1080/23312041.2019.1585406
25. Kisters AFM, Meyer FM, Znamensky SE, Seravkin IB, Ertl RGW, Kosarev AM (2000) Structural controls of lode-gold mineralization by mafic dykes in late-Paleozoic granitoids of the Kochkar district, southern Urals, Russia. Miner Deposita 35:157–168. https://doi.org/10.1007/s001260050013
26. Wilkinson JJ, Johnston JD (1996) Pressure fluctuations, phase separation, and gold precipitation during seismic fracture propagation. Geology 24:395–398. https://doi.org/10.1130/0091-7613(1996)024%3C0395:PFPSAG%3E2.3.CO;2
27. Goldfarb RJ, Groves DI (2015) Orogenic gold: common or evolving fluid and metal sources through time. Lithos 233:2–26. https://doi.org/10.1016/j.lithos.2015.07.011
28. Damien G (2019) Parameters for the formation of orogenic gold deposits. Appl Earth Sci (Trans Inst Min Metall B) 128(3):124–133. https://doi.org/10.1080/25726838.2019.1583310
29. Fiadonu EB, Hammond NQ, Dzigbodi-Adjimah K, Adomako-Ansah K (2023) Ore fluid characteristics and depositional processes of the hydrothermal gold mineralization at Damang, Southwestern Ghana. J Afr Earth Sc 197:104–141. https://doi.org/10.1016/j.jafrearsci.2022.104741
30. Streckeisen A (1974) Classification and nomenclature of plutonic rocks: recommendations of the IUGS subcommission on the systematics of igneous rocks. Geol Rundsch 63:773–786
31. Janoušek V, Farrow CM, Erban V (2006) Interpretation of whole-rock geochemical data in igneous geochemistry: introducing geochemical data toolkit (GCDkit). J Petrol 47:1255–1259. https://doi.org/10.1093/petrology/egl013
32. Large RR, Maslennikov VV, Robert F, Danyushevsky LV, Chang Z (2007) Multistage sedimentary and metamorphic origin of pyrite and gold in the giant sukhoi log deposit, Lena Gold Province, Russia. Econ Geol 102:1233–1267. https://doi.org/10.2113/gsecongeo.102.7.1233
33. Wu YF, Evans K, Hu SY, Fougerouse D, Zhou MF, Fisher LA, Guagliardo P, Li JW (2021) Decoupling of Au and As during rapid pyrite crystallization. Geology 49(7):827–831. https://doi.org/10.1130/G48443.1

34. Sun QF, Tang JX, Li SD, Pei QM, Sun XH, Chen C, Gao LL, Du XF, Xie JL (2024) Pyrite geochemistry and its implications for ore formation in reduced copper skarn systems: a case study of the Saibo deposit, Northwest Chian. Chem Geol 671:122478. https://doi.org/10.1016/j.chemgeo.2024.122478

Geology, Mineralogy, Geochemistry, Fluid Evolution and Genesis of Hydrothermal Gold Deposits in Ghana

Footprint of Hydrothermal Gold Mineralization in the Kawere Sandstone, Tarkwaian Group, Ghana: Findings from Two Drillhole Intersections at the Badukrom Prospect, Iduapriem

Kofi Adomako-Ansah, Charles Kusi-Manu, Michael Affam, Emmanuel Amekudzi, Yamoa Munko-Abo, Divine Kwapong, Kofi Poku Osei, and Justice Lawerh

1 Introduction

The global exploration of minerals has grown progressively more difficult as the majority of medium-depth and surface deposits have been identified. Consequently, the concurrent and cross-disciplinary use of knowledge derived from geological, geochemical, and mineralogical data is crucial for contemporary mineral exploration. Broadly, gold deposits in Ghana can be categorized into three primary types: (a) Hydrothermal gold deposits, previously believed to be exclusively linked to the Birimian Supergroup, have recently been discovered and are currently mined from the Tarkwaian Group as well [1–8]; (b) Paleo placer deposits, which are exclusively

K. Adomako-Ansah (✉) · C. Kusi-Manu · M. Affam · Y. Munko-Abo
Department of Geological Engineering, University of Mines and Technology, Tarkwa, Ghana
e-mail: kadomakoansah@umat.edu.gh

M. Affam
e-mail: maffam@umat.edu.gh

C. Kusi-Manu · D. Kwapong · K. P. Osei · J. Lawerh
Geology Department, AngloGold Ashanti Iduapriem Limited, Tarkwa, Ghana
e-mail: DKwapong@anglogoldashanti.com

K. P. Osei
e-mail: KPOsei@anglogoldashanti.com

J. Lawerh
e-mail: JLawerh@anglogoldashanti.com

E. Amekudzi
Department of Earth, Ocean and Atmospheric Science, University of British Columbia, Vancouver, Canada

E. D. Sunkari and S. Ali (eds.), *Geology and Gold Mineralization in Ghana*, Earth and Environmental Sciences Library, https://doi.org/10.1007/978-3-031-97163-1_5

associated with the Tarkwaian Group; and (c) Recent alluvial and eluvial deposits, primarily found in stream gravels and regarded as erosional products of the aforementioned geological groups [9]. Each of the aforementioned deposit types possesses distinct properties that render them suitable for exploration; thus, accurate classification of the mineralization style can enhance the discovery and extractability of these ore deposits.

For decades, AngloGold Ashanti Iduapriem Mine, Goldfields Ghana Tarkwa Mine, and Goldfields Ghana Damang Mine continue to exploit the paleoplacer gold mineralization of the Tarkwaian Group. In addition, the Damang gold deposit contains overprinting hydrothermal gold mineralization, which has garnered attention in recent studies [1–4, 6–8, 10] and ongoing operations by Gold Fields Ghana Mine. Currently, around 80% of gold mined at Damang is from the hydrothermal type. A similar hydrothermal gold mineralization has recently been intercepted at Badukrom within the Iduapriem concession of AngloGold Ashanti Iduapriem Ltd [5]. The concession is located approximately 30 km from the Damang deposit. In contrast to the Damang deposit, information regarding the geological and geochemical characteristics of the mineralization in Badukrom is scarce.

This study on the Badukrom prospect aims to describe the lithological and geochemical attributes of hydrothermal mineralization at Badukrom by:

(a) ascertaining rock types and stages of hydrothermal alteration based on field and petrographical features;
(b) correlating geochemical data from the drill holes to identify significant patterns and trends associated with the alteration haloes; and
(c) employing multivariate statistical analysis on the geochemical data to establish the elemental associations of the hydrothermal alteration and identify potential pathfinders to mineralization.

These data sets will be utilized to enhance the comprehension of hydrothermal mineralization and its implications for exploration targeting in the Badukrom area.

2 Materials and Methods

Description of Study Area

History of AngloGold Ashanti Iduapriem Limited

In Ghana, AngloGold Ashanti manages two mines: (a) Obuasi, which conducts both surface and underground mining, and (b) Iduapriem, which performs exclusively surface mining. Ashanti Goldfields Company and South Africa-based AngloGold merged in 2004, resulting in the formation of AngloGold Ashanti Ltd. The Iduapriem mine was founded by Golden Shamrock Mines (GSM) of Australia and commenced full-scale operations in 1992, with a designed processing rate of 2.4 Mtpa. It was referred to as Ghana Australia Goldfields (GAG). Subsequently, Ashanti Goldfields Company (AGC), which was managing the renowned Obuasi gold deposit, acquired GAG in early 1997 [11]. As of September 1997, Ashanti Goldfields Company (AGC) acquired exclusive ownership of the Iduapriem Mine [11]. The Iduapriem Mine is an open-pit gold mine consisting of Block 1–5, Teberebie (Block 7 and 8), Ajopa and Ajopa South West. The Iduapriem mine consists of the Iduapriem and Teberebie concessions, covering an area of approximately 140 km^2. The Iduapriem mine is situated in the Western Region of Ghana, approximately 70 km north of the coastal city of Takoradi. The mine is situated 10 km southwest of Tarkwa and 240 km west of Accra.

Geology

The Iduapriem Concession

The Iduapriem concession is located within the Tarkwaian Group of rocks which unconformably overly the metavolcanic rocks of the Birimian Supergroup (Figs. 1 and 2) [12]. The entire Tarkwaian Group has been folded into a broad syncline which is locally referred to as the Tarkwa Syncline (see Fig. 2) [11]. The Iduapriem concession is located along the south-western margin of the Tarkwa syncline (Fig. 2). The main lithologies in Iduapriem block include the Huni Group (phyllites, conglomerates, sandstones), Tarkwa Phyllites, Banket Group (conglomerates, sandstones), Kawere Group (sandstones, conglomerates) and basal Sefwi Group (mafic volcanics, volcaniclastics and granitoids) (Fig. 2). Gold mineralisation is confined to the Banket Reef Zone (BRZ), which comprises a sequence of individual beds of quartz pebble conglomerate, breccia conglomerate, quartzite and grit. The outcropping Banket Group in the mine lease area forms prominent curved ridges that extend southwards from Tarkwa, westwards through Iduapriem and northwards towards Teberebie.

The deposit consists of eight blocks (i.e. Blocks 1–8) [11]. At Blocks 7 and 8 (Teberebie deposit), the western limb of the syncline extends over 4 km on the property, with the eastern limb reaching the surface (outcropping) just beyond the eastern boundary of the concession. The western and the eastern limbs outcrop about 4 km apart with the mineralised horizons buried some 400 m below the surface at the centre of the syncline [11].

The Ajopa deposit (Block 6), comprise of a series of north-east south-west (NE-SW) trending ridges with dips generally between 50 and 60° W. The gold is fine-grained, particulate and free milling (i.e. not locked up with quartz or iron oxides). Mineralogical studies indicate that the grain size of native gold particles ranges between 2 and 500 μm and averages 130 μm. Sulphide mineralisation is present only at trace levels and is not associated with the gold. Haematite and magnetite are often extremely well developed on cross bed fore sets. Widespread distribution of hematite is often found in the conglomerate matrix with rare visible gold grains [11].

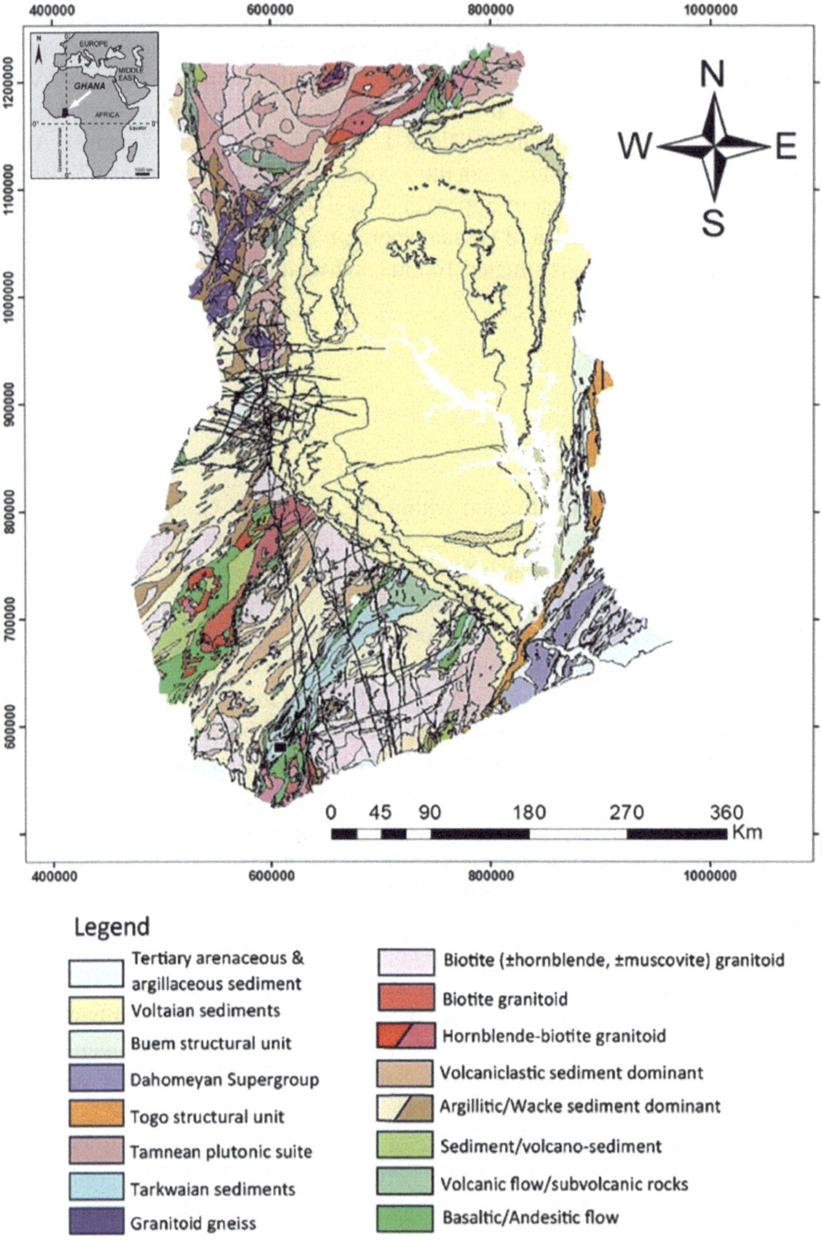

Fig. 1 Geological Map of Ghana, showing the location of the Tarkwa-Damang area (in black box) within the Tarkwaian Group of (meta) sediments. *Insert* Map of Africa showing the location of Ghana

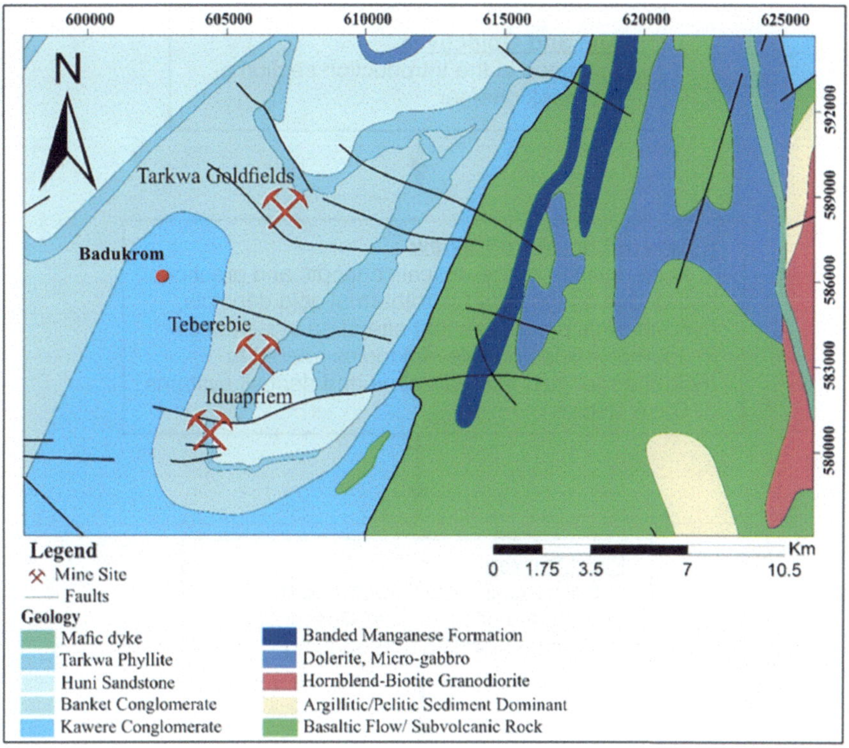

Fig. 2 Regional geology of the Tarkwa area, showing the location of the the Badukrom prospect (red dot)

The Badukrom Prospect

The hydrothermal target at Badukrom, a locality within the Iduapriem concession, sits within the Kawere Group at the base of the Tarkwaian Group (Fig. 2). Since the Badukrom prospect is not exposed at the surface, little information is known about the geological controls on the hydrothermal mineralisation. Osei et al. [5] conducted a correlative analysis and bivariate plots to determine that gold mineralisation, which is associated with pyrite + tourmaline ± carbonate alteration, is correlative with Sb, Fe, As, S and Te. It was also reported by Osei et al. [5] that the geochemical soil anomaly at the Badukrom prospect occur parallel to the regional scale anticline.

Data Collection and Acquisition

A flowchart of the methodology used for this study is shown in Fig. 3 and the details are outlined below

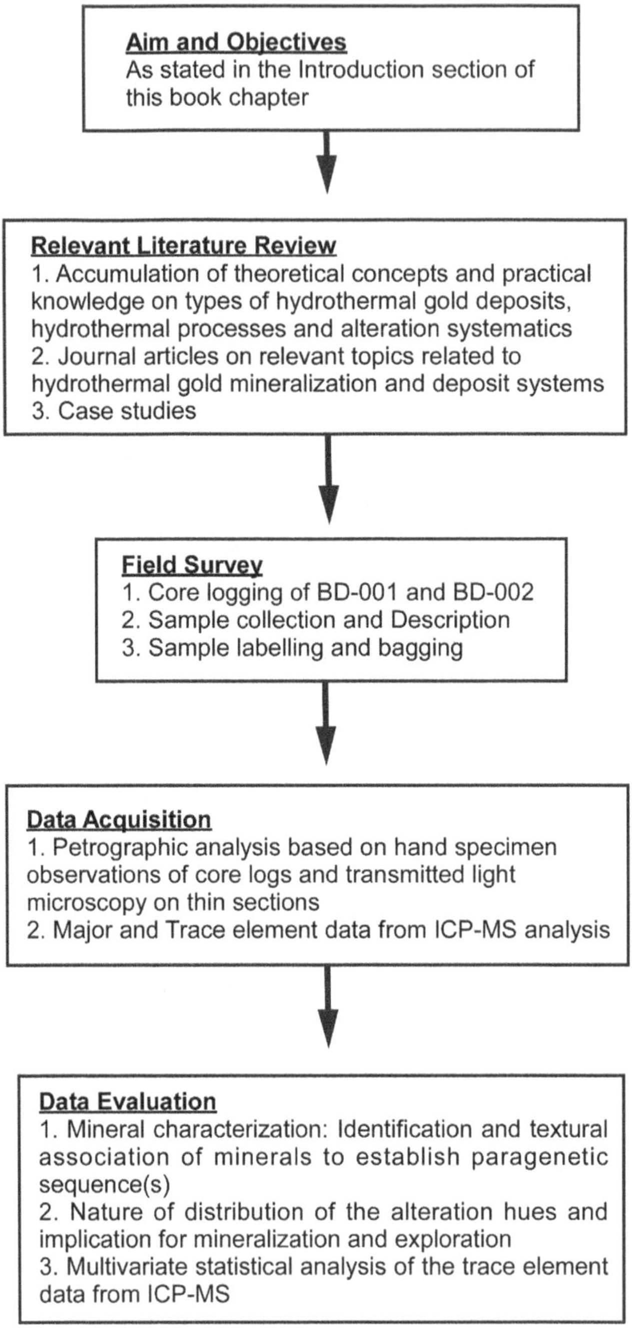

Fig. 3 Flowchart showing the methodology for conducting the studies on Badukrom Propsect

Selection of Samples

Due to the lack of field exposures only the drill cores were avaible for sampling at the time of conducting the field survey. Based on the aim and the objectives identified to tackle the challenge (see Sect. 1: Introduction), the study used 37 fresh samples collected from two-logged diamond drill holes (BD-001 and BD-002) from the Badukrom prospect to identify the hydrothermal characteristics of gold mineralization. This was executed in accordance with the sample acquisition protocol established by Adomako-Ansah et al. [13]. Fifteen (15) samples were chosen from areas exhibiting various color hues and mineral constituents (e.g. sulfides, tourmaline, carbonate and quartz veins) in the drill cores. Figures 4 and 5 display field images of the selected drill cores and grab samples.

Petrographic Examination

Nine samples (from BD 2–1 to BD 2–7A; Table 1) were selected for the preparation of thin and polished sections to identify the minerals, and textures associated with the hydrothermal gold mineralization in the Badukrom prospect. The Leica DM750P and Leica DM2700P petrographic microscopes, equipped with Leica MC 120HD digital cameras and linked to computers, were utilized for the analysis of the sections under transmitted and reflected light, respectively. The equipment is housed at the Petrology Laboratory of the Geological Engineering Department, University of Mines and Technology (UMaT), Tarkwa. Minerals were identified based on criteria that included color, isotropism/anisotropy, bireflectance, texture, twinning, extinction angle, cleavage, and pleochroism.

In hand specimen, classical description of core material was carried out, which included the identification of minerals, grain size, color, and microstructure for distinguishing between altered and unaltered rocks.

Geochemical Examination

For this study, twenty-five samples of quartz-tourmaline ± sulfide veins were selected for multi-element analysis [5]. The whole-rock multi-element analysis was conducted at SGS South Africa Pty Ltd in Johannesburg using four-acid digestion and subsequent analysis via ICP-OES and ICP-MS to obtain major and trace element geochemical data, respectively [14]. The four-acid digestion technique employed a mixture of nitric acid (HNO_3), hydrofluoric acid (HF), perchloric acid ($HClO_4$), and hydrochloric acid (HCl) to achieve near-total digestion [14]. Gold concentrations were determined by fire assay at SGS Tarkwa Ghana. Major elements are presented as wt% oxide, while trace elements are given as parts per million (ppm). The summary of the geochemical data is shown in Table 2.

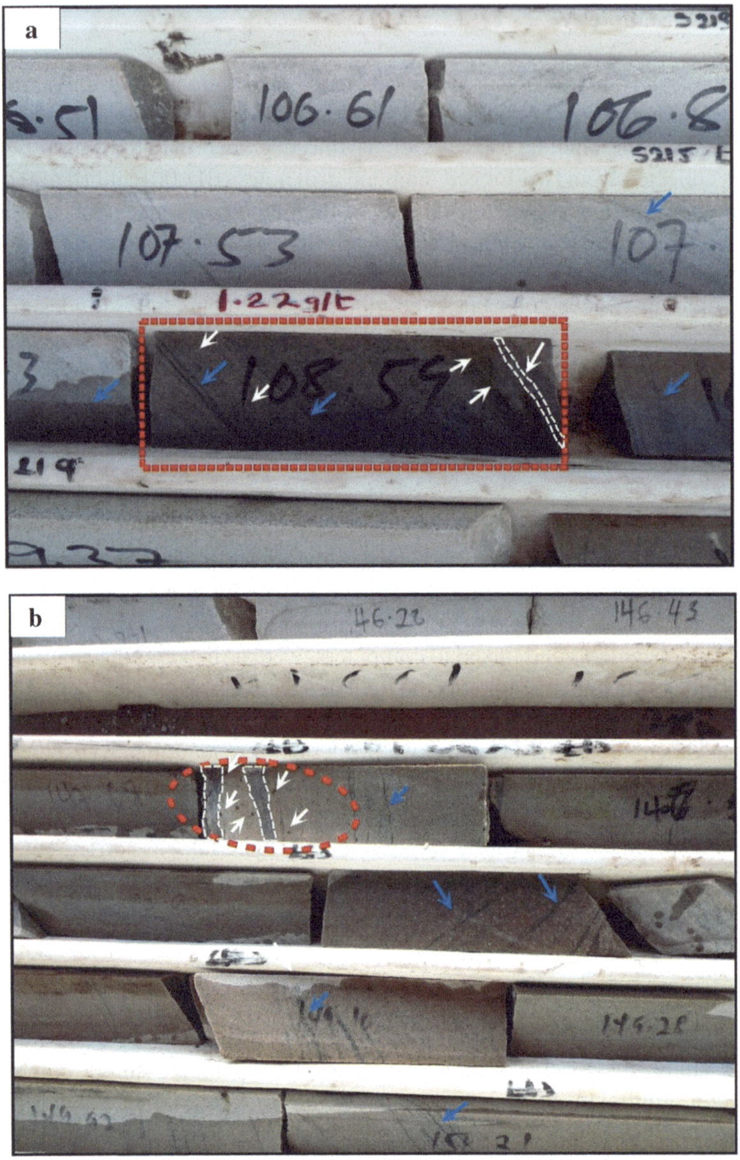

Fig. 4 Sections of drill hole BD-002 showing **a** light grey sandstone/quartzite **b** pink to light pink sandstone/quartzite. *Notes* Blue arrow = hematite ± magnetite along cross-bedded foresets; white arrow = disseminated sulfides (pyrite ± arsenopyrite); white dashed-line = quartz vein; Red square = disseminated sulfides in light grey sandstone/quartzite, with significant gold grade (1.22 g/t Au); Red circle = disseminated sulfides in light pink sandstone/quartzite but no significant Au concentrations

Fig. 5 Sulfide-tourmaline vein-rich samples from Badukrom. **a** Representative drill core sample showing quartz-tourmaline vein that is responsible for pyrite-arsenopyrite-gold mineralisation in phyllic- and propylitic-altered, light grey sandstone/quartzite. **b** Grab sample showing quartz-tourmaline vein which caused dominant tourmaline-sulfide-rich alteration. *Notes* qz—quartz, py—pyrite, apy—arsenopyrite, tur—tourmaline

Data Processing and Statistical Techniques

The geochemical data were examined by multivariate statistical analysis, which entails considering simultaneously two or more variables.

Table 1 Modal abundance of minerals in sandstone/quartzite samples

Sample No. (BD)	Modal abundance of minerals (%)									
	Quartz	Fine quartz	Plagioclase	Carbonate	Chlorite	Sulphide	Tourmaline	Sericite	Magnetite	Total
2–1A	30	30	5	15	–	–	–	20	–	100
2–5B	5	–	–	70	25	–	–	–	–	100
2–5C	–	30	10	40	–	10	10	–	–	100
2–5E	10	35	10	20	–	10	15	–	–	100
2–6	30	30	–	5	–	15	–	20	–	100
2–7A	25	5	20	20	5	10	5	10	–	100
2–7B	30	–	10	15	5	–	5	250	10	100
2–8A	20	25	15	15	–	–	15	10	–	100

Table 2 Summary statistics of the multi-element geochemical data obtained

	Range	Minimum	Maximum	Mean	Std. Deviation	Variance	Skewness	Kurtosis
Ag	1.00	0.30	1.30	0.52	0.26	0.07	1.41	1.83
Al	7.66	0.19	7.85	4.04	1.79	3.21	−0.34	0.36
As	9969.00	31.00	10,000.00	5687.52	4614.73	21,295,768.76	−0.29	−1.87
Au	29.02	0.08	29.10	2.82	5.77	33.31	4.27	19.65
Ba	1208.00	52.00	1260.00	641.76	337.59	113,966.19	−0.43	−0.60
Be	1.70	0.10	1.80	0.95	0.41	0.17	0.05	0.27
Bi	14.49	0.21	14.70	1.90	2.90	8.39	3.89	17.08
Ca	1.91	0.10	2.01	0.85	0.42	0.18	0.95	1.60
Cd	27.17	0.03	27.20	1.31	5.41	29.25	4.96	24.71
Ce	44.67	7.03	51.70	36.54	10.08	101.52	−0.91	1.67
Co	37.70	2.10	39.80	12.32	7.86	61.83	2.06	5.36
Cr	122.00	6.00	128.00	27.00	25.01	625.58	2.96	11.05
Cs	2.99	0.15	3.14	1.69	0.87	0.75	−0.56	−0.63
Cu	90.70	1.10	91.80	14.42	18.83	354.39	3.16	12.12
Eu	1.24	0.18	1.42	0.92	0.23	0.05	−0.98	3.92
K	2.45	0.04	2.49	1.19	0.80	0.64	−0.13	−1.23
La	22.00	3.60	25.60	17.57	5.13	26.30	−0.65	0.76
Li	64.00	2.00	66.00	20.52	18.28	334.09	1.07	0.18
Mg	0.84	0.03	0.87	0.36	0.17	0.03	1.11	3.05
Mn	543.00	66.00	609.00	348.84	145.60	21,198.56	−0.10	−0.89
Mo	3.46	0.91	4.37	1.63	0.74	0.54	2.47	7.40
Na	4.41	0.16	4.57	1.44	1.22	1.50	0.89	0.11
Nb	3.80	0.20	4.00	1.44	0.89	0.78	0.96	1.33
Nd	21.40	3.40	24.80	17.58	4.92	24.18	−0.80	1.44
Ni	33.00	3.50	36.50	16.68	6.69	44.76	0.91	2.13
P	1866.00	244.00	2110.00	736.28	366.19	134,093.46	2.72	8.72
Pb	240.20	6.80	247.00	25.81	46.57	2168.35	4.84	23.84
Pr	5.25	0.86	6.11	4.34	1.22	1.49	−0.76	1.27
Rb	75.20	2.90	78.10	41.36	22.75	517.77	−0.35	−0.83
S	4.99	0.01	5.00	0.93	1.01	1.03	2.90	10.79
Sb	16.13	0.37	16.50	3.29	3.29	10.84	3.02	10.92
Sc	18.40	1.30	19.70	9.48	4.65	21.66	0.91	0.66
Se	5.00	2.00	7.00	3.04	1.17	1.37	1.61	4.06
Sn	11.50	0.40	11.90	1.70	2.21	4.88	4.43	20.90
Sr	411.70	27.30	439.00	210.83	95.00	9024.60	0.29	0.32
Ta	0.59	0.05	0.64	0.34	0.16	0.03	0.11	−0.86

(continued)

Table 2 (continued)

	Range	Minimum	Maximum	Mean	Std. Deviation	Variance	Skewness	Kurtosis
Te	31.81	0.19	32.00	10.69	9.61	92.28	0.54	−0.74
Th	5.10	0.40	5.50	2.55	1.00	1.01	0.90	2.60
Ti	0.09	0.01	0.10	0.04	0.02	0.00	1.25	2.98
U	1.13	0.14	1.27	0.78	0.23	0.05	−0.31	1.75
V	251.00	13.00	264.00	63.92	50.16	2516.08	2.70	10.45
W	729.60	2.40	732.00	36.20	145.15	21,067.38	4.98	24.85
Y	7.40	0.90	8.30	4.44	1.32	1.75	0.24	3.57
Yb	0.90	0.10	1.00	0.71	0.19	0.03	−1.34	3.89
Zn	276.00	22.00	298.00	63.88	68.86	4741.28	2.82	7.54

P-P Plots

A potent graphical tool for stochastically comparing the magnitudes of two random variables is the P-P plot (also known as the probability–probability plot or percent–percent plot). It is a probability plot that is used to evaluate the degree of agreement between two data sets or the degree to which a dataset suits a specific model [15]. It functions by comparing the two cumulative distribution functions; if they are similar, the data will resemble a normal distribution and appear to be nearly a straight line. Conversely, the non-linearity of the distribution with respect to the straight line would indicate that the data set is not normally distributed [15, 16].

Centered-Log Ratio

Due to the frequent non-normal distribution of geochemical data, the range between the minimum and maximum values for each parameter is reduced to mitigate data abnormality [7, 17]. The formula used to achieve this purpose is shown as Eq. 1:

$$CLR(x) = \left(\log\left(\frac{x_1}{g(x)}\right), \ldots, \log\left(\frac{x_N}{g(x)}\right)\right) \tag{1}$$

where

CLR	Centered Log-Ratio (CLR)
x	Composition vector
$x_1, x_2, \ldots, x_N$	Single elements in the samples
$g(x)$	Geometric mean.

Spearman's Correlation

The Spearman's Rank correlation coefficient [18] quantifies the strength of a linear link between two variables. The Spearman correlation coefficient was utilized to calculate and present the correlation coefficient (r), which quantifies the extent of interrelations between gold and other elements. The Spearman correlation presupposes that the variables under investigation have a normal distribution in a graphical representation between gold and other elements. A correlation coefficient between 0.5 and 0.7 signifies moderate correlation, while a coefficient between 0.3 and 0.5 suggests low correlation. In addition, a correlation coefficient below 0.3 is considered negligible while above 0.7 is considered high. The highest possible correlation is indicated by $r = \pm 1$ and, a no correlation is indicated by $r = 0$. The correlation function is expressed as Eq. 2:

$$r = \frac{\sum (x_1 - \overline{x})(y_1 - \overline{y})}{\sum \sqrt{(x_i - x)^2 \sum (y_i - y)^2}} \tag{2}$$

where

- r Spearman's rank correlation coefficient
- x_i Values of the x-variable
- $\overline{x}$ Average of the x variable values
- y_I value of the y-variable
- $\overline{y}$ Values of the y-variable.

Factor and Hierarchical Cluster Analyses

Factor analysis utilizing principal component analysis for extraction and varimax rotation method and Hierarchical cluster analysis (HCA) were conducted in the SPSS environment to examine the dataset created to identify relationships between gold and the elements, and their possible origins.

The primary goal of factor analysis is to find hidden structures in multivariate data and to find fewer "factors" that can explain the variation in the data [17]. The number of components that needed to be removed was determined using the Kaiser criterion. Hence, only key components with eigenvalues of 1 or greater are extracted. Scree plots show the similarity index and covariance display the number of components [18].

In the case of the HCA, dendrograms were employed to illustrate the results of the hierarchical cluster analysis. The dendrogram displays the phases undertaken by the software to reach the solution, the multi-element clusters, and the distances between these clusters [19, 20]. Nude et al. [19] and Olise et al. [20] characterize the Hierarchical cluster analysis technique as the computation of similarity or dissimilarity among several objects. Subsequently, two components are grouped together with the condition that a specified agglomeration criterion will be diminished due to the clustering. The two items are in a class that is created by this clustering. Furthermore, two elements or elemental categories are grouped by assessing their similarities between the previously constructed class and the N-2 remaining components. This step is continued until all the components have been clustered.

3 Results and Discussion

Lithological Units

The stratigraphy and gold mineralization intercepted by the drill holes are illustrated in Fig. 3 The overall length of BD-001 and BD-002 is 260 m each, with the lithological units comprising sandstone and quartzite that are intermittently intruded by mafic dykes at various depths (Fig. 3). The sandstone and quartzite were recognized by the detrital characteristics of the mineral grains and matrix, along with the consistent, rhythmic presence of cross-bedded structures (Fig. 3). Coarse-to-fine, anhedral quartz and feldspar grains are prevalent in the least altered sandstone and quartzite, with the fine grains constituting the matrix. The quartz-feldspar grains typically measure between 0.25 and 0.5 mm (average 0.16 mm), categorizing them as medium sand to silt. In addition, the mineral grains in the quartzitic rocks are crystalline and have undergone recrystallization.

The hand specimen observations indicate that the grain size distribution in the sandstone-quartzite succession is as follows: Sand = 75%, Silt = 20%, Clay = 5%, suggesting that the sandstone/quartzite are composed of sandy silt material (Fig. 6a). Plotting the average modal abundances of the major minerals (Table 1) on a Quartz-Feldspar-Lithics (QFL) ternary diagram [21–23] reveals that all samples from the Kawere sandstone used in this study are subarkose and quartz arenite (Fig. 6b). Fine detrital grains of black hematite and magnetite were randomly dispersed within the sandstone and quartzite, occasionally appearing as narrow bands along the planes of cross-bedded foresets (Fig. 4). Magnetite and hematite were distinguished in the field using a magnetic pen. The magnetic pen was attracted to the magnetite-rich bands whereas the pen was unresponsive to the hematite-rich bands. Due to the resemblance in grain size distribution, pore structures, and cross-bedded structures, the quartzite is regarded as the metamorphic equivalent of the sandstone.

Interstitial mineral grains of fine- and patchy coarse-grained carbonate, chlorite and sericite within the matrix indicate that the alteration of carbonate and sericite occurred during the passage of hydrothermal fluids via pore spaces of the sandstone (Fig. 7a, b) and prior to the infiltration of ore-forming fluid. The presence of alteration-induced tourmaline and sulfides in the host rock matrix corroborates the interaction between the host rock and hydrothermal mineralizing fluid (Fig. 7c–f).

The sandstone and quartzitic units are intruded at various intervals by mafic dykes, at depths of 60 m, 160 m, 190.36 m, and 240 m in hole BD-001, and at 63 m, 73 m, and 197 m in hole BD-002. The lithological composition of these mafic dykes is predominantly diorite/basalt. 1–3 cm Quartz veins and veinlets are present within the sandstone at depths of 19 m, 46.3 m, 46.6 m, 101.3 m, 111 m, 120 m, etc. in BD001 and 83 m, 90 m, 102 m, etc. in BD002 (Fig. 3).

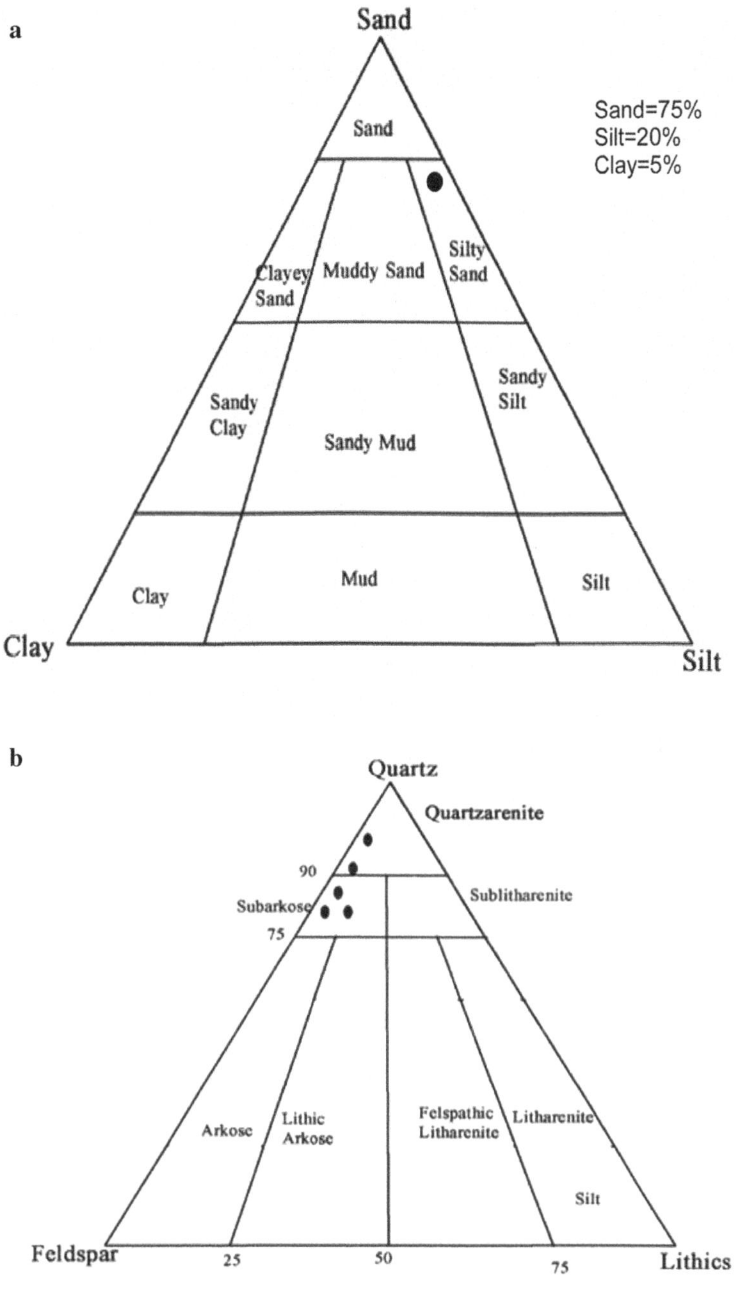

Fig. 6 Classification of the host sandstone/quartzite at Badukrom

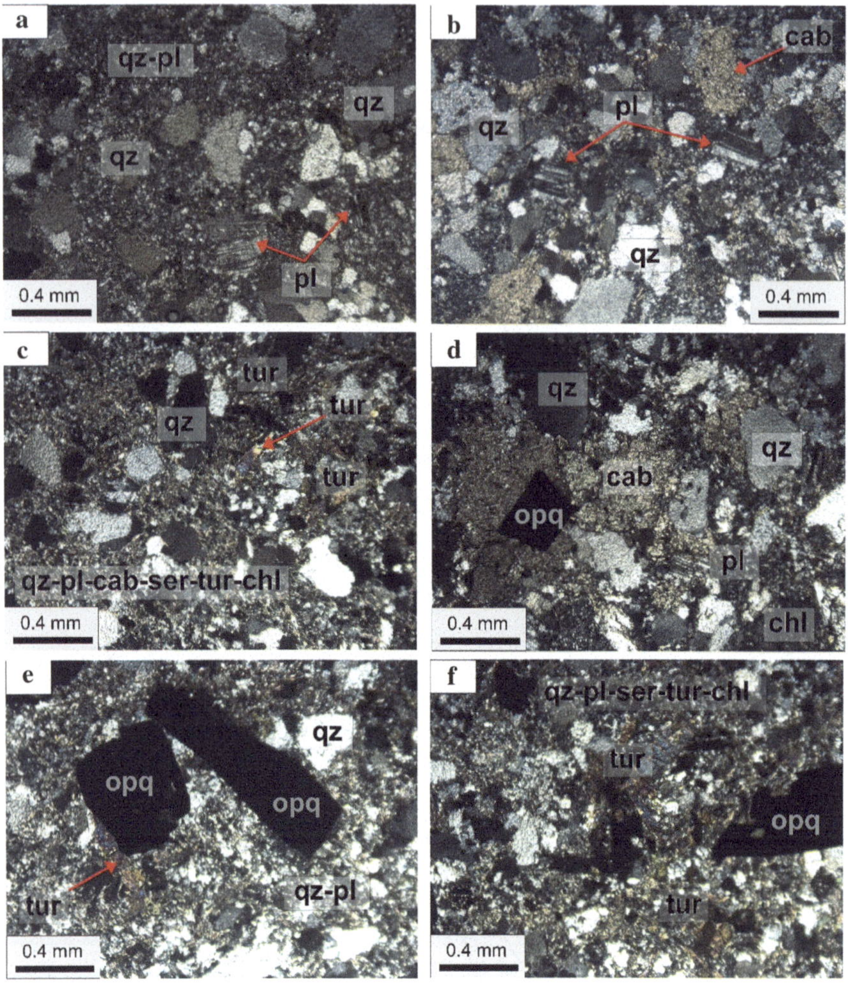

Fig. 7 Photomicrographs of the sandstone/quartzite samples from the Badukrom deposit. **a** Dark grey sandstone with minimal carbonate alteration. **b** Dark grey sandstone with patchy carbonate alteration. **c** Light grey sandstone with sulfide-tourmaline-carbonate-sericite alteration. **d** Light grey sandstone with sulfide-tourmaline-carbonate-sericite alteration. Opaque mineral (pyrite) forms euhedral shape in patchy carbonate. **e** & **f** Light grey sandstone with sulfide-tourmaline-sericite-chlorite alteration. *Notes* qz—quartz, pl—plagioclase, cab—carbonate, ser—sericite, py—pyrite, apy—arsenopyrite, tur—tourmaline

Alteration Hues: An Exploratory Indicator for Hydrothermal Gold Mineralization at Badukrom

The sandstone (and quartzite) displays two predominant colors: shades of grey (ranging from dark grey to light grey) and shades of pink (including pink and light pink). The dark grey sandstone is considered the original hue of the least altered sandstone unit before any geological events associated with alteration or metasomatism occurred in the area. Core logging reveals that the pink hue is unevenly distributed in the sandstone, indicating heterogeneous metasomatic activity within the rock. Pink hues observed in sedimentary and igneous rocks is commonly attributed to the albitization of plagioclase or the deposition of fine hematite within albite and plagioclase during fluid-rock interactions (e.g. [3, 24–26]). Notable instances of pink alteration of rocks have been recorded in the Damang gold deposit [1–3], the Chirano gold deposits [24], and the Luolo gold deposit [25]. The petrographic analysis of the Badukrom prospect did not indicate any distinctive textural variations between the prevalent mineral grains (i.e., quartz and plagioclase) in the dark grey sandstone and the pink sandstone. Therefore, the pink sandstone most probably formed during the albitization of plagioclase minerals in the sandstone.

Based on petrographic observations and analogous descriptions of the pink alteration in the Damang deposit, it is hypothesized that the pink alteration observed at Badukrom, could be a metasomatic event associated with mafic intrusions [3, 26] or the effect of Na-alteration by saline paleo-water that permeated the sandstone. Any of these geological activities would have transpired before the hydrothermal alteration related to gold mineralization. During the gold mineralization event, hydrothermal ore-forming fluid permeated the rocks and transformed dark grey sandstone into light grey sandstone while the pink sandstone was converted into light pink sandstone (Fig. 5). Consequently, the light grey and light pink hues signify the hydrothermally-altered (meta)sedimentary rock formations in the Badukrom area.

The mafic dykes are dark green in color and occasionally display speckled white patches of plagioclase phenocrysts. The hydrothermally altered mafic dykes exhibit a bleached appearance, rendering them also as prime candidates for exploration targeting.

Gold Mineralization and Hydrothermal Minerals Associated with the Ore-Forming Fluid

Major ore and associated minerals related to the hydrothermal system at Badukrom have been identified and discussed below:

(a) The distribution of hydrothermal sulfides (i.e. pyrite and arsenopyrite), which occurred as cubes, rhombuses, and/or bands within the altered sandstone, quartzite, and bleached mafic dykes, is associated with hydrothermal alteration during the gold mineralization event (Fig. 7d, f).

(b) The distribution of hydrothermal tourmaline, in the form of fine needles and massive ores, inside the altered sandstone and quartzite, was associated with hydrothermal alteration during the gold mineralization event (Figs. 5 and 8c, e, f).

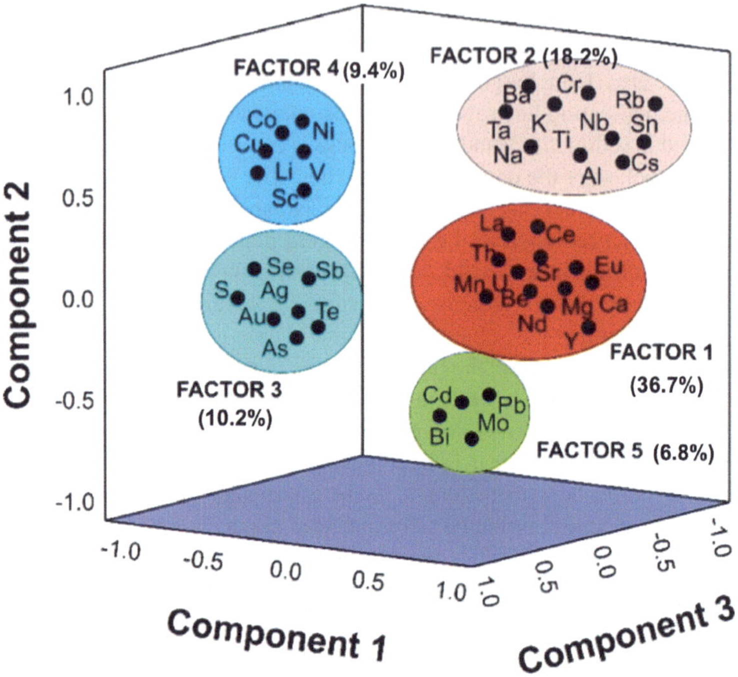

Fig. 8 Rotated component matrix showing the factors and total variances in percentages

(c) Carbonate and chlorite minerals appeared as fine- to coarse-grained patches, surrounding the sulfides and tourmaline within the altered sandstone and quartzite. Sericite manifested as fine-grained minerals coupled with the carbonates inside the matrix of the sandstone (Fig. 7c, e, f). The presence of carbonates and chlorite patches in the dark grey sandstone suggests that some of the carbonate and chlorite precipitated prior to the formation of the ore minerals (Fig. 7a, b). In the bleached mafic dykes, carbonate alteration predominates over chlorite.

(d) Nearly all significant gold concentrations are located within altered rock units, particularly the light grey sandstones/quartzites (e.g., Figs. 4a and 8d–f) and the bleached mafic dykes. This signifies a major overprinting hydrothermal system was involved in the gold mineralization.

(e) No substantial gold concentrations were recorded in the dark grey, pink, or light pink sandstone/quartzite. Although the light-pink sandstone is hydrothermally altered and may locally contain sulfides, no substantial gold amounts were documented in them (e.g., Fig. 4b).

(f) The occurrence of sulfides (pyrite and arsenopyrite), tourmaline, sericite and fine chlorite, with or without carbonate, correlates with substantial gold concentrations in the light grey sandstones (e.g. Figs. 5, 6 and 8e–f). This assemblage is directly linked to the hydrothermal alteration.

Multivariate Statistical Analysis and Its Implications for Ore Genesis and Exploration

The multi-element geochemistry of hydrothermal system at the Badukrom area is summarized in Table 2. The mean concentrations for the elements are respectively given as follows: As (5687.5) > P (736.3) > Ba (641.8) > Mn (348.8) > Sr (210.8) > V (63.92) > Zn (63.88) > Rb (41.36) > Ce (36.54) > W (36.2) > Cr (27.00) > Pb (25.81) > Li (20.52) > Nd (17.58) > La (17.57) > Ni (16.68) > Cu (14.42) > Co (12.32) > Te (10.69) > Sc (9.48) > Y (4.44) > Pr (4.34) > Al (4.04) > Sb (3.29) > Se (3.04) > Au (2.82) > Th (2.548) > Bi (1.9) > Sn (1.70) > Cs (1.69) > Mo (1.63) > Na (1.45) > Nb (1.44) > Cd (1.31) > K (1.19) > Be (0.95) > S (0.93) > Eu (0.92) > Ca (0.84) > U (0.78) > Yb (0.71) > Ag (0.52) > Mg (0.36) > Ta (0.34) > Ti (0.04). Maximum and minimum values aided in determining the range for each element with Ti (0.09) being the minimum range and As (9969) being the maximum range. The values for asymmetry and kurtosis between −2 and +2 and −7 and + 7 are considered acceptable in order to prove normal univariate distribution [27]. Skewness values (−1.62–4.98) and Kurtosis values of (−1.87–24.85) shows that some of the elements have their dataset within the acceptable range for skewness and kurtosis as proposed by George and Mallery [27] and some elements have their dataset exceeding this value. This shows that the dataset needs to be transformed before any parametric statistics are done.

Probability (P-P) plots for Au, Ti, Mg, V, Al, and Ni, indicated that the geochemical data had anomalous outliers and therefore did not exhibit a normal distribution. It also indicates that the distribution of the above elements and gold are weakly normal. This agrees with the notion that geochemical data of mineral deposits frequently do not follow a normal distribution because the data of an element vary from its relationships with other elements due to the diverse composition of the crust and its differing elemental concentrations [28, 29]. Hence, due to the non-normal distribution in the dataset, a centered log- transformation was employed to reduce the distances between the minimum and maximum values and, hence make the dataset fall within the acceptable range for a normally distributed data [27].

Elemental Signatures of the Hydrothermal System at Badukrom

To evaluate the correlation between the elements of interests, Spearman's correlation was used (Table 3). This method was selected because the dataset used in this current paper was log-transformed. For Au, Spearman's correlation revealed strong positive to low positive correlations among Au and the following elements: S ($r = 0.47$), Ag ($r = 0.56$), As ($r = 0.5$), Sb ($r = 0.47$), Se ($r = 0.39$), Te ($r = 0.45$), U (0.34), and V (0.31). These set of elements can be recognized as pathfinder elements of Au ([5]). The pair of the rest of the elements with Au had either a positive weak or a negative weak to poor correlation.

Table 3 Spearman's correlation analysis ($r \geq 0.50$ indicated in bold)

	Au	Ba	Ca	Cr	Cu	K	Li	Mg	Mn	Na	P	S	Sr	V
Au	1.00													
Ba	0.16	1.00												
Ca	−0.36	0.11	1.00											
Cr	0.15	0.54	−0.14	1.00										
Cu	−0.20	−0.27	−0.15	0.15	1.00									
K	0.13	0.92	0.10	0.61	−0.13	1.00								
Li	0.02	0.13	0.00	0.12	0.44	0.02	1.00							
Mg	−0.16	0.21	0.71	0.21	0.23	0.24	0.48	1.00						
Mn	−0.12	0.29	0.65	0.39	0.24	0.41	0.26	0.78	1.00					
Na	−0.24	−0.62	0.18	−0.62	−0.15	−0.73	−0.04	−0.13	−0.30	1.00				
P	0.18	−0.24	0.16	−0.17	−0.23	−0.26	0.06	−0.03	0.10	0.30	1.00			
S	**0.47**	−0.14	−0.21	0.14	−0.17	−0.10	0.09	0.09	−0.14	−0.06	0.28	1.00		
Sr	−0.03	0.58	0.59	0.20	−0.29	0.60	−0.04	0.55	0.42	−0.17	−0.28	−0.02	1.00	
V	0.31	0.35	0.03	0.40	0.16	0.31	0.64	0.50	0.32	−0.32	0.05	0.53	0.30	1.00
Zn	0.03	−0.71	−0.08	−0.23	0.36	−0.70	0.13	0.09	−0.02	0.44	0.39	0.22	−0.53	−0.11
Ag	**0.56**	−0.26	−0.59	0.02	−0.19	−0.19	−0.23	−0.40	−0.34	0.03	0.27	0.52	−0.38	−0.06
As	**0.50**	−0.08	−0.12	−0.17	−0.45	−0.12	0.07	−0.06	−0.20	0.04	0.36	0.75	−0.02	0.39
Bi	0.16	−0.31	−0.42	0.18	−0.08	−0.19	−0.30	−0.29	−0.24	0.13	0.11	0.39	−0.19	−0.09
Co	0.29	−0.20	0.00	−0.04	0.26	−0.20	0.48	0.35	0.07	0.06	0.20	0.70	0.03	0.71
Eu	0.08	0.45	0.20	0.47	−0.32	0.47	−0.23	0.13	0.21	−0.16	0.08	0.33	0.46	0.21
Mo	0.00	−0.25	−0.42	0.18	0.05	−0.18	−0.22	−0.34	−0.27	0.15	0.02	0.11	−0.23	−0.22
Ni	0.29	0.04	0.10	0.23	0.11	−0.03	0.63	0.50	0.25	−0.01	0.28	0.68	0.18	0.84

(continued)

Table 3 (continued)

	Au	Ba	Ca	Cr	Cu	K	Li	Mg	Mn	Na	P	S	Sr	V
Pb	0.06	−0.36	−0.06	−0.17	−0.17	−0.30	−0.17	0.00	−0.17	0.28	0.31	0.40	−0.07	−0.06
Sb	**0.47**	−0.04	−0.06	0.06	−0.19	−0.08	0.36	0.16	0.03	0.01	0.37	0.67	0.01	0.62
Se	**0.39**	−0.33	−0.07	−0.19	−0.39	−0.35	0.09	0.01	−0.16	0.20	0.53	0.76	−0.11	0.32
Te	**0.45**	−0.11	−0.08	−0.07	−0.48	−0.10	−0.12	0.02	−0.20	0.03	0.34	0.84	0.10	0.32
W	0.19	0.58	−0.11	0.48	−0.21	0.62	0.21	0.03	0.31	−0.47	0.09	0.04	0.20	0.38

	Zn	Ag	As	Bi	Co	Eu	Mo	Ni	Pb	Pr	Sb	Se	Te	W
Au														
Ba														
Ca														
Cr														
Cu														
K														
Li														
Mg														
Mn														
Na														
P														
S														
Sr														
V														
Zn	1.00													
Ag	0.40	1.00												

(continued)

Table 3 (continued)

	Zn	Ag	As	Bi	Co	Eu	Mo	Ni	Pb	Pr	Sb	Se	Te	W
As	−0.05	0.40	1.00											
Bi	0.30	0.74	0.26	1.00										
Co	0.30	0.12	0.43	0.02	1.00									
Eu	−0.36	0.03	0.15	0.24	0.04	1.00								
Mo	0.22	0.48	−0.08	0.75	−0.02	0.09	1.00							
Ni	0.18	0.13	0.49	0.13	0.82	0.26	−0.04	1.00						
Pb	0.51	0.57	0.31	0.66	0.23	0.00	0.52	0.16	1.00					
Sb	0.02	0.28	0.82	0.18	0.58	0.12	−0.15	0.73	0.20	0.18	1.00			
Se	0.31	0.44	0.80	0.24	0.53	0.03	−0.03	0.53	0.48	0.14	0.70	1.00		
Te	0.12	0.59	0.84	0.49	0.44	0.23	0.11	0.51	0.55	0.29	0.67	0.78	1.00	
W	−0.51	−0.03	0.22	−0.02	−0.10	0.37	−0.10	0.17	−0.26	0.39	0.31	0.04	−0.03	1.00

In the case of factor analysis (FA), the minimum factor loading criteria was set to 0.50. The communality of the scale, which indicates the amount of variance in each dimension, was also assessed to ensure acceptable levels of explanation. The results show that all communalities except W, Zn, and P were over 0.50. Results for total variance explained is presented in Table 4. In the initial FA, four items (W, Zn, P, and Mn) did not load onto any of the factors, so the four items were removed from further analysis. The results for the new analysis confirmed five dimensional structures or components (Table 4). The five dimensions explained 81.28% of the variance among the elements in the study (Table 5). Factor 1 accounts for 28.24% of the total variance and includes Al–Ca–Mg–Mn–Sr–Be–Ce–Eu–La–Nd–Th–U–Y–Yb multi-element association (Table 5 and Fig. 8). These elements had moderate to perfect correlation ($r = 0.599$ to $r = 0.967$) with factor 1. Factor 2 accounts for 20.38% of the total variance and includes Ba–Cr–K–Na–Ti–Cs–Nb–Rb–Sn–Ta–Yb multi-elements association (Table 5 and Fig. 8). This element had a low to perfect correlation ($r = 0.580$ to $r = 0.869$) with Factor 2 except for Na, which had a strong negative correlation with Factor 2. Factor 4 accounts for 10.94% of the total variance and included Au–S–Ag–As–Sb–Se–Te multi-element association (Table 5 and Fig. 8). Element on this factor had a moderate to strong with a correlation coefficient ($r = 0.64$ to $r = 0.84$) between elements and the Factor 3. Factor Three accounts for 14.16% of the total variance and included Cu–Li–V–Co–Ni–Sc multi-element association (Table 5 and Fig. 8). This element had a moderate to perfect positive correlation ($r = 0.5$ to $r = 0.91$) with Factor 4. Factor 5 accounts for 10.48% of the total variance and included Bi–Cd–Pb–Mo multi-element association (Table 5 and Fig. 7). The elements loaded strongly on this factor with correlation coefficient between $r > 0.89$.

For the cluster analysis, five clusters representing multi-element associations could be observed in the dendrogram derived from the dataset (Fig. 9). Cluster 1 has a multi-element cluster of As–Te–S–Se–Au. Cluster 2 has multi-element association of Nb–Ta–Cs–Rb–Ba–K–Be–Al–Ti–Sn–Cr–W. Cluster 3 has multi-element association of Nd–Ce–La–Eu–Th–U–Yb–Y–Mg–Sc–Mn–Ca–Sr–P–Na. Cluster 4 has a multi-element association of Co–Ni–Sb–V–Li–Cu. Cluster 5 has a multi-element association of Cd–Pb–Bi–Mo–Ag–Zn.

Elemental Vectors to Hydrothermal Gold Mineralization at Badukrom

In geochemical investigation of mineral deposits, very high or low element concentrations indicate various processes or anomalies associated with mineralization [17, 30]. A strong correlation among components indicates that analogous processes contributed to their formation in the hard rock. In the correlation analysis, factor analysis (Factor 3), and cluster analysis (Cluster 1) of the transformed geochemical data, gold is consistently associated with Au–S–Ag–As–Sb–Se–Te. This elemental association is reflective of sulfidation and tourmalinization associated with the gold mineralisation. Cluster 4 and Factor 4 reveal a Co–Ni–Sb–V–Li–Cu association that can also be attributed to sulfidation and the abundant tourmaline associated with

Table 4 Rotated component matrix showing the distribution of elements into factors

Element	Factor				
	1	2	3	4	5
Au			0.801		
Al	0.595				
Ba		0.681			
Ca	0.767				
Cr		0.768			
Cu				0.589	
K		0.869			
Li				0.913	
Mg	0.741			0.501	
Mn	0.599				
Na		−0.790			
S			0.844		
Sr	0.808				
Ti		0.580		0.550	
V				0.707	
Ag			0.692		0.515
As			0.704		
Be	0.719				
Bi					0.851
Cd					0.856
Ce	0.890				
Co				0.720	
Cs		0.754			
Eu	0.967				
La	0.860				
Mo					0.842
Nb		0.757			
Nd	0.909				
Ni	0.575			0.652	
Pb					0.855
Pr	0.901				
Rb		0.772			
Sb			0.643		
Sc	0.576			0.720	
Se			0.778		
Sn		0.634			0.506

(continued)

Table 4 (continued)

Element	Factor				
	1	2	3	4	5
Ta	0.586	0.647			
Te			0.790		
Th	0.784				
U	0.824				
Y	0.910				
Yb	0.965				

the quartz veins in the deposit [30]. Kalliomäki et al. [30] asserts that all these elemental components can be integrated into the lattice structure of tourmaline. The good elemental correlation indicates that the hydrothermal alteration of the Kawere sandstone/quartzite led to fixation of these elements in the sulfides and tourmaline. Clusters 2 and 3 and Factors 1 and 2 are related to carbonatization, chloritization, and sericitization of the sandstone/quartzite during hydrothermal fluid rock interaction.

4 Implication for Hydrothermal Gold Systems in the Tarkwaian Group

Before the discovery of hydrothermal-gold-mineralization overprints on the placer gold deposit at Damang, it was often believed that the Tarkwaian Group of rocks solely contained placer gold. The interception of hydrothermal gold mineralization in the Kawere sandstones at the Kobeda and Badukrom prospects, respectively, suggest that hydrothermal gold mineralization in the Tarkwaian Group could be more widespread than previously believed. The Damang deposit and the Badukrom prospect are situated within the Tarkwaian Group of sedimentary rocks and are intruded by mafic dykes/sills. The Damang hydrothermal gold mineralisation is found in all the four stratigraphic units of the Tarkwaian Group: Kawere conglomerate, Banket conglomerate and sandstone/quartzite, Tarkwa Phyllite, and Huni sandstone [1, 2]; Exploration geologists, Damang, pers comm. 2022), whereas the Badukrom deposit is currently intercepted within the Kawere sandstone/quartzite unit [5]. Consequently, the two deposits are also fundamentally situated within the Tarkwaian Group of (meta)sediments. The Au–Ag–Te–Sb–S–As–Se association identified for the hydrothermal system at Badukrom in the present study is somewhat similar to that of the hydrothermal system at Damang (Au–Ag–Te–Sb–S–Bi–W; [8] or Au–Ag–Te; [7]). This elemental association is likely caused by any of the following factors: (a) identical hydrothermal ore-forming fluids in the Damang-Tarkwa region interacted with various rock units to generate distinct elemental associations, or (b) disparate fluids from diverse sources interacted with the rocks of the region and resulted in

Table 5 Variance explained for transformed dataset

Total variance explained									
Component	Initial eigenvalues			Extraction sums of squared loadings			Rotation sums of squared loadings		
	Total	% of variance	Cumulative %	Total	% of variance	Cumulative %	Total	% of variance	Cumulative %
1	16.493	36.652	36.652	16.493	36.652	36.652	12.708	28.240	28.240
2	8.188	18.196	54.848	8.188	18.196	54.848	7.855	17.454	45.695
3	4.606	10.235	65.083	4.606	10.235	65.083	6.372	14.160	59.855
4	4.242	9.427	74.510	4.242	9.427	74.510	4.924	10.942	70.797
5	3.046	6.769	81.279	3.046	6.769	81.279	4.717	10.482	81.279

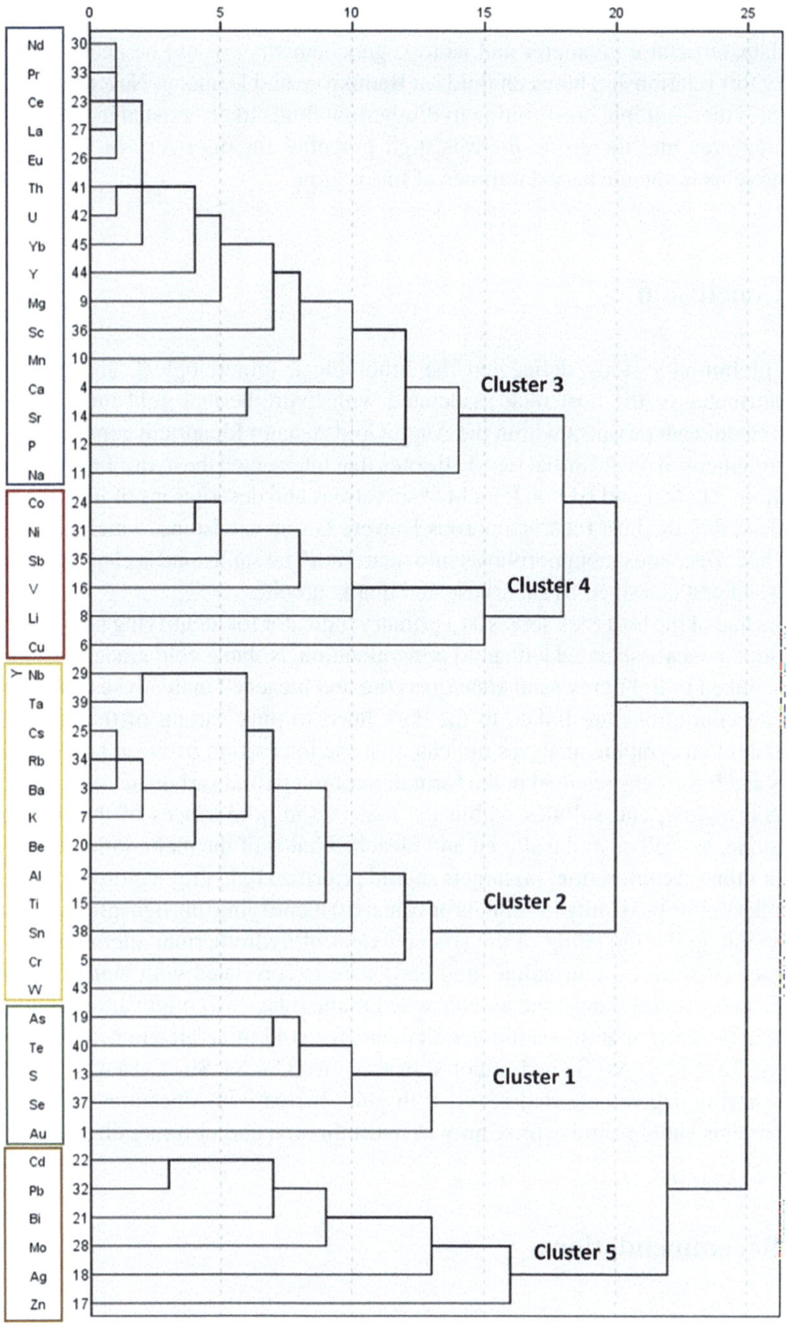

Fig. 9 Dendrogram from hierarchical cluster analysis showing five (5) elemental clusters

the varied elemental associations. Further studies (e.g. full complement of geochemical data, structural geometry and isotope geochemistry) would be needed to fully clarify this relationship between fluids at Badukrom and Damang. Nevertheless, this indicates that multiple ore-forming hydrothermal fluids likely existed in the Tarkwa-Damang area and therefore, there is high potential for discovery of gold-related quartz veins in the uncharted terranes of this region.

5 Conclusion

This preliminary study delineated the lithological, mineralogical, and geochemical attributes of the host rock associated with hydrothermal gold mineralization at the Badukrom prospect within the AngloGold Ashanti Iduapriem concession. The research focused on the initial two drill cores that intersected the hydrothermal mineralization: BD-001 and BD-002. Field observations and descriptions of the drill cores indicated that the host rock is a porous Kawere Group sandstone, which has undergone heterogeneous metamorphosis into quartzite. The sandstone is characterized as silty sand and classified as subarkose and quartz arenite.

The hue of the host rock serves as a primary indicator for identifying hydrothermal alteration zones associated with gold mineralization. Notable gold grades are consistently linked to light grey sandstone/quartzite and bleached mafic dykes. Negligible gold concentrations are linked to the dark-hued to pink variant of the sandstone/quartzite. Petrographic analyses indicate that the interaction between hydrothermal fluids and host rocks resulted in the formation of interstitial carbonate, sericite, chlorite, tourmaline, and sulfides within the matrix and pore spaces of the light grey sandstone, as well as in the altered and bleached halos of the mafic intrusive rocks.

For mineral exploration, (a) targets should prioritize light grey-colored sandstone, regardless of its proximity to veinlets or veins; (b) identifying this light grey sandstone will facilitate the modeling of the spatial extent of hydrothermal alteration; (c) the presence of sulfides, tourmaline, and carbonate is correlated with significant gold grades in light grey sandstone and bleached mafic dykes; (d) multivariate statistical analysis indicates that the pathfinder elements for gold mineralization at Badukrom are Ag, Te, Sb, S, As, Se and minor signatures for Co–Ni–Sb–V–Li–Cu, based on their varying degrees of relationship with gold. Increase in concentrations of these components should signify proximity to hydrothermal and/or mineralization zones.

6 Recommendations

Based on the above, we can state the following recommendations:

(a) Field mapping strategies must be devised to identify the color variations of the host rocks for delineation of alteration zones linked to the hydrothermal system in the Badukrom area. This will facilitate the identification of the spatial distribution of the hydrothermal system associated with the light-colored sandstone/quartzite of the Kware Group and the bleached mafic dykes.
(b) A total of eleven holes were drilled to intercept and identify the hydrothermal mineralization at Badukrom. The BD-001 and BD-002 boreholes reported in this book chapter are the first two holes. Consequently, similar investigations ought to be performed on the remaining drill holes to thoroughly understand and clarify the characteristics of the gold mineralization at Badukrom.
(c) This investigation acquired geochemical data from quartz-tourmaline veins linked to gold mineralization. Future studies should acquire geochemical data from both unaltered and altered host rocks to assess the impact of host rock interaction with hydrothermal fluids and identify any similarities or differences in the modeling outcomes associated with this fluid-rock interaction at Badukrom and the surrounding region.
(d) Comprehensive structural, geochemical, and isotopic analyses must be conducted on all cores to formulate an ore genetic model for the Badukrom deposit, which will also have broad exploratory significance for analogous deposit types both regionally and globally.

Acknowledgements The Management of the AngloGold Ashanti Iduapriem Ltd including the Geology Department, is gratefully acknowledged for their invaluable support, insightful discussions and permission to publish the results. The Editors are duly acknowledged for their constructive comments and suggestions.

References

1. Pigois J, Groves D, Fletcher I, McNaughton N, Snee L (2003) Age constraints on Tarkwaian palaeoplacer and lode-gold formation in the Tarkwa-Damang district, SW Ghana. Mineralium Deposita 38:695–714. https://doi.org/10.1007/s00126-003-0360-5
2. Tunks JA, Selley D, Rogers J, Brabham G (2004) Vein mineralization at the damang gold mine, Ghana: controls on mineralization. J Struct Geol 26(6–7):1257–1273. https://doi.org/10.1016/j.jsg.2003.11.005
3. White AJR, Waters DJ, Robb LJ (2015) Exhumation-driven devolatilization as a fluid source for orogenic gold mineralization at the damang deposit, Ghana. Economic Geology 110:1009–1025. https://doi.org/10.2113/econgeo.110.4.1009
4. Goldfarb RJ, André-Mayer AS, Jowitt SM, Mudd GM (2017) West Africa: the world's premier Paleoproterozoic gold province. Econ Geol 112(1):123–143. https://doi.org/10.2113/econgeo.112.1.123
5. Osei KP, Affam M, Kusi-Manu C, Kwapong DT (2021) Multivariate statistical evaluation of geochemical data from quartz vein-associated gold mineralization at the badukrom prospect within the Tarkwaian Base. Ghana Mining Journal 21(1):22–30
6. Fiadonu EB, Hammond NQ, Dzigbodi-Adjimah K, Adomako-Ansah K (2023) Ore fluid characteristics and depositional processes of the hydrothermal gold mineralization at Damang,

Southwestern Ghana. J Afr Earth Sc 197:104–141. https://doi.org/10.1016/j.jafrearsci.2022.104741

7. Sunkari ED, Gan AN, Fiadonu EB (2024) Geochemical vectors for targeting hydrothermal gold mineralization in the Damang area, southwestern Ghana: constraints from petrography, multi-element geochemistry and multivariate statistical analysis. Ore Energy Resour Geol. https://doi.org/10.1016/j.oreoa.2024.100071
8. Adomako-Ansah K, Osei-Somiah I (2024) Trace element characteristics in ore sulfides and their records of fluid conditions during over printing hydrothermal gold mineralization in the Damang Placer Gold deposit, SW Ghana. Book Chapter, in press
9. Groves DI, Goldfarb RJ, Robert F, Hart CJ (2003) Gold deposits in metamorphic belts: overview of current understanding, outstanding problems, future research, and exploration significance. Econ Geol 98(1):1–29
10. Le Mignot E, Reisberg L, André-Mayer AS, Bourassa Y, Fontaine A, Miller J (2017) Re-Os geochronological evidence for multiple Paleoproterozoic gold events at the scale of the West African craton. Econ Geol 112(1):145–168. https://doi.org/10.2113/econgeo.112.1.145
11. AngloGold Ashanti Iduapriem Ltd Unpublished Annual Report (2005) Iduapriem 2006 Business Plan Geology Report, Iduapriem, p 49
12. Griffis RJ, Barning K, Agezo FL, Akosah FK (2002) Gold deposits of Ghana. Minerals Commission, Accra, p 430
13. Adomako-Ansah K, Ishiyama D, Allen R (2018) Origin of Zn–Pb–Sb–Au mineralization adjacent to the paleoproterozoic boliden Au-Rich VMS deposit, Sweden: evidence from petrographic and oxygen isotope characteristics. Miner Deposita 53(5):683–700
14. SGS Geochemistry Guide 2024, accessed on November 27, 2024
15. Sinclair AJ (1976) Application of probability graphs in mineral exploration. Association of Exploration Geochemists, Special Volume, p 95
16. Gao Y, Chik ABR (2013) A multiple regression analysis on influencing factors of urban services growth in China". Technol Invest 4:1–5
17. Zhao J, Chen S, Zuo R (2016) Identifying geochemical anomalies associated with Au–Cu mineralization using multifractal and artificial neural network models in the Ningqiang district, Shaanxi, China. J Geochem Explor 164:54–64. https://doi.org/10.1016/j.gexplo.2015.06.018
18. Spearman C (1987) The proof and measurement of association between two things. Am J Psychol 100(3/4):441–471. https://doi.org/10.2307/142268
19. Nude PM, Asigri JM, Yidana SM, Arhin E, Foli G, Kutu JM (2012) Identifying pathfinder elements for gold in multi-element soil geochemical data from the Wa-Lawra belt, Northwest Ghana: a multivariate statistical approach. Int J Geosci 3(1):62–70. https://doi.org/10.4236/ijg.2012.31008
20. Olise FS, Adeojo SA, Owoade OK, Oketayo OO, Adekola SA Akinlua A (2019) Geochemical characterization of soil samples from gold mining area using PIXE spectroscopy. Environ Sci Pollut Res 26(5):4924–4937
21. Folk RL, Ward WC (1957) Brazos river bar [Texas]; a study in the significance of grain size parameters. J Sediment Res 27(1):3–26. https://doi.org/10.1306/74D70646-2B21-11D7-8648000102C1865D
22. Johnson MR (1994) Thin section grain size analysis revisited. Sedimentology 41(5):985–999. https://doi.org/10.1111/j.1365-3091.1994.tb01436.x
23. Li DL, Zhao WW (2013) Diagenesis and its effect on reservoir properties of Chang 2 oil reservoir set in longzuigou area of xiasiwan, ordos basin. Lithologic Reservoirs 25(3):77–81. [In Chinese] Allibone A, Teasdale J, Cameron G, Etheridge M, Uttley P, Soboh A, Appiah-Kubi L, Adanu A, Arthur R, Mamphey J, Odoom B, Zuta J, Tsikata A, Pataye F, Famiyeh S, Lamb E (2002) Timing and structural controls on gold mineralization at the Bogoso gold mine, Ghana, West Africa. Econ Geol 97:949–969
24. Allibone A, Hayden P, Cameron G, Duku F (2004) Paleoproterozoic gold deposits hosted by albite- and carbonate-altered tonalite in the Chirano district, Ghana, West Africa. Econ Geol 99:479–497

25. Lawrence DM, Treloar PJ, Rankin AH, Boyce A, Harbidge P (2013) A fluid inclusion and stable isotope study at the Loulo mining district, Mali, West Africa: implications for multifluid sources in the generation of orogenic gold deposits. Econ Geol 108(2):229–257. https://doi.org/10.2113/econgeo.108.2.199
26. Sharpe R (2005) Damang regional prospectivity analysis: stratigraphic & structural framework of hydrothermal & paleoplacer gold deposits in the Damang District—constraints to target additional resources, Volume 1 of 2, Report Prepared for Abosso Goldfields Limited, Ghana Goldfields Limited & Gyata Exploration Limited
27. George D, Mallery P (2016) IBM SPSS statistics 23 step by step: a simple guide and reference (14th ed.). Routledge, New York. https://doi.org/10.4324/9781315545899
28. Stracke A (2012) Earth's heterogeneous mantle: a product of convection-driven interaction between crust and mantle. Chem Geol 330:274–299
29. Zuo R (2011) Identifying geochemical anomalies associated with Cu and Pb–Zn skarn mineralization using principal component analysis and spectrum–area fractal modeling in the Gangdese Belt, Tibet (China). J Geochem Expl 111(2):13–22
30. Kalliomäki H, Wagner T, Fusswinkel T, Sakellaris G (2017) Major and trace element geochemistry of tourmalines from archean orogenic gold deposits: proxies for the origin of gold mineralizing fluids? Ore Geol Rev 91:906–927

Trace Element Characteristics in Ore Sulfides and Their Records of Fluid Conditions During Overprinting Hydrothermal Gold Mineralization in the Damang Placer Gold Deposit, SW Ghana

Isaac Osei-Somiah and Kofi Adomako-Ansah

1 Introduction

Pyrite is one of the most ubiquitous minerals associated with gold. Commonly known as 'fools gold', recent advances in mineral microanalysis (e.g. LA-ICPMS, SIMS, SEM, EPMA etc.) have revealed that pyrite is 'no fool at all' and can host gold and other trace elements in its crystal lattice [1]. Previous studies have shown that, as long as complete recrystallization did not occur, pyrite can record changes in fluid chemistry during the evolution of the hydrothermal system from which it has precipitated [2–5]. Differences in the composition of fluid and/or conditions of precipitation generally lead to textural and geochemical variations [2, 6]. Some of the textural forms that may influence the chemical variations include euhedral pyrite, anhedral pyrite, porous pyrite, massive pyrite, rim-to-core pyrite, and core-to-rim pyrite [7–9]. Thus, detailed studies involving in situ analyses (spots, lines, or maps) of pyrite can detect these subtle variations and provide critical information about pyrite growth history and physicochemical conditions of ore formation [2, 9]. Many researchers have developed various stringent approaches that use SIMS and LA-ICP-MS multi-element database on well-characterized deposits to improve our understanding of mineral systems and the controls on incorporation of trace element metals in pyrite

I. Osei-Somiah · K. Adomako-Ansah (✉)
Department of Geological Engineering, University of Mines and Technology, Tarkwa, Ghana
e-mail: kadomakoansah@umat.edu.gh

I. Osei-Somiah
e-mail: isaac.osei-somiah@mincom.gov.gh

I. Osei-Somiah
Eastern Regional Office, Minerals Commission, Koforidua, Ghana

E. D. Sunkari and S. Ali (eds.), *Geology and Gold Mineralization in Ghana*, Earth and Environmental Sciences Library, https://doi.org/10.1007/978-3-031-97163-1_6

and other sulfides [2–13]. This database has been helpful in clarifying, to a greater extent, the evolution and/or origin of pyrite during magmatic, hydrothermal, sedimentary or diagenetic mineralization. According to Feng et al. [2], these approaches have been successfully applied for several gold [3, 10, 11], porphyry [5, 12], and uranium [13] deposits.

The paleoplacer gold deposit located at Damang, SW Ghana, is overprinted by hydrothermal gold-forming fluids that formed quartz-carbonate veins related to the hydrothermal gold and associated sulfides [14–20]. The quartz-carbonate veins cut across all the lithologies of the Tarkwaian Group and intrusive rocks in the deposit [14–17], in addition to wall rock alteration and phase separation [15, 18, 20]. Fiadonu et al. [20] also indicated from fluid inclusions and mineral thermometric studies that a decrease from high to low temperatures was another factor for ore precipitation at Damang. Sharpe [16] suggested that two types of gold-forming fluids exist in the Damang area: (a) pyrite ore in the conglomerate was formed by oxidized magmatic-hydrothermal fluids emanating from the crystallizing dykes or intrusive rocks, while (b) the pyrrhotite ore in the intrusive rocks were caused by metamorphic-hydrothermal fluids. Other workers suggested only a metamorphic fluid source for the hydrothermal gold mineralization based on thermal modelling [19] and fluid inclusion and stable isotope studies [20]. Most recently, Sunkari et al. [21] used multivariate statistical analysis to establish that the pathfinder elements of gold in Damang deposit is Bi, Te and Ag. However, none of these previous studies considered the implication of mineralogical and textural controls on the distribution of trace metals in the hydrothermal sulfides and its insight into the ore-forming process in the deposit. Using LA-ICP-MS technique, the current study obtained trace element data on the hydrothermal sulfides (i.e., pyrite and pyrrhotite) related to the quartz veins in the Damang deposit. Since the hydrothermal ore minerals are one of the direct pieces of evidence, the textural and trace element associations in these sulfides were evaluated to improve our understanding of the conditions and evolution of the ore-forming fluid that led to the overprint of hydrothermal gold mineralization on a paleoplacer deposit.

2 Materials and Methods

Regional Geology

The Paleoproterozoic supracrustal rocks that directly cover the Man Shield of West Africa [21] are known as the Birimian Supergroup and in Ghana, they occur at the southwestern through northwestern parts of the country (Fig. 1a). These subparallel, >100 km long, NE-SW trending volcano-sedimentary greenstone belts and intervening sedimentary basins are now widely considered contemporaneously developed during a 2200–2000 Ma phase of crustal expansion [22–27]. The Tarkwaian Group sporadically overlie the volcanic belts and comprise a broadly fining-upward sequence of clastic sediments that are interpreted as infills of a tensional rift basin within the volcanic belts at approximately 2135–2100 Ma [27–31]. Both the Birimian and Tarkwaian rocks were then affected by a 2130–1980 Ma Eburnean orogeny.

In terms of gold mineralization, there are generally two distinct styles observed in Ghana. First, paleoplacer-style gold mineralization contained in extensive quartz-pebble conglomerate horizons of the Banket Conglomerate of the Tarkwaian Group (e.g., at the Tarkwa, Iduapriem, Teberebie, and Damang mines; [15, 22, 32, 33]). Numerous studies suggest that provenance of these paleoplacer deposits were sourced from pre-Eburnean gold-bearing rocks that are presently unidentified [27, 34, 35]. Secondly, orogenic-style gold mineralization occurred later in the Eburnean orogenic cycle, and therefore formed after the paleoplacer mineralization [27, 31, 34]. The orogenic gold mineralization is more widely distributed throughout the entire Birimian terrane whereas the paleoplacer deposit is primarily restricted to the Tarkwaian Group in the Tarkwa-Damang area. Numerous researchers agree that

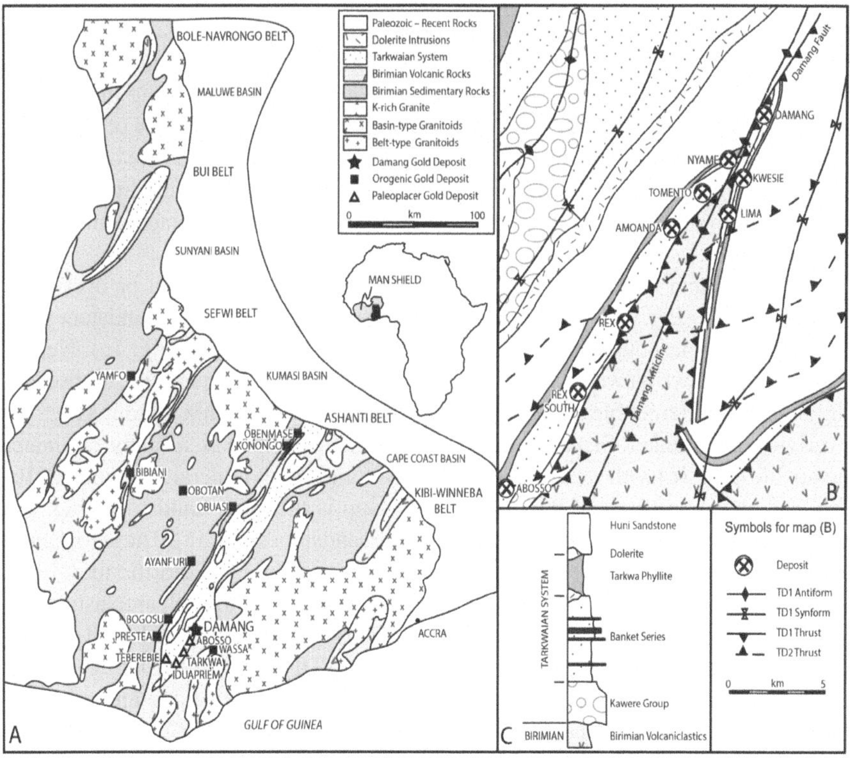

Fig. 1 Simplified geologic maps of the Birimian terrane of Ghana and the Damang area (obtained from [62]) (**a**). Simplified geology of southwest Ghana, showing the location of major gold deposits, including the Damang deposit (modified from [51]) (**b**). Simplified geologic map and stratigraphic column (**c**) of the Damang deposit (modified from [60]), illustrating the regional structure and location of the Damang deposit and its satellite deposits. Fold and thrust terminology from [60]

low salinity, H_2O–CO_2–(CH_4)–(N_2)-rich metamorphic-hydrothermal fluids are the major contributors to the orogenic ore-forming systems, and Ghana is no exception (e.g., [18, 20, 34, 36–40]). However, studies on granitoid-hosted gold mineralization in Ghana [41] and other parts of the Birimian of West Africa [42–47] suggest that contributions from magmatic sources cannot be totally ruled out. Alluvial gold deposits are often regarded as a third style of gold mineralization in Ghana but in terms of origin, we consider them to be younger versions of the paleoplacer-style deposits in Tarkwa and its environs.

Local Geology

The Damang gold mine is situated in southwest Ghana, roughly 30 km northeast of Tarkwa (Fig. 1a, b). On the basis of extensive structural studies conducted on the deposit, Tunks et al. [17] revealed that regional deformation within the Tarkwaian involved an initial NW–SE directed shortening (D_1) and the formation of a major NNE-trending F_1 anticline that hosts the Damang orebody (Figs. 1b and 2). Broadly N–S shortening during D_2 resulted in the formation of E–W-trending thrusts with small displacements. The D_3 shortening direction was similar to that of D_1. Steep D_1 faults were reactivated and a new set of low angle thrusts and associated flat-lying extension quartz-carbonate veins were formed. The bulk of the hydrothermal mineralization observed at Damang is associated with the low displacement D_3 fault fracture mesh.

The Tarkwaian Group of rocks in the deposit are faulted against or overlie the Birimian rocks (Fig. 1c). At the base of the Tarkwaian Group are sandstones and polymictic pebble-boulder conglomerates of the Kawere Group (Fig. 1c). Next in the sequence above Kawere Group are cross- to planar-bedded quartzite, arkose, and oligomictic quartz-pebble conglomerates that comprise the Banket Group (Fig. 1c). Overlying the Banket Group is the Tarkwa Phyllite, which is a finely laminated metapelite (Fig. 1c) with well-developed amphibolite facies mineral assemblage [16–18]. The uppermost unit of the Tarkwaian Group is the Huni Sandstone (Fig. 1c), which is a thick sequence of massive feldspathic sandstones [16, 48]. These rock units of the Tarkwaian Group are then intruded by sills and dykes of dioritic to doleritic composition. According to White [49], U–Pb zircon ages for the volcaniclastic rocks and the diorite are 2178.0 ± 9.3 and 2164.6 ± 8.0 Ma, respectively. Based on thermodynamic modeling of the Tarkwa Phyllite and U–Th–Pb chemical dating of metamorphic monazite, White et al. [18] also concluded peak regional metamorphism in the Damang area to have reached 595 °C and 5.5 kbars at 2005 ± 26 Ma.

The paleoplacer-style gold mineralization of the Damang deposit is restricted to four quartz-pebble conglomeratic layers of the Banket Group, which is the same for other placer deposits in the Tarkwaian Group elsewhere in the region (e.g., Abosso, Tarkwa, Iduapriem, and Tebrebe; Fig. 1). The placer gold is associated with other heavy minerals like magnetite along bedding planes and crossbedded foresets. On the other hand, the hydrothermal gold mineralization overprints the

Fig. 2 Saddle Pit of the Damang deposit (facing north) showing the Damang Fault and the quartz veins on the left side of the fault (western limb). Absence of quartz veins on the right side of the Fault is due to throw-down movement. Insert photo is shown in Fig. 4

Tarkwaian Group and associated intrusive rocks. It has been observed at Damang (Gold Fields Damang Mine), Kobeda (Gold Fields Tarkwa Mine) and Badukrom (AngloGold Ashanti Iduapriem Mine). Whether or not the hydrothermal systems in these localities are connected is beyond the scope of this study. The hydrothermal gold is related to crosscutting, subhorizontal, extensional quartz vein arrays that were created during regional compression [17, 19]. At Damang, the hydrothermal gold mineralization is found in the eastern and western limbs of the Damang anticline [17]. The western limb hosts the bulk of the hydrothermal gold mineralization (Fig. 2; [17,

20]). Gold is free-milled both inside the quartz veins and in close proximity to sulfide minerals (pyrite and pyrrhotite) in the hydrothermal alteration halos that surround the quartz-carbonate veins [14, 16, 50].

Methodology

A flowchart of the methodology used for this study is shown in Fig. 3 and the details are outlined below.

Field Survey, Core Logging and Sampling

Based on the aim and the objectives of the project, a field survey of the Damang Mine was carried out in the Saddle Pit to appreciate the geological and structural context of the two types of gold mineralization in the Damang area (Figs. 2, 4 and 5), as described by Tunks et al. [17]. Subhorizontal quartz veins, which are representations of the ore-forming fluid, were observed to crosscut and cause hydrothermal gold mineralization in all the lithologies in the Saddle Pit, including the Banket Conglomerate, the Tarkwa Phyllite and the dioritic intrusive rocks (e.g., Fig. 6). From the field survey, a total of 80 samples were taken from the pit and logged drilled cores, respectively, for petrographic analysis. The sample selection process on the drill cores was prioritized using the results of the metallurgical assay obtained from Gold Fields Damang Mine. An electric core cutter was used to cut the chosen core along the orientation lines. The half-core samples were placed in properly labeled sample bags, sealed and sent to various laboratories for further analyses.

Hand Specimen and Microscopic Analyses

The lithology, color, grain sizes, grain shapes, ore mineralogy, alteration, textural fabrics and structural aspects of the core samples were noted and described. The identification of the minerals were aided by the use of a hand lens with a ($\times$ 10) magnification. The reaction of carbonate minerals with HCl and the use of a magnetic pen were additional methods employed to distinguish some of the minerals. From core samples, a total of 62 thin and polished block sections were made at the Petrology laboratory of the University of Mines and Technology (UMaT), Tarkwa. The sections were examined under LEICA DM750P and LEICA DM2700P petrographic microscopes with Leica MC 120HD digital cameras affixed and connected to a computer in order to identify both rock-forming and ore minerals and their textural relationships. Various optical characteristics, including color, isotropy/anisotropy, bireferingence,

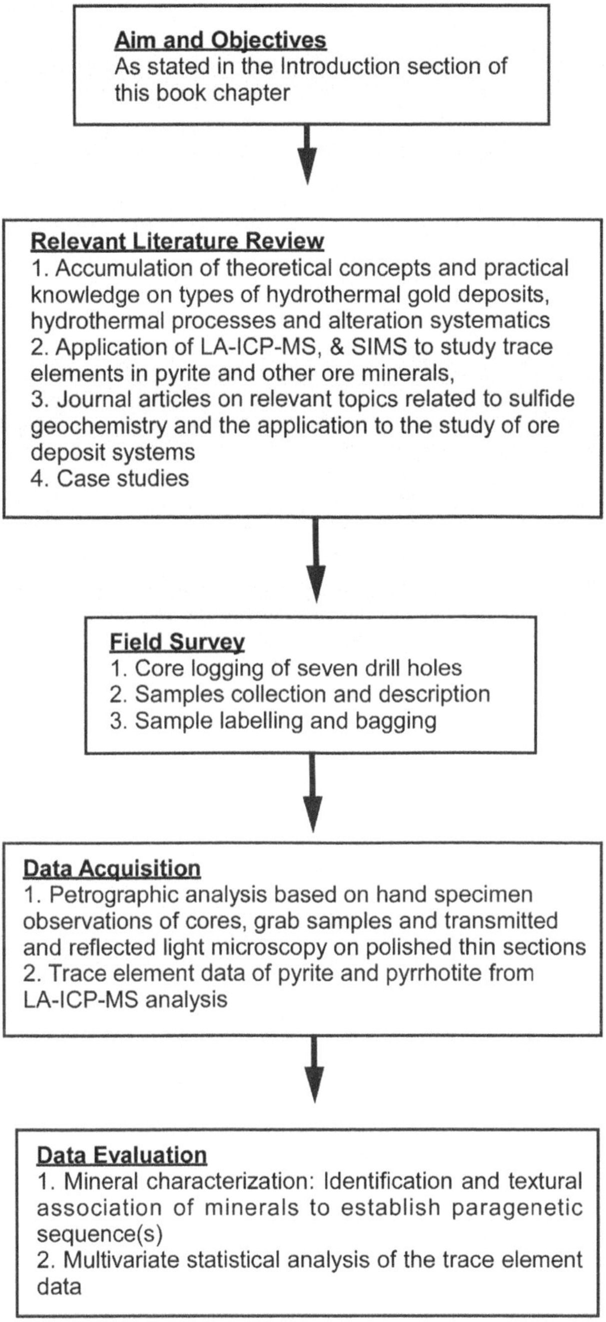

Fig. 3 Flowchart showing the methodology for conducting the studies on Damang deposit

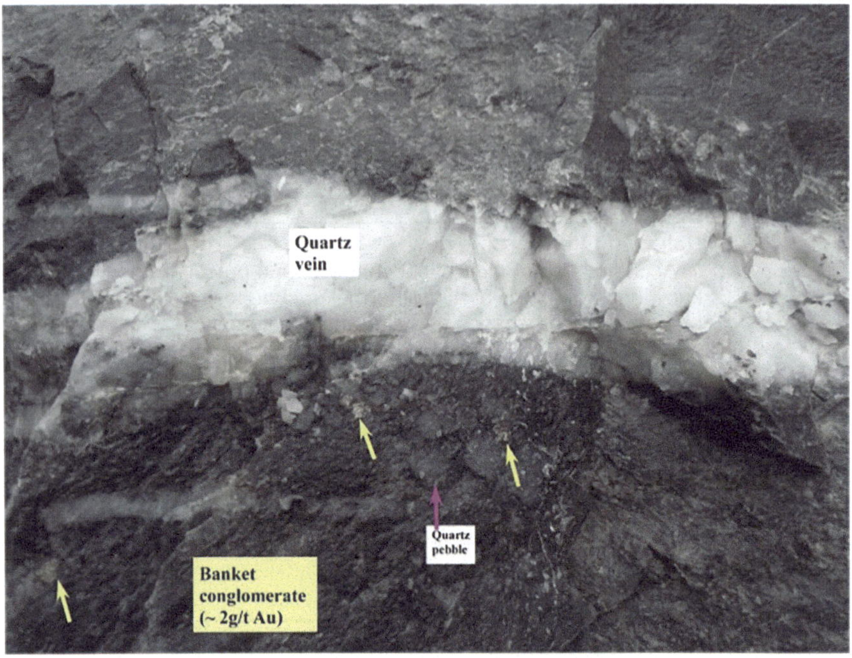

Fig. 4 Insert from Fig. 2 showing a close-up photograph of quartz veins cross-cutting the Banket conglomeratic unit exposed in the mine pit. Pyrite grains are indicated by the yellow arrows. The thickness of the quartz vein is 5 cm

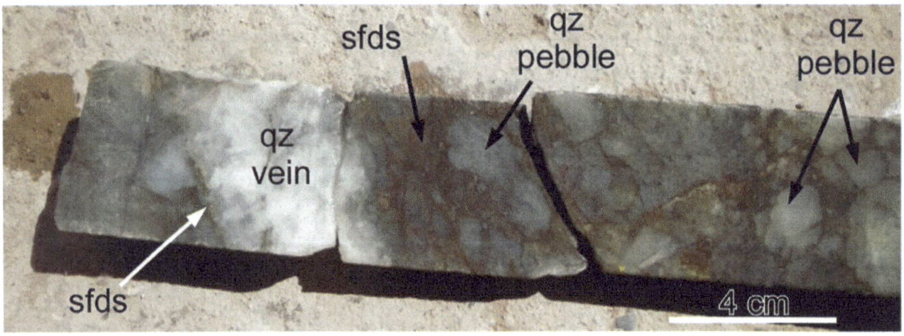

Fig. 5 A drill core showing cross-cutting quartz veins that are related to abundant sulfides (*sfds*) and hydrothermal gold mineralization in the Banket Conglomerate

texture, twining, extinction angle, cleavage, pleochroism, bireflectance, and reflectivity were appropriately used to discriminate between the rock-forming and ore minerals in the samples [51].

Fig. 6 Placer gold-bearing Banket Conglomerate (left) intruded by diorite (right). Both lithological units are hydrothermally mineralized with gold by the cross-cutting quartz vein(s). The yellow dash line is along the contact between the two lithological units. Pyrite grains are indicated by yellow arrows. The pen in the photo is used as scale

LA-ICP-MS Analysis

After logging the eleven diamond drill cores, five (5) samples that include dioritic intrusive rocks or dykes (196, 198-9A and 207-4) and conglomerate (216-13A and 218-10) were selected based on the very high Au assay values recorded in these rocks (unpublished company metallurgical report) and the occurrence of abundant hydrothermal sulfides. The selected samples were prepared as 25-mm round grain mounts for the LA-ICP-MS analysis at CODES, University of Tasmania. Pyrite and pyrrhotite, determined from microscopy in the samples, were analyzed for trace elements using a RESOlution 193 nm excimer laser ablation system coupled with an Agilent 8900 ICP-MS (Table 1). A range of elements and isotopes were analyzed and the following are reported: ^{57}Fe, ^{59}Co, ^{60}Ni, ^{65}Cu, ^{66}Zn, ^{75}As, ^{77}Se, ^{95}Mo, ^{107}Ag, ^{111}Cd, ^{118}Sn, ^{121}Sb, ^{125}Te, ^{197}Au, ^{205}Tl, ^{208}Pb and ^{209}Bi. The internal standard elements used for quantification is Fe. Analyses are normalized to 100 wt.% total. A spot size of 29 μm was used for all mineral analyses. For each analysis, the background signal is recorded for 30 s, then the laser is turned on and the targeted mineral is ablated while the ICP-MS collects data for each element for ~60 s. During spot analysis, the material analyzed is typically dominated by the targeted mineral. Element signals that show no changes, gradual smooth changes, or changes consistent with chemical zonation are interpreted to be chemically bound into the target mineral structure. However, element signals may also show no or gradual change when the element is present in evenly distributed inclusions within the targeted mineral. Such signals are indistinguishable from 'true' chemical substitution into the mineral structure. Both types of occurrences are referred to as refractory. Elements that have signals with discrete, sharp changes in the laser signal unrelated to sharp chemical zoning, are interpreted as inclusions. When inclusions are small, they do not decrease element signals from the targeted mineral. Such inclusions are referred to as micro-inclusions. To calculate concentrations, the average of the signal during the time interval of interest within the sample grain is calibrated against reference standards, i.e., STDGL3 (in-house sulfur-rich glass), USGS SRM GSD-1G, and in-house pyrite.

Multivariate Statistical Analysis

Probability–Probability Plots

In order to test for the normality in the data set, probability–probability plots (or P-P plots) were used. The P-P plots operate by comparing two cumulative distribution functions. If the distribution functions are similar, the data will resemble a normal distribution and appear to be nearly a straight line. Conversely, the non-linearity of the distribution with respect to the straight line would indicate that the data set is not normally distributed [21, 52–55].

Centred-Log Ratio

Commonly, chemical data acquired from geological formations are statistically not normally distributed due to the heterogeneity of the crust, which consequently influences the nature of the distribution, concentration and enrichment of the elements/metals in the crust. Thus, it becomes imperative to reduce the range between the minimum and maximum values for each parameter in order to mitigate data abnormality [21, 54]. The formula used to achieve this purpose is shown in Eq. 1:

$$\mathrm{CLR} = \log\left(\frac{x_1}{g(x)}\right), \ldots, \log\left(\frac{x_n}{g(x)}\right) \tag{1}$$

where

CLR	Centered Log-Ratio (CLR)
x	composition vector
$x_1, x_2, \ldots, x_n$	individual elements in the samples
$g(x)$	geometric mean.

Table 1 LA-ICP-MS data on hydrothermal sulfides from the Banket Conglomerate and Diorite

Quartz-veined, altered rock	Sample analyte	Spot size (μm)	Mineral	Au	Ag	As	Te	Sb	Bi	Se	Cd	Ni
Banket Conglomerate	218-10_x1	29.00	pyrite	0.02	0.02	1.64	17.53	0.80	0.13	25.37	0.01	572.07
Banket Conglomerate	218-10_x2	29.00	pyrite	0.03	0.12	1.65	8.71	1.04	0.16	22.57	0.01	405.68
Banket Conglomerate	218-10_x3	29.00	pyrite	0.60	3.77	3.18	14.90	16.88	3.42	24.43	0.07	306.85
Banket Conglomerate	218-10_x4	29.00	pyrite	0.01	1.73	1.20	0.32	0.85	1.44	26.82	0.02	329.73
Banket Conglomerate	218-10_x5	29.00	pyrite	0.00	0.19	1.10	0.29	0.05	0.01	17.67	0.01	347.61
Banket Conglomerate	216-13A_x6	29.00	pyrite	0.00	0.01	16.87	0.46	0.03	0.07	5.42	0.01	26.36
Banket Conglomerate	216-13A_x7	29.00	pyrite	6.82	1.07	11.63	23.63	6.09	14.75	15.97	0.01	106.26
Banket Conglomerate	216013A_x8	29.00	pyrite	2.34	0.70	10.01	21.54	17.56	13.37	16.51	0.02	66.01
Banket Conglomerate	216-13A_x9	29.00	pyrite	0.00	0.02	12.18	0.17	0.02	0.02	30.80	0.09	260.73
Diorite	207-10_x10	29.00	pyrite	0.00	0.01	38.89	7.80	0.02	0.02	31.09	0.01	1.69
Diorite	207-10_x11	29.00	pyrrhotite	0.00	0.26	271.51	1.23	0.09	0.52	38.75	0.03	276.03
Diorite	207-4_x12	29.00	pyrrhotite	0.00	0.02	2.59	1.30	0.07	0.03	20.30	0.02	59.40
Diorite	207-4_x14	29.00	pyrite	2.21	1.74	6.16	290.80	3.71	9.47	8.65	0.02	21.92
Diorite	207-4_4	29.00	pyrite	0.64	0.30	36.96	16.13	0.16	1.23	8.69	0.01	1.17

(continued)

Table 1 (continued)

Quartz-veined, altered rock	Sample analyte	Spot size (μm)	Mineral	Au	Ag	As	Te	Sb	Bi	Se	Cd	Ni
Diorite	207-4_x13	29.00	pyrrhotite	0.00	0.03	2.43	1.25	1.82	0.03	25.57	0.03	101.88
Diorite	207-4_x15	29.00	pyrrhotite	0.04	0.16	2.43	9.32	1.77	12.73	32.61	0.23	90.51

Spearman's Correlation

To measure the extent of interrelationships between gold and other elements, the Spearman correlation coefficient was applied [56]. The Spearman correlation assumes that the variables under examination have a normal distribution. The assumed values typically fall between −1 and +1, where ±1 indicates the strongest possible correlation and 0 indicates no correlation. A correlation coefficient that falls between 0.3 and 0.5 would indicate a low level of correlation, while a correlation coefficient that falls between 0.5 and 0.7 would indicate a moderate correlation. Correlation coefficient below 0.3 is considered negligible while above 0.7 is considered high. The equation that represents the Spearman correlation function is presented in Eq. (2).

$$r = \frac{(\Sigma(x_1 - \overline{x})(y_1 - \overline{y}))}{\left(\Sigma\sqrt{(x_i - x)^2\Sigma(y_i - y)^2}\right)} \tag{2}$$

where

r Spearman's rank correlation coefficient
x_i values of the x-variable
$\overline{x}$ average of the x variable values
y_1 value of the y-variable
$\overline{y}$ values of the y-variable.

Factor and Hierarchical Cluster Analyses

To analyze the geochemical dataset and find links between gold and the other elements of interest, we used the IBM-SPSS software to carry out factor analysis with principal component analysis for extraction, the varimax rotation, and hierarchical cluster analysis. Using multivariate data, the factor analysis sought to uncover latent patterns and reduce the number of "factors" that account for data variation (54). We used the Kaiser criterion to figure out how many components needed to be removed. Therefore, we only extracted critical components whose eigenvalues were 1 or above. Covariance shows how many components there are, while scree plots reveal the similarity index [57].

Hierarchical cluster analysis is defined by Olise et al. [58] as the process of calculating the degree of similarity or dissimilarity between many objects. For the analysis, we made use of dendograms to depict hierarchical cluster analytical data. (e.g., [21, 58]). Next, two parts are clustered together under the assumption that a given agglomeration criterion will be reduced because of the grouping. This clustering creates a class that includes the two entries. In addition, the similarities between the previously built class and the N-2 remaining components are evaluated in order to group two elements or elemental categories. After each component has been clustered, the process is repeated (e.g. in [21]).

3 Results

Petrography and Ore Textures

The Banket Group conglomerate comprised oligomictic milky to grey, coarse-grained quartz pebbles that have been compacted together in a sandy matrix constituted by abundant fine quartz, plagioclase, K-feldspar, sericite, chlorite, calcite and detrital magnetite and hematite (Fig. 7a, b). The detrital magnetite and hematite commonly formed crossbedded foresets (Fig. 7c, d), which suggest their deposition in a stream-dominant paleoenvironment. On the other hand, detrital and/or metamorphic pyrite was not observed in the samples studied, although there are some reports of detrital pyrite in the Huni Sandstone (e.g., [14, 16, 19]). Many sections of the Banket Conglomerate that are crosscut by the hydrothermal quartz veins exhibited sulfidation and carbonatization as the main alteration types surrounding the quartz veins (Figs. 4 and 5). Sulfidation is dominantly represented by hydrothermal pyrite (py) and pyrrhotite (po).

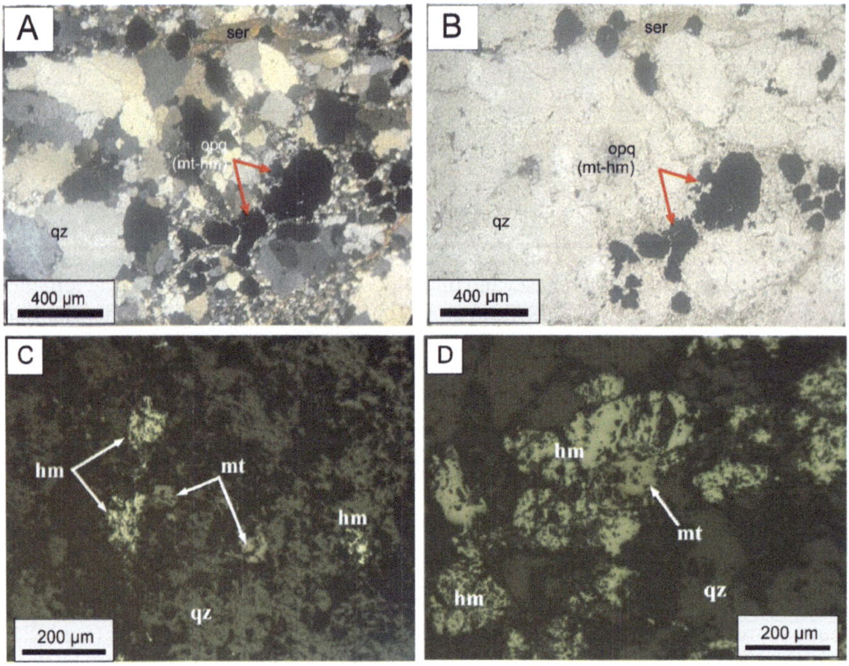

Fig. 7 **a–d** Photomicrographs of least altered conglomerate units showing occurrences of quartz, sericite, magnetite and hematite. **a** In transmitted light under crossed polars **b** in transmitted light under plane polars **c** and **d** in reflected light. *Note* qz—quartz; ser—sericite; mt—magnetite, hm—hematite

Disseminated fine pyrite is commonly anhedral, while coarse pyrite is anhedral through euhedral (Fig. 8a–d). At least two major types of pyrite were observed: (a) anhedral and porous pyrite, py1 (Fig. 8a–c,) and (b) anhedral to subhedral py2, with a porous core py2a and a massive rim, py2b) (Fig. 8d). In both types of pyrite, the porous sections (i.e., py1 and py2a) contained inclusions of gangue minerals that are part of the host rock mineralogy (Fig. 8d). It was noticed from this study that pyrrhotite, po1, mostly occurred as rounded to sub-rounded inclusions within pyrite (Fig. 8b–d).

The mafic intrusive rocks are medium to coarse-grained, dark green to almost black lithologies (Fig. 6), and regionally, considered hypabyssal intrusive rocks into the Tarkwaian Group [17, 19]. They occur as dykes and sills. The ore mineralogy of altered intrusive rocks is constituted by dominantly large, anhedral pyrrhotite (po2), with subordinate subhedral to euhedral pyrite, py3 (Fig. 8a). The textural relations of the hydrothermal assemblage show that py3 forms mutual boundary textural relationship with po2 or as euhedral pyrite grains surrounded by po2 (Fig. 9b–d).

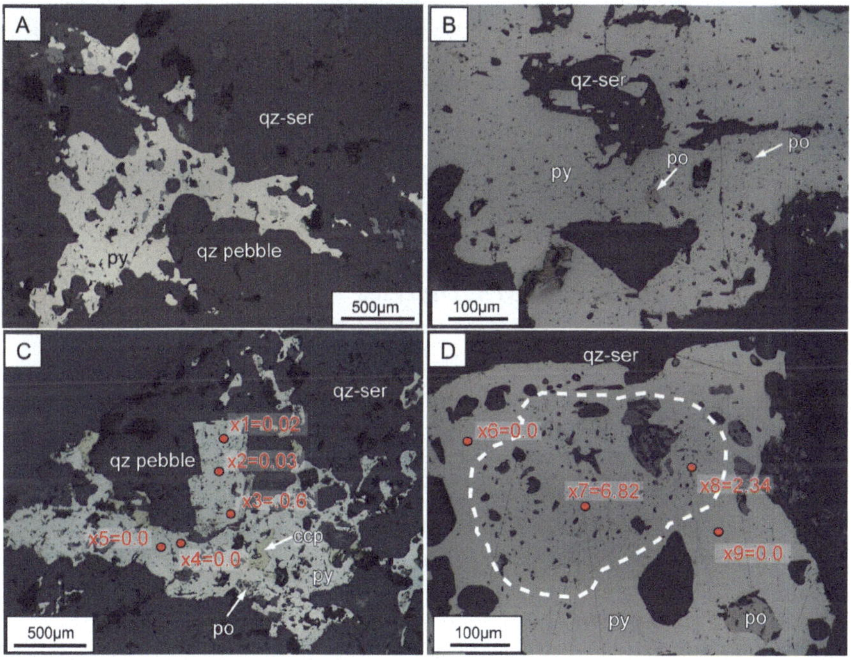

Fig. 8 **a–d** Hydrothermal ore sulfides in the Banket Conglomerate showing location of LA-ICP-MS ablated points in (**c**) and (**d**). The gold concentrations in ppm are shown for each ablated point. See Table 1 for more information. **d** Pyrite exhbiting porous core (py2a) and massive rim (py2b), with inclusions of pyrrhotite (po1). *Note* qz—quartz; ser—sericite; py—pyrite; po—pyrrhotite; ccp-chalcopyrite; X1-X9—LA-ICP-MS analyte points

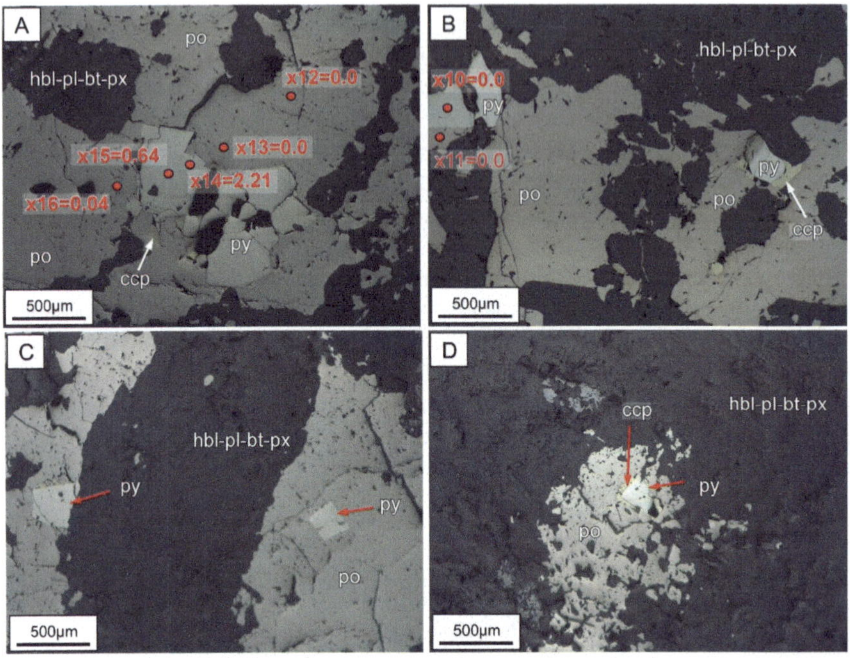

Fig. 9 a–d Hydrothermal ore sulfides in diorite showing the location of LA-ICP-MS ablated points in (**a**) and (**b**). The gold concentrations in ppm are shown for each ablated point. See Table 1 for more information. *Note* qz—quartz; px—pyroxene; pl—plagioclase; bt—biotite; ser—sericite; py—pyrite; po—pyrrhotite; ccp—chalcopyrite; X10-X16—LA-ICP-MS analyte points

Constraints from LA-ICP-MS Analyses

The LA-ICP-MS data obtained for this study is shown in Table 1. The range of concentrations (in ppm) for elements in pyrite (py1 and py2) from the Banket Conglomerate are given as follows:

Au (0.00–6.82), Ag (0.01–3.77), Bi (0.01–14.75), Te (0.17–23.63), Sb (0.02–17.56), Pb (0.02–47.95), Cu (0.89–16,262.48), Sn (0.01–2.75), Zn (0.14–200.76), Co (8.14–1473), Ni (26.36–572.07), Mo (0.00–0.13) and Se (5.42–30.80).

The elemental concentrations (in ppm) for pyrite (py3) from the intrusive rocks are given as follows:

Au (0.00–2.21), Ag (0.01–1.74), Bi (0.02–9.47), Te (7.80–290.8), Sb (0.02–3.71), Pb (0.04–29.62), Cu (0.19–0.77), Sn (0.14–0.15), Zn (0.10–0.36), Co (3682–15,930), Ni (1.17–21.92), Mo (0.00–0.01) and Se (8.65–31.09).

The range of concentrations (in ppm) for elements in po2 from the intrusive rocks are Au (0.01–0.04), Ag (0.02–0.26), Bi (0.03–12.73), Te (1.23–9.32), Sb (0.07–1.82), Pb (0.05–23.62), Cu (0.33–20,983), Sn (0.27–0.64), Zn (0.16–2.19), Co (90.89–312.61), Ni (59.40–276.03), Mo (0.02–0.07) and Se (20.3–38.75). Due to its small size, po1 could not be analyzed to obtain any meaningful data for processing and interpretation.

The distribution of Co and Ni in the pyrites from both rock types reveal some important information during the sequential precipitation of the hydrothermal sulfides. Comparatively, Ni and Co contents in pyrite from the Banket Conglomerate are enriched and depleted, respectively whereas, for pyrite from the intrusive rocks, Ni is depleted, and Co is highly enriched.

Multi-Element Analysis

Conglomerate

For the Banket Conglomerate, Spearman correlation reveals a significant medium to strong positive correlation between Au and Bi (0.95), Au and Ag (0.53), Au and Te (0.92), Au and Sb (0.88), Au and Zn (0.62), Au and Sn (0.92) and Au and Pb (0.80) (Table 2). The pair of the rest of the elements with Au either shows a positive or negative negligible to low correlation.

The results for the factor analysis (FA) confirmed four-dimensional structures. The four dimensions explained 93.67% of the total variance among the elements in the study with Eigen values greater than one. The summary of the results of the FA for the Banket Conglomerate is shown in Table 3 and illustrated in Fig. 10. Factor one accounts for 40.866% of the total variance and include Au–Ag–Te–Sb–Bi–Zn–Sn–Pb multi-element association (Fig. 10). These elements had a strong to perfect correlation with factor one. Factor two accounts for 25.96% (Table 3) of the total variance and include As–Se–Ni–Co multi-element association (Fig. 10). These elements had both strong to perfect negative and positive correlation. Co and As had strong negative correlation with factor two. Factor three accounts for 15.88% (Table 3) of the total variance and included Mo-W multi-element association (Fig. 10). Elements in this factor had a moderate to strong positive correlation coefficient between elements and the factor three. Factor four accounts for 10.971% (Table 3) of the total variance and loaded on Cu (Fig. 10). This element had a strong correlation with factor four.

For the hierarchical cluster analysis (HCA), four (4) clusters representing multi-element associations were observed in the dendrogram derived from the dataset (Fig. 11). Cluster one has a multi-element cluster of Au–Bi–Sb–Zn–Sn–Te–Ag–Pb–Cu. Cluster two has multi-element associations of As–Co. Cluster three has multi-element associations of Se–Ni. Cluster four has a multi-element association of Mo–W.

Diorite

For Diorite, Spearman correlation reveals a significant medium to strong positive correlation between Au and Bi (0.883), Au and Ag (0.96), Au and Te (0.67), Au and

Table 2 Spearman correlation of LA-ICP-MS data for sulfides from the Banket Conglomerate

	S	Fe	Co	Ni	Cu	Zn	As	Se	Ag	Sn	Sb	Te	W	Au	Pb	Bi
S	1.00															
Fe	0.88	1.00														
Co	−0.53	−0.42	1.00													
Ni	0.42	0.33	−0.93	1.00												
Cu	−0.07	−0.32	−0.07	−0.03	1.00											
Zn	−0.77	−0.72	0.40	−0.52	0.30	1.00										
As	−0.05	0.10	0.77	−0.80	0.00	0.08	1.00									
Se	0.27	0.18	−0.43	0.55	0.40	−0.48	−0.35	1.00								
Ag	−0.64	−0.74	−0.11	0.04	0.54	0.77	−0.39	0.02	1.00							
Sn	−0.55	−0.27	0.27	−0.25	0.03	0.45	0.38	−0.25	0.31	1.00						
Sb	−0.67	−0.63	0.02	−0.08	0.32	0.72	−0.12	−0.25	0.76	0.72	1.00					
Te	−0.67	−0.45	0.22	−0.15	−0.15	0.50	0.15	−0.47	0.38	0.88	0.77	1.00				
W	−0.33	−0.24	0.21	−0.18	−0.14	0.39	−0.28	−0.11	0.19	−0.35	−0.15	−0.23	1.00			
Au	−0.65	−0.50	0.20	−0.20	0.23	0.62	0.20	−0.32	0.54	0.92	0.88	0.92	−0.28	1.00		
Pb	−0.53	−0.50	−0.13	0.00	0.47	0.75	−0.17	−0.13	0.86	0.65	0.92	0.62	−0.09	0.80	1.00	
Bi	−0.78	−0.70	0.30	−0.28	0.40	0.75	0.15	−0.25	0.70	0.80	0.88	0.82	−0.08	0.95	0.82	1.00

Table 3 Total variance explained for transformed dataset for pyrite from Banket Conglomerate

Component	Initial Eigenvalues			Extraction sums of squared loadings			Rotation sums of squared loadings		
	Total	% of variance	Cumulative %	Total	% of variance	Cumulative %	Total	% of variance	Cumulative %
1	6.732	44.882	44.882	6.732	44.882	44.882	6.130	40.866	40.866
2	4.150	27.669	72.551	4.150	27.669	72.551	3.893	25.955	66.821
3	2.113	14.088	86.639	2.113	14.088	86.639	2.382	15.881	82.702
4	1.055	7.034	93.673	1.055	7.034	93.673	1.646	10.971	93.673

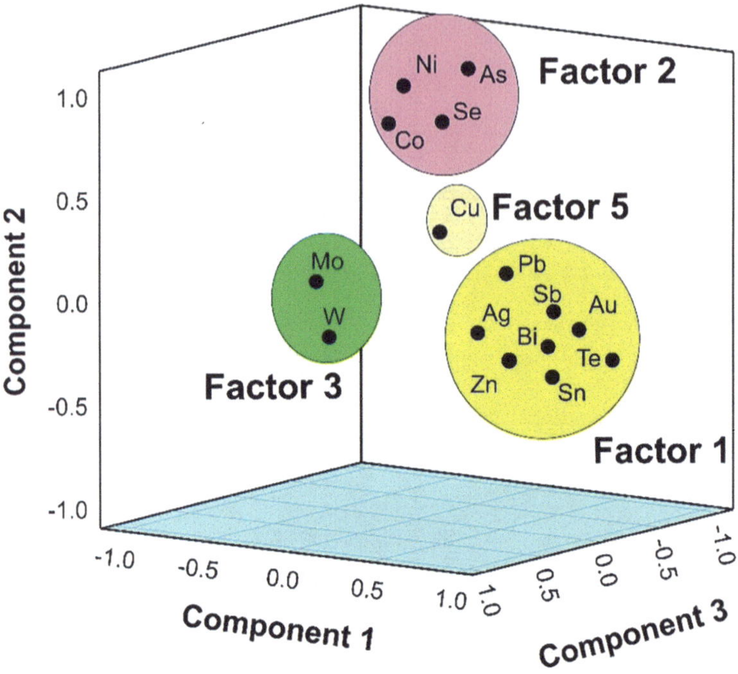

Fig. 10 Rotated component matrix showing the factors for trace element association in hydrothermal pyrite obtained from the Banket Conglomerate

Sb (0.69), Au and W (0.5), and Au and Pb (0.88) (Table 4). The pair of the rest of the elements with Au either shows a positive or negative negligible to low correlation.

For the factor analysis, four-dimensional components were also confirmed (Fig. 12). The four dimensions explained 94.65% of the total variance among the elements in the study with Eigen values greater than one. Factor one accounts for 34.37% (Table 5) of the total variance and includes Au–Ag–Te–Sb–Bi–W–Pb multi-element association (Fig. 12). This element had a moderate to perfect correlation with factor one. Factor two accounts for 29.79% (Table 5) of the total variance and includes Sn–Mo–Ni–Co multi-element association (Fig. 12). These elements had both very good to strong negative and positive correlation. Co had strong negative correlation with factor two. Factor three accounts for 19.34% (Table 5) of the total variance and included Se–Cu–Zn multi-element association (Fig. 12). Elements on this factor had a moderate to strong positive correlation coefficient with factor three. Factor four accounts for 11.16% (Table 5) of the total variance and loaded on As (Fig. 12). This element had a strong correlation with factor four.

For the HCA, four clusters representing multi-element associations are observed in the dendrogram derived from the dataset (Fig. 13). Cluster one has a multi-element

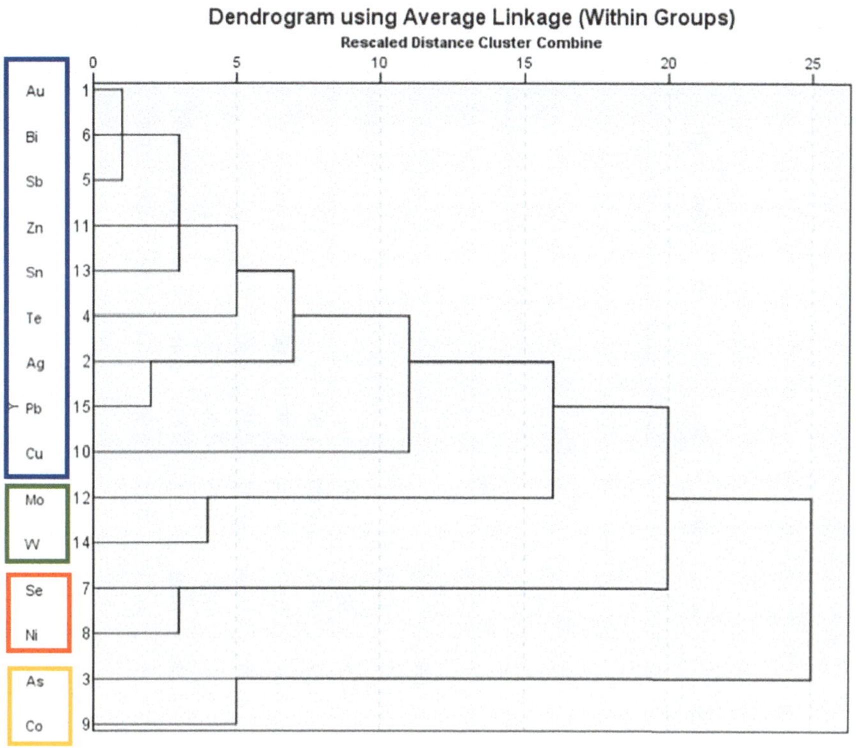

Fig. 11 Dendrogram from hierarchical cluster analysis showing four (4) elemental clusters of trace element data in hydrothermal pyrite obtained from the Banket Conglomerate

cluster of Au–W–Te–Co. Cluster two has multi-element associations of Cu–Zn–Pb–Bi–Ag–Sb. Cluster three loaded on As. Cluster four has a multi-element association of Ni–Mo–Sn–Se.

4 Discussion

Ore Petrography: Insight for the Nature of Hydrothermal Mineralization at Damang

All the previous works and theses on the hydrothermal gold mineralization at Damang have indicated that pyrite and pyrrhotite are the dominant ore minerals with subordinate ilmenite, chalcopyrite, and rare pentlandite [15–21, 48, 50, 59, 60]. However, the controls of the textural relationships and ore paragenetic sequence on the sulfide trace element distribution have not been studied. In this study, the occurrence of

Table 4 Spearman correlation of LA-ICP-MS data for sulfides from Diorite

	S	Fe	Co	Ni	Cu	Zn	As	Se	Ag	Sn	Sb	Te	W	Au	Pb	Bi
S	1.00															
Fe	−0.50	1.00														
Co	0.50	−1.00	1.00													
Ni	−0.79	0.71	−0.71	1.00												
Cu	−0.75	0.25	−0.25	0.75	1.00											
Zn	−0.52	0.23	−0.23	0.43	0.78	1.00										
As	0.21	−0.54	0.54	−0.21	−0.21	−0.65	1.00									
Se	−0.79	0.21	−0.21	0.64	0.50	0.02	0.21	1.00								
Ag	0.32	−0.18	0.18	−0.07	0.25	0.13	0.14	−0.39	1.00							
Sn	−0.64	0.86	−0.86	0.75	0.43	0.36	−0.29	0.25	−0.11	1.00						
Sb	0.14	0.14	−0.14	0.18	0.43	0.58	−0.54	−0.39	0.68	0.00	1.00					
Te	0.57	−0.68	0.68	−0.75	−0.18	0.16	−0.07	−0.68	0.50	−0.64	0.39	1.00				
W	0.74	−0.82	0.82	−0.85	−0.41	−0.26	0.26	−0.56	0.48	−0.85	0.19	0.89	1.00			
Au	0.33	−0.45	0.45	−0.30	0.22	0.19	0.15	−0.41	0.93	−0.33	0.59	0.74	0.69	1.00		
Pb	−0.14	−0.14	0.14	0.14	0.64	0.56	0.00	−0.14	0.82	0.11	0.61	0.46	0.26	0.85	1.00	
Bi	−0.07	−0.33	0.33	−0.07	0.52	0.45	−0.04	−0.07	0.74	−0.22	0.56	0.63	0.50	0.89	0.89	1.00

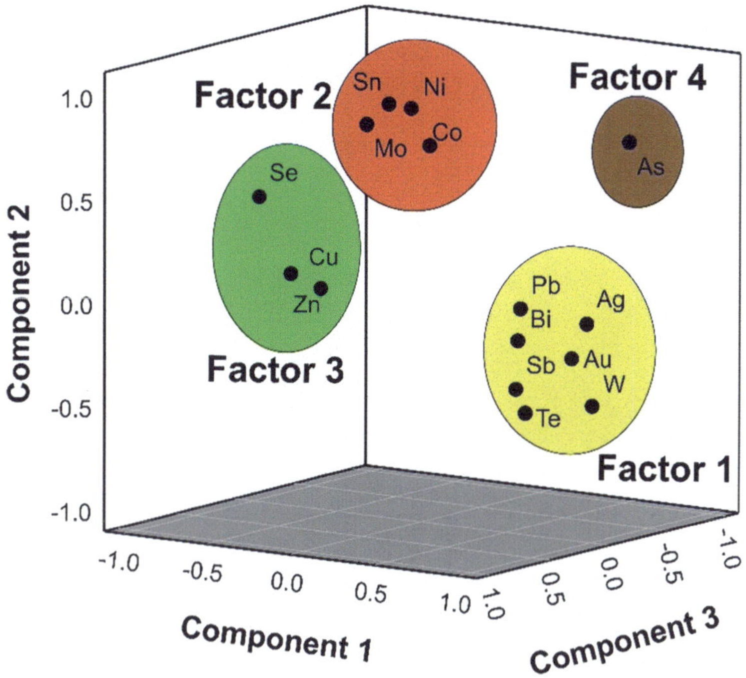

Fig. 12 Rotated component matrix showing the factors for trace element association in hydrothermal pyrite from diorite (i.e. the intrusive rock)

pyrrhotite and pyrite were observed with rare chalcopyrite in both the quartz-veined, hydrothermally altered Banket Conglomerate and the intrusive rocks. All the aforementioned sulfides occur as anhedral through euhedral in form and ranged in grain size from finely disseminated to coarse.

In the Banket Conglomerate, the occurrence of po1 as rounded to sub-rounded inclusions within py1 and py2 (Fig. 8b–d), suggests that the precipitation of po1 was contemporaneous with the precipitation of py1 and py2. The matrix, which supports the quartz pebbles in the Banket Conglomerate, contains Fe-rich minerals such as fine-grained detrital magnetite and hematite (Fig. 6), implying an oxidized nature for the Banket Conglomerate prior to water–rock interaction. Since the quartz pebbles are essentially composed of SiO_2 and therefore unreactive to the fluid, most of the hydrothermal pyrites precipitated within the matrix (Fig. 8a–c). The replacement of some of the Fe-oxides by pyrite suggest sulfidation processes during water–rock interaction (e.g., [16, 19, 20]). The textural occurrence of pyrrhotite inclusions (po1) within the hydrothermal pyrites (Fig. 8b–d) suggests that the fluid was initially reduced. However, the oxidized nature of the sedimentary Banket Group unit caused an oxidation of the fluid to gradually change from a pyrrhotite-stable fluid (po1 inclusions) through pyrite-stable fluid (py1 and py2). The absence of sulphate minerals that normally accompany highly oxidized gold mineralization processes suggests the

Table 5 Total variance explained for transformed dataset for pyrite from diorite

Component	Initial Eigenvalues			Extraction sums of squared loadings			Rotation sums of squared loadings		
	Total	% of variance	Cumulative %	Total	% of variance	Cumulative %	Total	% of variance	Cumulative %
1	6.562	43.748	43.748	6.562	43.748	43.748	5.155	34.369	34.369
2	4.813	32.084	75.833	4.813	32.084	75.833	4.468	29.786	64.154
3	1.448	9.655	85.488	1.448	9.655	85.488	2.901	19.340	83.494
4	1.375	9.164	94.652	1.375	9.164	94.652	1.674	11.158	94.652

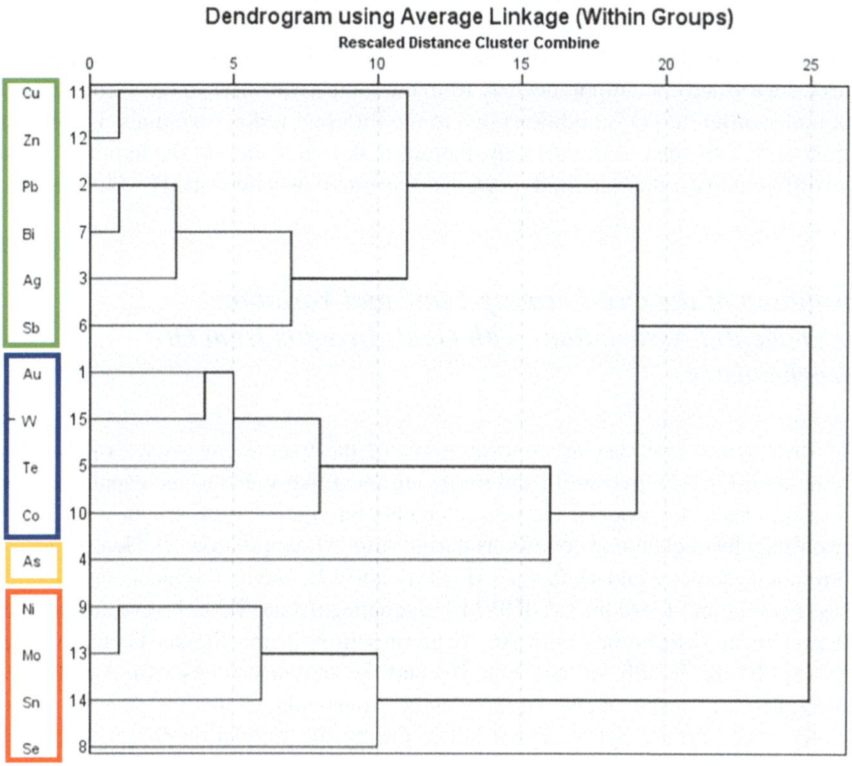

Fig. 13 Dendrogram from hierarchical cluster analysis showing four (4) elemental clusters of trace element data in hydrothermal pyrite obtained from diorite (i.e. the intrusive rock)

change in redox condition of the hydrothermal fluid to have remained in the pyrite-pyrrhotite stability field (e.g., [60]). This explains the py >> po association in the Banket Group within the Damang deposit.

In the case of the intrusive rocks, ore textural relations (Fig. 9) suggest that pyrrhotite (po2) was deposited with predominantly euhedral pyrite (py3) (e.g., [52]). This precipitation sequence also mirrors the sequence of sulfide precipitation (i.e., py1 and py2 with po1) in the Banket Conglomerate (Fig. 9). These sequences of sulfide precipitation suggest that both mineralization processes in the Banket Conglomerate and the intrusive rocks were contemporaneously caused by the same low fO_2 hydrothermal ore-forming fluid. The differences in textural and modal abundances of sulfide precipitation (i.e. (py1 + py2) > po1 and py3 $\leq$ po2) only reflected the nature of the host rocks with which the fluid interacted: oxidized conglomeratic sedimentary rocks versus relatively reduced medium-grained mafic intrusive rocks, respectively.

The similar sequence of sulfide precipitation in both rock types can be related to the field evidence by the observance of the same orientation (subhorizontal vein sets in similar D3 deformational trends; [17]; Figs. 2 and 5) for the causative hydrothermal quartz veins that cut across all the lithologies in the Damang deposit (Figs. 2 and 6). This clearly indicates that the hydrothermal gold-sulfide mineralization in the lithologies were formed contemporaneously from the same hydrothermal ore-forming fluid but under different fO_2 conditions due to the different redox potentials of the host lithologies. This idea from the study, therefore, does not support the hypothesis for two different hydrothermal fluids at Damang as previously described by Sharpe [16].

Evolution of the Ore-Forming Fluid and Variation in Elemental Associations with Gold: Insights from Ore Geochemistry

The distribution and varying concentrations of the trace elements with particular ore textural types in pyrrhotite and pyrite are most likely due to the capabilities of these sulfides to incorporate the trace elements into their lattice structure or by the prevailing physiochemical conditions during sulfide precipitation [9]. Generally, the correlation (Tables 2 and 4), factor 1 (Figs. 10 and 12), and hierarchical cluster analyses (Figs. 11 and 13) of the LA-ICP-MS geochemical data (Table 1) revealed similar trace element associations with Au for pyrites in both the Banket Conglomerate (Au–Ag–Bi–Te–Sb–Pb–Zn–Sn; Fig. 10) and the intrusive rocks (Au–Ag–Bi–Te–Sb–W; Fig. 12), respectively. All these associations indicate that the elements were incorporated into the pyrite crystal lattice during the sulfidation process. Sunkari et al. [21] indicated a Au–Ag–Bi–Te association at Damang, which was also derived from multivariate statistical evaluation of whole-rock geochemical data. This clearly shows that the hydrothermal signature exhibited by the associations of these metals was controlled by the ore minerals themselves. However, elemental associations from factor 2 (Ni–Co $\pm$ Se $\pm$ Mo; Figs. 10 and 12) for both rock types were indicative of the controls on the sulfide precipitation by redox conditions and nature of the host rock.

To explore the effect of temperature, wall rock alteration and fO_2 conditions on ore-formation in the Damang deposit, the Co–Ni relationship was used because there are enough research and experimental works on many deposits that utilized Co and Ni in pyrite to understand the aforementioned conditions of ore-forming processes (e.g., [3, 61–68]). It is expected that high temperature ($T \geq 300$ °C) hydrothermal fluids would transport Co and Ni. From LA-ICP-MS data, Gregory et al. [65] and Large et al. [63] expressed the strong dependence of temperature for controlling Co and Ni concentrations in pyrite and concluded that low Ni/Co suggest high temperatures of mineralization while high Ni/Co implies low-temperature conditions. According to the thermometric works of White et al. [19] and Fiadonu et al. [20] on the Damang mineralization, the temperature of the hydrothermal ore-forming fluid is high: ~300 to 420 °C. Therefore, the high Ni/Co ratio in pyrite from the Banket Conglomerate (Fig. 14) could not be due to low temperature conditions but rather the increase in fluid fO_2 conditions from pyrrhotite stability to pyrite stability. This increase in fO_2

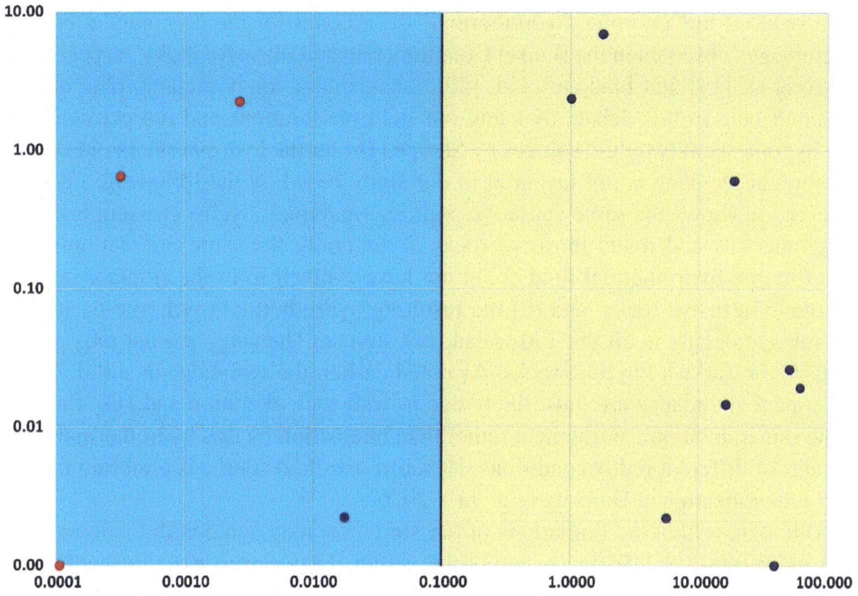

Fig. 14 Plots of Ni/Co ratios (abscissa axis) verses Au (ordinate axis) for hydrothermal pyrite from diorite (red dots) and Banket Conglomerate (black dots). The blue range is low temperature region (Ni/Co < 0.1) and yellow range is high temperature region (Ni/Co > 0.1) (modified after [34]). However, in the case of the hydrothermal system at Damang, the consistent high temperature condition during hydrothermal gold mineralization (T > 300 °C), makes the Ni/Co ratios to rather reflect the redox environment of mineralization: Blue range represents low fO_2 for pyrite from diorite (py ≤ po) and yellow range represents high fO_2 for pyrite from Banket conglomerate (py > po)

condition during fluid-rock interaction caused the increase in Ni concentrations to be accommodated in the pyrite crystal structure during hydrothermal gold mineralization in the Banket Conglomerate. The reverse in Ni concentrations is true for the low fO_2 conditions envisaged in the mafic intrusives. This means that wall rock interaction and its attendant changes in the fO_2 conditions at high temperatures were dominant over temperature decrease during the mineralization process at Damang. This supports the statement of Fiadonu et al. [20] that ore and silicate alteration assemblage associated with the mineralization was initiated at high temperatures (325 ± 20 °C) while the prolonged precipitation of quartz ± carbonates in the veins continued to low temperatures (~210 °C).

Origin of the Hydrothermal Ore-Forming Fluid at Damang

The origin of the hydrothermal ore-forming fluid at Damang is debatable. Sharpe [16] proposed two separate origins: (a) an oxidized magmatic-hydrothermal and

(b) a reduced metamorphic-hydrothermal—to account for the dominant alteration assemblages observed in the Banket Conglomerate and Intrusive rocks, respectively. White et al. [19] and Fiadonu et al. [20] considered a solely metamorphic origin. We contribute to this debate by using our field observations and ore petrography. The hypothesis of two fluid sources by Sharpe [16] for the hydrothermal gold mineralization at Damang is not favoured in our study based on the following: (a) field observation shows the same causative hydrothermal quartz veins crosscut both the conglomerates and mafic intrusive rocks at generally the same orientations, and therefore the hydrothermal fluid could not have evolved from the emplacement of the mafic intrusive rocks, and (b) the resulting hydrothermal pyrrhotite-pyrite ore assemblage occurs in all the Tarkwaian rock units at Damang, and not only in the intrusive or Tarkwa Phyllitic rocks. As noted earlier, the variations in sulfide types and modal abundance are most likely due to wall rock alteration and fO_2 changes in the mineralization environment rather than interaction by two hydrothermal fluid systems of different redox conditions. Hence, a one-fluid source is preferred for the gold mineralization at Damang (e.g., [19, 20]).

Although, within the limitations of our study, we may agree with Fiadonu et al. [20] and White et al. [19] on a metamorphic origin, the occurrence of rare pentlandite [16, 68] and Au–Te ores [59] associated with gold mineralization in the deposit as well as the extensional and non-sheared nature of the related-hydrothermal quartz veins [17, 19] may support some contribution of magmatic-hydrothermal fluid at Damang (e.g., [53, 69, 70]). Moreover, the hydrothermal fluid composition (i.e., low salinity, CO_2–CH_4–H_2O–NaCl and O-isotope characteristics) presented by Fiadonue et al. [20] for their metamorphic-hydrothermal model of the gold mineralization at Damang, can also be invoked for a magmatic-hydrothermal origin (e.g., [43, 46, 71–74]). Therefore, it will not be surprising to have an equivocal source of metamorphic and/or magmatic-hydrothermal nature for the Damang deposit, as has been shown for many orogenic gold deposits worldwide [40, 47, 71, 73]. However, the non-discovery of igneous intrusions in the Damang area that postdate the mineralized quartz-veined mafic intrusive rocks makes the invocation of a magmatic-hydrothermally-related gold mineralization [16] in this orogenic setting a bit challenging for the Damang deposit (e.g., [71, 73]).

5 Conclusion

This study employed LA-ICP-MS technique to establish trace element concentrations in hydrothermal sulfides related to gold-associated quartz-carbonate veins crosscutting the paleoplacer gold deposit at Damang. The textural and trace element distributions in these sulfides improved our understanding of the conditions and evolution of the ore-forming fluid that led to the overprinting hydrothermal gold mineralization. From our study,

(a) Both pyrite (py1, py2, py3) and pyrrhotite (po1, po2) formed contemporaneously from the same hydrothermal fluid (i.e. the one-fluid source model) and not by two different hydrothermal systems of different redox conditions (i.e. the two-fluid source model).
(b) Multivariate statistical analyses of the LA-ICP-MS data for pyrite in the Banket Conglomerate show Au–Ag–Bi–Te–Sb association, while in the mafic intrusive rocks, Au–Ag–Bi–Te–Sb–W association is registered, indicating similar incorporation of ore metals in the pyrite lattice during sulfidation.
(c) Variations in Co and Ni ratios of the pyrite showed that during fluid-rock interactions, the hydrothermal fluid, possibly of metamorphic ± magmatic origin, evolved from low fO_2 conditions in the mafic intrusives to high fO_2 conditions in the Banket Conglomerate. These changes in the fO_2 conditions resulted in the py $\leq$ po and py > po ore assemblages in the mafic intrusives and Banket Conglomerate, respectively. The relative oxidation of the ore-forming fluid by the Banket conglomeratic unit is likely due to the abundant Fe-rich oxide minerals such as magnetite and hematite, some of which occurred as cross-bedded foresets. This means that in addition to phase separation, wall rock alteration and its attendant changes in the fO_2 conditions at high temperatures ($T > 300$ °C) were dominant over temperature decrease during the hydrothermal gold mineralization process at Damang.

6 Recommendations

It is recommended that additional LA-ICP-MS data on the ore minerals should be acquired to:

(a) establish consistency with the current data set reported herein; and
(b) compare with the large database in the literature that have utilized trace element plots of pyrites to clarify the origins of hydrothermal ore-forming fluids.

Acknowledgements This work forms part of the first author's MPhil thesis at the University of Mines and Technology (UMaT) Tarkwa, Ghana, and it was supported by the Chamber of Mines Postgraduate Grant (GCM-TEF_UMaT-Gh-Pg/006/21) to the first author. The Management of the Damang Mine including the Mine Geology and Exploration Departments, is gratefully acknowledged for their invaluable support, insightful discussions on the Damang mineralization and, permission to carry out this work and publish the results. The presentation and conclusion for this book chapter is solely the idea of the authors and do not necessarily reflect the idea of the Company. The Editors are duly acknowledged for their critical constructive suggestions.

References

1. Barrie CD (2010) Pyrite—'fool's gold' or misunderstood mineral? Geol Today 26(1):28–33. https://doi.org/10.1111/j.1365-2451.2010.00741

2. Feng Y, Zhang Y, Xie Y, Shao Y, Lai C (2020) Pyrite geochemistry and metallogenic implications of Gutaishan Au deposit in Jiangnan Orogen, South China. Ore Geol Rev 117:103298. https://doi.org/10.1016/j.oregeorev.2019.103298
3. Large RR, Danyushevsky L, Hollit C, Maslennikov V, Meffre S, Gilbert S, Bull S, Scott R, Emsbo P, Thomas H, Singh B (2009) Gold and trace element zonation in pyrite using a laser imaging technique: implications for the timing of gold in orogenic and Carlin-style sediment-hosted deposits. Econ Geol 104(5):635–668. https://doi.org/10.2113/gsecongeo.104.5.635
4. Agangi A, Hofmann A, Wohlgemuth-Ueberwasser CC (2013) Pyrite zoning as a record of mineralization in the Ventersdorp Contact Reef, Witwatersrand Basin, South Africa. Econ Geol 108(6):1243–1272. https://doi.org/10.2113/econgeo.108.6.1243
5. Reich M, Deditius A, Chryssoulis S, Li JW, Ma CQ, Parada MA, Barra F, Mittermay F (2013) Pyrite as a record of hydrothermal fluid evolution in a porphyry copper system: a SIMS/EMPA trace element study. Geochemica Cosmochimica Acta 104:42–62. https://doi.org/10.1016/j.gca.2012.11.006
6. Steadman JA, Large RR, Olin PH, Danyushevsky LV, Meffre S, Huston D, Fabris A, Lisitsin V, Wells T (2021) Pyrite trace element behavior in magmatic-hydrothermal environments: an LA-ICPMS imaging study. Ore Geol Rev 128:103–878. https://doi.org/10.1016/j.oregeorev.2020.103878
7. Asadi HH, Voncken JHL, Hale M (1999) Invisible gold at Zarshuran, Iran. Econ Geol 94(8):1367–1374. https://doi.org/10.2113/gsecongeo.94.8.1367
8. Liu Z, Mao X, Jedemann A, Bayless RC, Deng H, Chen J, Xiao K (2021) Evolution of pyrite compositions at the Sizhuang Gold Deposit, Jiaodong Peninsula, Eastern China: implications for the genesis of jiaodong-type orogenic gold mineralization. Minerals 11:23. https://doi.org/10.3390/min11040344
9. Berthier C, Perret J, Eglinger A, André-Mayer AS, Feneyrol J, Voinot A, Teitler Y, Bosc R (2023) Pyrite as a microtextural and geochemical tracer of ore-forming processes, central zone orogenic Gold Deposit, Gabgaba District, Sudan. Econ Geol 118(5):1031–1053. https://doi.org/10.5382/econgeo.5001
10. Thomas HV, Large RR, Bull SW, Masslennikov V, Berry RF, Fraser R, Froud S, Moye R (2011) Pyrite and pyrrhotite textures and composition in sediments, laminated quartz veins, and reefs at Bendigo gold mine, Australia: insights for ore genesis. Econ Geol 106:1–31. https://doi.org/10.2113/econgeo.106.1.1
11. Hazarika P, Mishra B, Chinnasamy SS, Bernhardt HJ (2013) Multi-Stage growth and invisible gold distribution in pyrite from the Kundarkocha sediment-hosted gold deposit, eastern India. Ore Geol Rev 55:134–145. https://doi.org/10.1016/j.oregeorev.2013.05.006
12. Franchini M, McFarlane C, Maydagán L, Reich M, Lentz DR, Meinert L, Bouhier V (2015) Trace metals in pyrite and marcasite from the Agua Rica porphyry-high sulfidation epithermal deposit, Catamarca, Argentina: textural features and metal zoning at the porphyry to epithermal transition. Ore Geol Rev 66:366–387. https://doi.org/10.1016/j.oregeorev.2014.10.022
13. Ingham ES, Cook NJ, Cliff J, Ciobanu CL, Huddleston A (2014) A combined chemical, isotopic and microstructural study of pyrite from roll-front uranium deposits, Lake Eyre Basin, South Australia. Geochim Cosmochim Acta 125:440–465. https://doi.org/10.1016/j.gca.2013.10.017
14. Brabham G (1998) The regional geological setting and nature of the Damang stockwork gold deposit—a new type of gold deposit in Ghana. Master's thesis, Department of Geology and Geophysics, Univ. of Western Australia, p 101
15. Pigois J, Groves D, Fletcher I, McNaughton N, Snee L (2003) Age constraints on Tarkwaian palaeoplacer and lode-gold formation in the Tarkwa-Damang district, SW Ghana. Mineralium Deposita 38:695–714. https://doi.org/10.1007/s00126-003-0360-5
16. Sharpe R (2005) Damang regional prospectivity analysis: stratigraphic & structural framework of hydrothermal & paleoplacer Gold Deposits in the Damang District—constraints to target additional resources, Volume 1 of 2, Report Prepared for Abosso Goldfields Limited, Ghana Goldfields Limited & Gyata Exploration Limited

17. Tunks JA, Selley D, Rogers J, Brabham G (2004) Vein mineralization at the Damang Gold Mine, Ghana: controls on mineralization. J Struct Geol 26(6–7):1257–1273. https://doi.org/10.1016/j.jsg.2003.11.005
18. White AJR, Waters DJ, Robb LJ (2013) The application of P–T–X (CO_2) modelling in constraining metamorphism and hydrothermal alteration at the Damang gold deposit, Ghana. J Metamorphic Geol 31:937–961. https://doi.org/10.1111/jmg.12051
19. White AJR (2015) Exhumation-driven devolatilization as a fluid source for orogenic gold mineralization at the Damang deposit, Ghana. Econ Geol 110:1009–1025. https://doi.org/10.2113/econgeo.110.4.1009
20. Fiadonu EB, Hammond NQ, Dzigbodi-Adjimah K, Adomako-Ansah K (2023) Ore fluid characteristics and depositional processes of the hydrothermal gold mineralization at Damang, Southwestern Ghana. J Afr Earth Sc 197:104–141. https://doi.org/10.1016/j.jafrearsci.2022.104741
21. Sunkari ED, Gan AN, Fiadonu EB (2024) Geochemical vectors for targeting hydrothermal gold mineralization in the Damang area, southwestern Ghana: constraints from petrography, multi-element geochemistry and multivariate statistical analysis. Ore Energy Resour Geol. https://doi.org/10.1016/j.oreoa.2024.100071
22. Ennih N, Liégeois JP (2008) The boundaries of the West African craton, with special reference to the basement of the Moroccan metacratonic Anti-Atlas belt. Geol Soc, London, Special Publications 297(1):1–17. https://doi.org/10.1144/SP297.1
23. Leube A, Hirdes W, Mauer R, Kesse GO (1990) The early Proterozoic Birimian supergroup of Ghana and some aspects of its associated gold mineralization. Precambr Res 46:139–165. https://doi.org/10.1016/0301-9268(90)90070-7
24. Abouchami W, Boher M, Michard A, Albarede F (1990) A major 2.1 Ga event of mafic magmatism in West Africa: an early stage of crustal accretion. J Geophys Res: Solid Earth 95(B11):17605–17629. https://doi.org/10.1029/JB095iB11p17605
25. Boher M, Abouchami W, Michard A, Albarede F, Arndt NT (1992) Crustal growth in west Africa at 2.1 Ga. J Geophys Res Solid Earth 97(B1):345–369. https://doi.org/10.1029/91JB01640
26. Hirdes W, Davis DW, Eisenlohr BN (1992) Reassessment of Proterozoic granitoid ages in Ghana on the basis of U/Pb zircon and monazite dating. Precambr Res 56(1–2):89–96. https://doi.org/10.1016/0301-9268(92)90085-3
27. Dampare SB, Shibata T, Asiedu DK, Osae S, Banoeng-Yakubo B (2008) Geochemistry of Paleoproterozoic metavolcanic rocks from the southern Ashanti volcanic belt, Ghana: petrogenetic and tectonic setting implications. Precambr Res 162(3–4):403–423. https://doi.org/10.1016/j.precamres.2007.10.001
28. Perrouty S, Aillères L, Jessell MW, Baratoux L, Bourassa Y, Crawford B (2012) Revised Eburnean geodynamic evolution of the gold-rich southern Ashanti Belt, Ghana, with new field and geophysical evidence of pre-Tarkwaian deformations. Precambr Res 204:12–39. https://doi.org/10.1016/j.precamres.2012.01.003
29. Davis DW, Hirdes W, Schaltegger U, Nunoo EA (1994) U-Pb age constraints on deposition and provenance of Birimian and gold-bearing Tarkwaian sediments in Ghana, West Africa. Precambrian Research 67(1–2):89–107. https://doi.org/10.1016/0301-9268(94)90006-X
30. Ledru P, Johan V, Milési JP, Tegyey M (1994) Markers of the last stages of the Palaeoproterozoic collision: evidence for a 2 Ga continent involving circum-South Atlantic provinces. Precambr Res 69(1–4):169–191. https://doi.org/10.1016/0301-9268(94)90085-X
31. Feybesse JL, Billa M, Guerrot C, Duguey E, Lescuyer JL, Milesi JP, Bouchot V (2006) The paleoproterozoic Ghanaian province: geodynamic model and ore controls, including regional stress modeling. Precambr Res 149(3–4):149–196. https://doi.org/10.1016/j.precamres.2006.06.003
32. Chudasama B, Porwal A, Kreuzer OP, Butera K (2016) Geology, geodynamics and orogenic gold prospectivity modelling of the Paleoproterozoic Kumasi Basin, Ghana, West Africa. Ore Geol Rev 78:692–711. https://doi.org/10.1016/j.oregeorev.2015.08.012

33. Eisenlohr BN (1992) Conflicting evidence on the timing of mesothermal and paleoplacer gold mineralization in early Proterozoic rocks from southwest Ghana, West Africa. Miner Deposita 27:23–29. https://doi.org/10.1007/BF00209263
34. Milési JP, Ledru P, Ankrah P, Johan V, Marcoux E, Vinchon C (1991) The metallogenic relationship between Birimian and Tarkwaian gold deposits in Ghana. Miner Deposita 26:228–238. https://doi.org/10.1007/BF00209263
35. Oberthürr T, Mumm AS, Vetter U, Simon K, Amanor JA (1996) Gold mineralization in the Ashanti Belt of Ghana; genetic constraints of the stable isotope geochemistry. Econ Geol 91(2):289–301. https://doi.org/10.2113/gsecongeo.91.2.289
36. Perrouty S, Jessell MW, Bourassa Y, Miller J, Apau D, Siebenaller L, Velásquez G, Baratoux L, Aillères L, Béziat D, Salvi S (2015) The Wassa deposit: a poly- deformed orogenic gold system in Southwest Ghana—implications for regional exploration. J Afr Earth Sci. https://doi.org/10.1016/j.jafrearsci.2015.03.003
37. Mumin AH, Fleet ME (1995) Evolution of gold mineralization in the Ashanti Gold Belt, Ghana: evidence from carbonate compositions and parageneses. Mineral Petrol 55(4):265–280. https://doi.org/10.2113/gsecongeo.91.1.135
38. Schmidt Mumm A, Oberthür T, Vetter U, Blenkinsop TG (1997) High CO_2 content of fluid inclusions in gold mineralizations in the Ashanti Belt, Ghana: a new category of ore forming fluids? Miner Deposita 32:107–118. https://doi.org/10.1007/s001260050078
39. Yao Y, Murphy PJ, Robb LJ (2001) Fluid characteristics of granitoid-hosted gold deposits in the Birimian Terrane of Ghana: a fluid inclusion microthermometric and Raman spectroscopic study. Econ Geol 96(7):1611–1643. https://doi.org/10.1007/s00126-003-0380-1
40. Wille SE, Klemd R (2004) Fluid inclusion studies of the Abawso gold prospect, near the Ashanti Belt, Ghana. Miner Deposita 39:31–45. https://doi.org/10.1007/s00126-003-0380-1
41. Goldfarb RJ, André-Mayer AS, Jowitt SM, Mudd GM (2017) West Africa: the world's premier Paleoproterozoic gold province. Econ Geol 112(1):123–143. https://doi.org/10.2113/econgeo.112.1.123
42. Agbenyezi TK, Adomako-Ansah K, Foli G, Sunkari ED (2024) A review of intrusion-hosted gold systems of the Palaeoproterozoic Birimian terrane of Ghana. Int J Earth Sci. https://doi.org/10.1007/s00531-024-02455-1
43. Masurel Q, Thébaud N, Allibone A, André-Mayer AS, Hein K AA, Reisberg L, Bruguier O, Eglinger A, Miller J (2019) Intrusion-related affinity and orogenic gold overprint at the Paleoproterozoic Bonikro Au–(Mo) deposit (Côte d'Ivoire, West African Craton). Mineralium Deposita. https://doi.org/10.1007/s00126-019-00888-2
44. Treloar PJ, Lawrence DM, Senghor D, Boyce A, Harbidge P (2015) The Massawa gold deposit, Eastern Senegal, West Africa: an orogenic gold deposit sourced from magmatically derived fluids. Ore Deposits in an Evolving Earth. Geol Soc, London, Special Publications 393:135–160. https://doi.org/10.1144/SP393.12
45. Lawrence DM, Treloar PJ, Rankin AH, Boyce A, Harbidge P (2013) A fluid inclusion and stable isotope study at the Loulo mining district, Mali, West Africa: implications for multifluid sources in the generation of orogenic gold deposits. Econ Geol 108(2):229–257. https://doi.org/10.2113/econgeo.108.2.229
46. Hammond NQ, Shimazaki H (1994) Geology and geochemical aspects of ore formation at the Prestea mesothermal vein gold deposit in the Birimian system of Ghana. Int Geol Rev 36(8):715–731. https://doi.org/10.1016/j.oregeorev.2011.03.002
47. McFarlane CR, Mavrogenes J, Lentz D, King K, Allibone A, Holcombe R (2011) Geology and intrusion-related affinity of the Morila gold mine, southeast Mali. Econ Geol 106(5):727–750. https://doi.org/10.2113/econgeo.106.5.727
48. Groves DI, Santosh M, Deng J, Wang Q, Yang L, Zhang L (2020) A holistic model for the origin of orogenic gold deposits and its implications for exploration. Miner Deposita 55:275–292. https://doi.org/10.1007/s00126-019-00877-5
49. White AJR (2011) The nature and origin of gold mineralization at Damang mine, Ghana. Ph.D. thesis, University of Oxford, p 314

50. White AJR, Robb VM, Robb LJ, Waters DJ (2010) Portable infrared spectroscopy as a tool for the exploration of gold deposits in tropical terrains: a case study at the Damang deposit, Ghana. Soc Econ Geol Special Publication 15:67–84. https://doi.org/10.5382/SP.15.1.06
51. Spry PG, Fuhrmann GD (1993) Fluid mixing during deposition of bedded-replacement (BR) deposits in the Illinois-Kentucky fluorspar district. Geol Soc America, Abstracts with Programs; (United States), 25(CONF-9303210-)
52. Sinclair AJ (1976) Application of probability graphs in mineral exploration. Association of Exploration Geochemists, Special Volume, p 95
53. Gao Y, Chik ABR (2013) A multiple regression analysis on influencing factors of urban services growth in China". Technol Invest 4:1–5
54. Zhao J, Chen S, Zuo R (2016) Identifying geochemical anomalies associated with Au–Cu mineralization using multifractal and artificial neural network models in the Ningqiang district, Shaanxi, China. J Geochem Explor 164:54–64. https://doi.org/10.1016/j.gexplo.2015.06.018
55. Sunkari ED, Tetteh SEK, Sumbobo M, Demir Y (2022) Identifying pathfinder elements of gold in the Wa-Lawra Belt, NW Ghana: constraints from petrography, multi-element geochemistry, and multivariate statistical analysis. In: Proceedings of the 7th UMaT Biennial International Mining and Mineral Conference, Tarkwa, Ghana, pp 1–13
56. Spearman C (1987) The proof and measurement of association between two things. Am J Psychol 100(3/4):441–471. https://doi.org/10.2307/142268
57. Nude PM, Asigri JM, Yidana SM, Arhin E, Foli G, Kutu JM (2012) Identifying pathfinder elements for gold in multi-element soil geochemical data from the Wa-Lawra belt, Northwest Ghana: a multivariate statistical approach. Int J Geosci 3(1):62–70. https://doi.org/10.4236/ijg.2012.31008
58. Olise FS, Adeojo SA, Owoade OK, Oketayo OO, Adekola SA Akinlua A (2019), Geochemical characterization of soil samples from gold mining area using PIXE spectroscopy. Environ Sci Pollut Rese 26(5):4924–4937
59. Le Mignot E, Reisberg L, André-Mayer AS, Bourassa Y, Fontaine A, Miller J (2017) Re-Os geochronological evidence for multiple Paleoproterozoic gold events at the scale of the West African craton. Econ Geol 112(1):145–168. https://doi.org/10.2113/econgeo.112.1.145
60. Adomako-Ansah K, Mizuta T, Ishiyama D, Hammond NQ (2017) Nature of ore-forming fluid and formation conditions of BIF-hosted gold mineralization in the Archean Amalia Greenstone Belt, South Africa: constraints from fluid inclusion and stable isotope studies. Ore Geol Rev 89:609–626. https://doi.org/10.5382/SP.15.1.06
61. Jansson NF, Liu W (2020) Controls on cobalt and nickel distribution in hydrothermal sulphide deposits in Bergslagen. Sweden—Constraints from Solubility Modelling, GFF. https://doi.org/10.1080/11035897.2020.1751270
62. Li R, Chen H, Large RR, Zhao L, Liu Y, Jiao J, Xia XP, Yang Q (2020) Ore-forming fluid source of the orogenic gold deposit: implications from a combined pyrite texture and geochemistry study. Chem Geol 552:119–781. https://doi.org/10.1016/j.chemgeo.2020.119781
63. Large RR, Maslennikov VV (2020) Invisible gold paragenesis and geochemistry in pyrite from orogenic and sediment-hosted gold deposits. Minerals 10(4):339. https://doi.org/10.3390/min10040339
64. Jiang J, Zhu Y (2017) Geology and geochemistry of the Jianchaling hydrothermal nickel deposit: T–pH–fO2–fS2 conditions and nickel precipitation mechanism. Ore Geol Rev. https://doi.org/10.1016/j.oregeorev.2017.10.005
65. Gregory DD, Large RR, Halpin JA, Baturina EL, Lyons TW, Wu S, Danyushevsky L, Sack PJ, Chappaz A, Maslennikov VV, Bull SW (2015) Trace element content of sedimentary pyrite in black shales. Econ Geol 110(6):1389–1410. https://doi.org/10.2113/econgeo.110.6.1389
66. Bralia A, Sabatini G, Troja F (1979) A revaluation of the Co/Ni ratio in pyrite as geochemical tool in ore genesis problems: evidences from southern Tuscany pyritic deposits. Miner Deposita 14:353–374. https://doi.org/10.1007/BF00206365
67. Bajwah ZU, Seccombe PK, Offler R (1987) Trace element distribution Co: Ni ratios and genesis of the Big Cadia iron-copper deposit, New South Wales, Australia. Miner Deposita 22:292–300. https://doi.org/10.1007/BF00204522

68. Fiadonu E (2019) The geology and physico-chemical characteristics of the hydrothermal gold mineralization at the Damang gold deposit, Ghana (Doctoral dissertation). University of Mines and Technology, Tarkwa
69. Cameron EM, Hattori K (1987) Archean gold mineralization and oxidized hydrothermal fluids. Econ Geol 82(5):1177–1191. https://doi.org/10.2113/gsecongeo.82.5.1177
70. Dzigbodi-Adjimah K (1993) Geology and geochemical patterns of the Birimian gold deposits, Ghana, West Africa. J Geochem Explor 47(1–3):305–320. https://doi.org/10.1016/0375-6742(93)90073-U
71. Baker T (2000) Emplacement depth and carbon dioxide-rich fluid inclusions in intrusion-related gold deposits. Econ Geol 97:1111–1117. https://doi.org/10.2113/gsecongeo.97.5.1111
72. De Ronde CE, Kamo SL (2000) An Archean arc-arc collisional event: a short-lived (ca 3 Myr) episode, Weltevreden area, Barberton greenstone belt, South Africa. J Afr Earth Sc 30(2):219–248. https://doi.org/10.1016/S0899-5362(00)00017-8
73. Tomkins AG (2013) On the source of orogenic gold. Geology 41(12):1255–1256. https://doi.org/10.1130/focus122013.1
74. Lawrence DM, Treloar PJ, Rankin AH, Boyce A, Harbidge P (2013) A fluid inclusion and stable isotope study at the Loulo mining district, Mali, West Africa: implications for multifluid sources in the generation of orogenic gold deposits. Econ Geol 108(2):229–257. https://doi.org/10.2113/econgeo.108.2.199

Constraints on Fluid Composition and Physico-Chemical Conditions for the Birimian Gold Deposits in Ghana

Johnson Manu

1 Introduction

Till the 1970s, most literature on gold occurrences in the Birimian had been in the form of geological reports of the Geological Survey Department and one such report is on the geology of the Ashanti mine (the largest gold mine in West Africa) [1] in which the source of the Birimian gold was attributed to hydrothermal fluid from deep seated granitic rocks. The first comprehensive work on the geological background of the gold deposits of Ghana was also published [2], which was re-issued in 1973 that summarized the history of gold mining and documented the geology and nature of the mineralization in the active mining areas at that time.

Another possible source for the Birimian gold was put forward [3] that suggested the gold was formed through (a) concentration of the gold in gold-bearing rocks during sedimentation and (b) mobilization and concentration of gold in silicate minerals of the basic rocks from the volcanic activity through volcanic processes. Based on comprehensive geochemical and petrographic work on Birimian lode gold deposits, [4] proposed that the auriferous fluid was metamorphic, which formed by dehydration reactions, during discharge on a sea floor located near a boundary between a basin and a volcanic belt. On account of the broad similarities on structural, lithological and depositional controls on the Birimian greenstone gold deposits, there is the widely accepted genetic model that involved metamorphic fluids with possible role of magmatic fluid [5–8].

There exist numerous literatures on gold mineralization and deformation in the Lower Proterozoic greenstone belts of West Africa [9–20] and on lithological and wall rock alterations [4, 6, 21–25].

J. Manu (✉)
Department of Earth Science, School of Physical and Mathematical Sciences, University of Ghana, Legon-Accra, Ghana
e-mail: jmanu@ug.edu.gh

E. D. Sunkari and S. Ali (eds.), *Geology and Gold Mineralization in Ghana*, Earth and Environmental Sciences Library, https://doi.org/10.1007/978-3-031-97163-1_7

Studies on metamorphic grades in this geological setting include [4, 6, 11, 16, 21, 26–29] and on fluid characteristics from the Birimian gold deposits in Ghana, Burkina Faso, Mali and Senegal [5–8, 30, 31] (Fig. 1).

Based on integrated structural, mineralogical, lithological, and fluid inclusion data, exhibiting trends broadly analogous to those observed in Archaean greenstone belts, despite inherent geological differences, multiple geodynamic models have been proposed to account for the spatial distribution of gold mineralization. This chapter presents a novel, regionally contextualized model that has not previously been considered, emphasizing the role of fluid composition and physico-chemical constraints in elucidating the origin and evolution of the hydrothermal fluids responsible for gold deposition. To illustrate the nature of such gold deposits to ensure maximum benefit from the depositional control dataset, larger sampling was done in Ghana from two major gold mines (Ashanti mine and Prestea mine) and four abandoned gold mines (termed in this study as prospects) such that outlying values are less influential. Furthermore, the alteration style, relevant rock type as determined petrographically or stratigraphically was considered.

2 Geological Setting

During the period from 2100 to 1950 Ma [32], the early Proterozoic rocks of the West African Craton were deformed, regionally metamorphosed and tectonically stabilized, and this period of time is referred to as the Eburnean tectonic event which is widespread on the African continent. In West Africa, it is recognized by the presence of linear zones of metasediments and metavolcanics which are intruded by syntectonic and late-tectonic suites of granitoids. These north and northeastern striking belts of the sedimentary-volcanic rocks are referred to as the Birimian lithological unit. The Birimian terrane which is exposed in Ghana, Burkina Faso and Ivory Coast and is characterized by large sedimentary basins and volcanic belts and can be subdivided into the Lower Birimian and Upper Birimian which represent mainly metasediments and metavolcanics respectively (Fig. 1). The economic mineral deposits in the Birimian in Ghana are gold and manganese and the lode gold type deposit in the Birimian (i.e. gold in the consolidated Birimian rock type) occur as massive auriferous quartz veins mainly in the metasediments or as auriferous sulphides in the wall rocks, particularly tuffaceous bodies (termed as disseminated sulphide ore type).

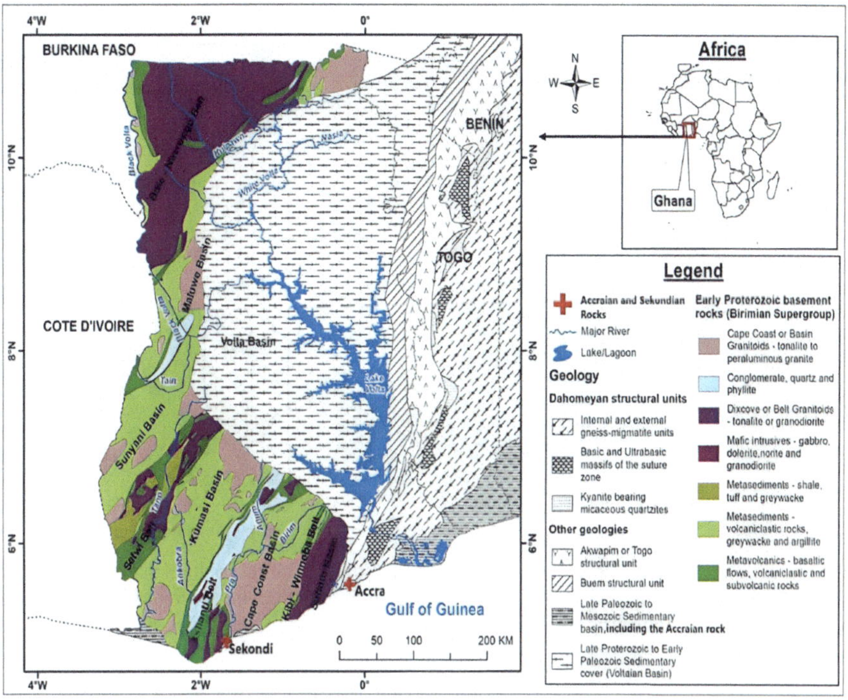

Fig. 1 Modified geology of central and western Ghana (excluding post-Togo Series), simplified after [33]

3 Materials and Methods

Sample Collection Points

Fig. 2a is a flowchart illustrating the methods used in achieving the objectives of this study. Sampling of wall rock types was limited to the deposits at Obuasi and Prestea mines because fresh samples could only be collected from the underground workings there, as against Mpesatia, Oboum, Akoko and Kubekrom prospects, where country rocks have been weathered leaving only the resistant quartz reefs. Nevertheless, to help set up the fundamentals of the auriferous fluid in the Birimian, samples were taken in different geological settings: (i) at various depths, gold-bearing quartz rocks and Birimian rocks (metasediments, metavolcanics and in the metasediments/metavolcanics contact) at the Prestea and Obuasi mines and (ii) quartz reefs from abandoned mines (termed in this study as prospects) at Obuom, Mpesatia, Akoko and Kubekrom (Fig. 2b).

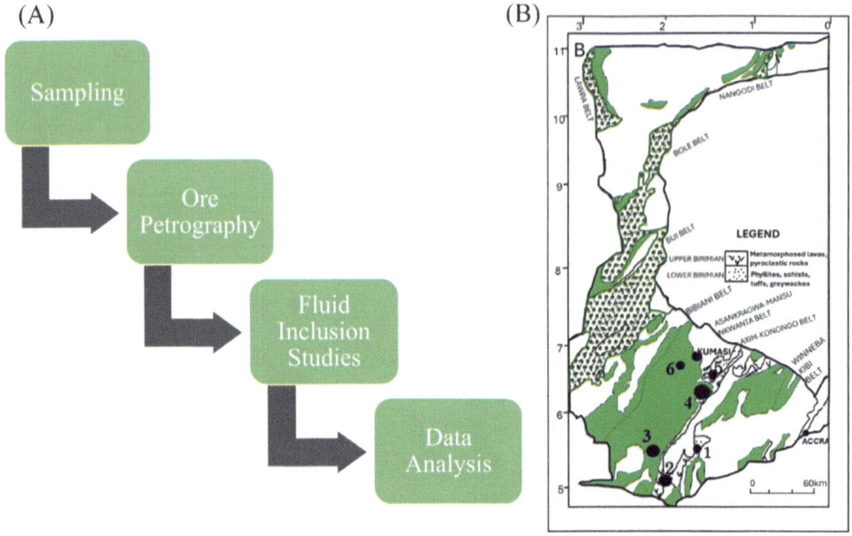

Fig. 2 **a** A flowchart of the general methodology of the study. **b**. Lithostructural map of central and western Ghana (excluding post-Birimian cover), with gold belts simplified after [34]. Also shown are places of sample collection. 1 = Kubekrom prospect; 2 = Akoko prospect; 3 = Prestea mine; 4 = Ashanti mine; 5 = Obuom prospect; 6 = Mpesatia prospect

4 Mineralization

Metasediments/Metavolcanics

Since more extensive and detailed information exists on ore petrography in the Birimian, only a summary together with information obtained during the study of 20 polished sections are presented here. Even though detailed petrographic work in the past has focused mainly on the two prominent mines at Obuasi and Prestea, similar parageneses also occur in other mines or gold occurrences in Ghana [24, 35, 36]. Ore microscopy at Obuasi mine has been conducted by researchers like [6, 27, 35, 37] and at Prestea mine [6, 38–40].

The mineral assemblages in the metasediments are ilmenite-(magnetite?)-pyrrhotite-arsenopyrite-rutile-pyrite-gold. Rutile occurs after ilmenite and titano-magnetite whilst TiO_2 mineral aggregates as leucoxene and rare anatase mineral exist. Visible gold is rarely observed in the arsenopyrite, and it is generally assumed that most of the gold is locked up as submicroscopic grains in the arsenopyrite crystals. Figures 3 and 4 show the paragenetic diagram of the minerals and their major chemical components in the metasediments, respectively.

In the metavolcanics, as also in the metasediments, magnetite-ilmenite-rutile transformations have been observed and the ore minerals in the metavolcanics are magnetite, ilmenite, rutile, pyrrhotite, pentlandite, arsenopyrite, pyrite, marcasite,

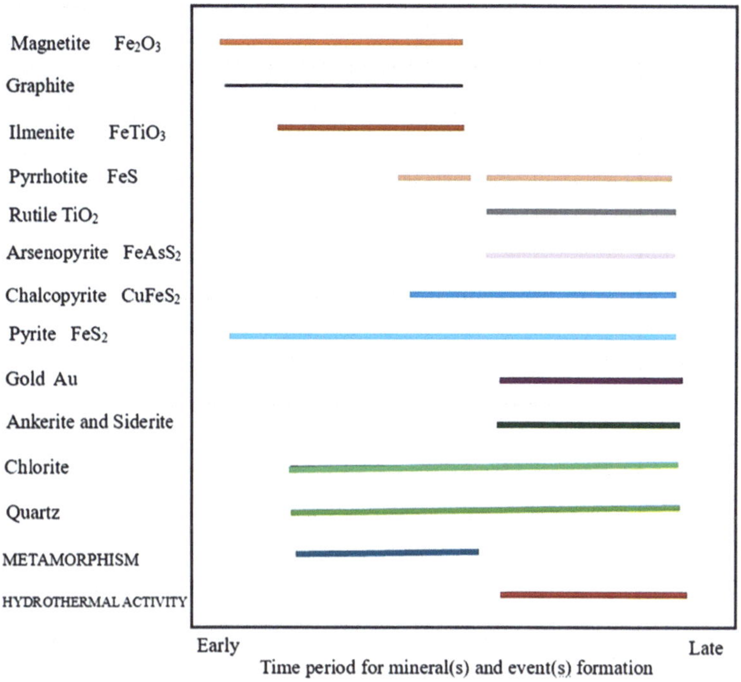

Fig. 3 Paragenetic diagram of minerals (showing metamorphism and hydrothermal activity) in the metasediments

chalcopyrite, sphalerite and gold. Two types of pyrrhotite occur: one with blebs of pentlandite and twinned chalcopyrite and this has been suggested [24, 38] to be of a magmatic origin whilst the other with untwined chalcopyrite is of hydrothermal source [38].

Generally, the pyrrhotite shows alteration into marcasite and pyrite and visible gold occurs occasionally in arsenopyrite and in the metasediments, most of the gold is present sub-microscopically. Figures 5 and 6 show the paragenetic diagram of the minerals and their chemical components for the metavolcanics respectively.

In both the metavolcanics and metasediments, arsenopyrite is the dominant ore mineral lying within the foliation and also across the foliation. They occur together with the other sulphides. However, since most of the gold in the country rocks is within or intergrown with the arsenopyrites (which are in the foliation), it can be concluded that the gold deposition is also post-metamorphic. That is, gold as well as the sulphides were deposited in the existing foliation planes or pathways in the rocks.

Gold-TiO_2 oxides-sulphides assemblages are typical for the country rocks though some differences exist in the assemblages between the metasediments and the metavolcanics. The assemblage in the metavolcanics is complex and shows extensive

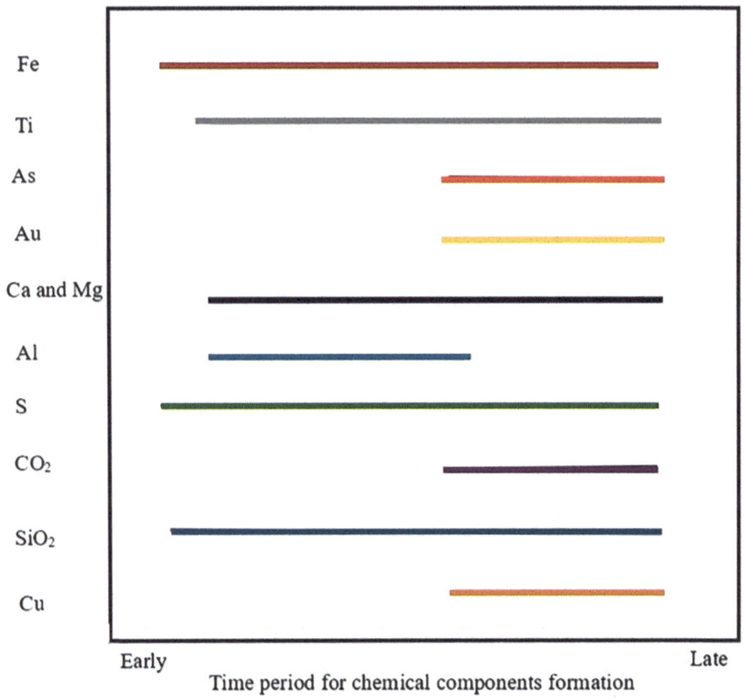

Fig. 4 Paragenetic diagram in terms of the major chemical components that make up the minerals in the metasediments

mineral replacements compared to the simple mineral associations in the metasediments and textural evidence suggests that the gold was formed in the country rocks during hydrothermal alteration.

Translating the paragenetic diagram of the minerals in terms of their radicals shows a strong positive correlation of As, Cu and CO_3 with Au in the metasediments and in the metavolcanics, a strong correlation with Au exists for Cu, Zn, As, S and CO_3.

5 Quartz Veins

The interlocked quartz grains may show effects such as undulose extinction, deformation lamellae, recrystallisation or granulation depending on the degree of deformation and accessories are tourmaline, ankerite, sericite, muscovite, apatite, chlorite and opaques. The presence of such fluid-containing minerals suggests that the quartz reefs have formed in approximate equilibrium with the country rock alteration.

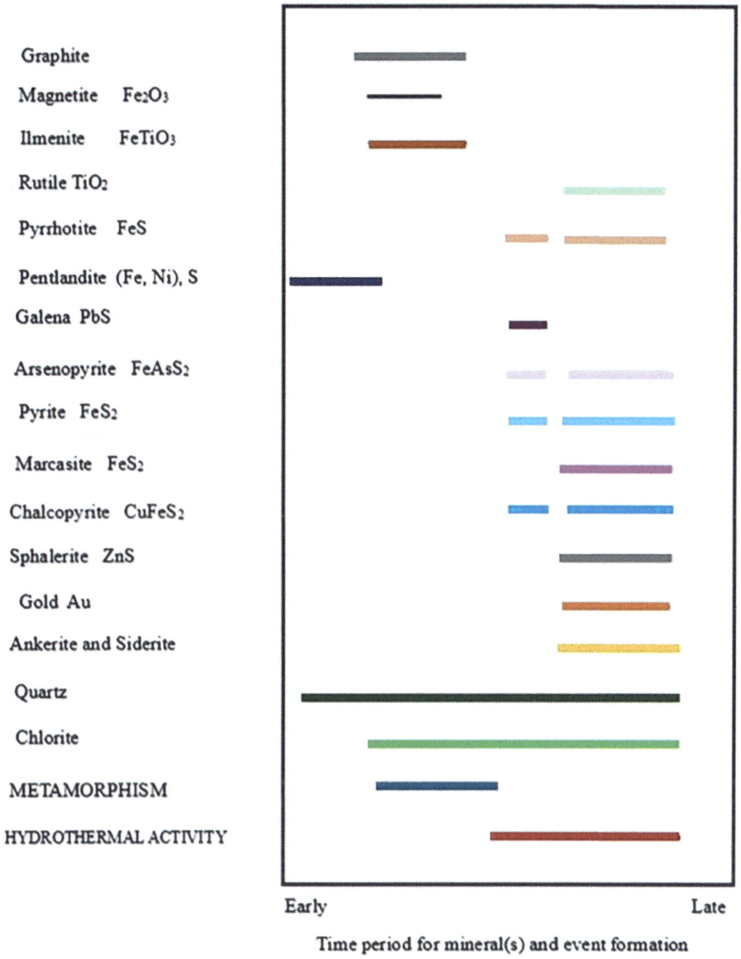

Fig. 5 Paragenetic diagram of minerals in the metavolcanics. Metamorphism and hydrothermal activity are also shown

Texturally, the gold-bearing quartz veins are characterized by ribbon texture which is induced by phyllitic wall rock graphitic partings or relatively insoluble phases in the quartz.

The ore minerals in the quartz veins are gold + sulphides (pyrite, arsenopyrite, chalcopyrite, pyrrhotite, bornite, galena, sphalerite, covellite) + sulphosalts (tetrahedrite, bournonite, boulangerite, jamesonite) and ± tellurides (altaite, nagyagite, petzite, sylvanite, calaverite, kostovite). Figures 7, 8 and 9 show gold in the quartz veins.

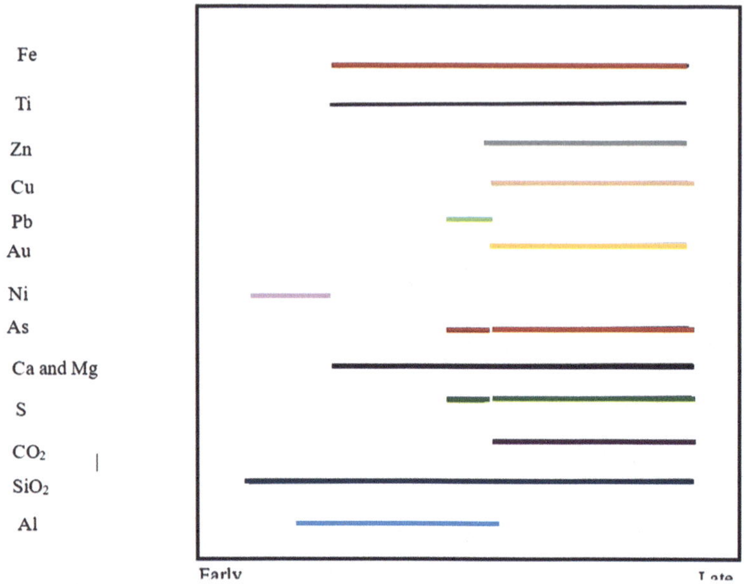

Fig. 6 Paragenetic diagram in terms of the major chemical components that make up the ore in the metavolcanics

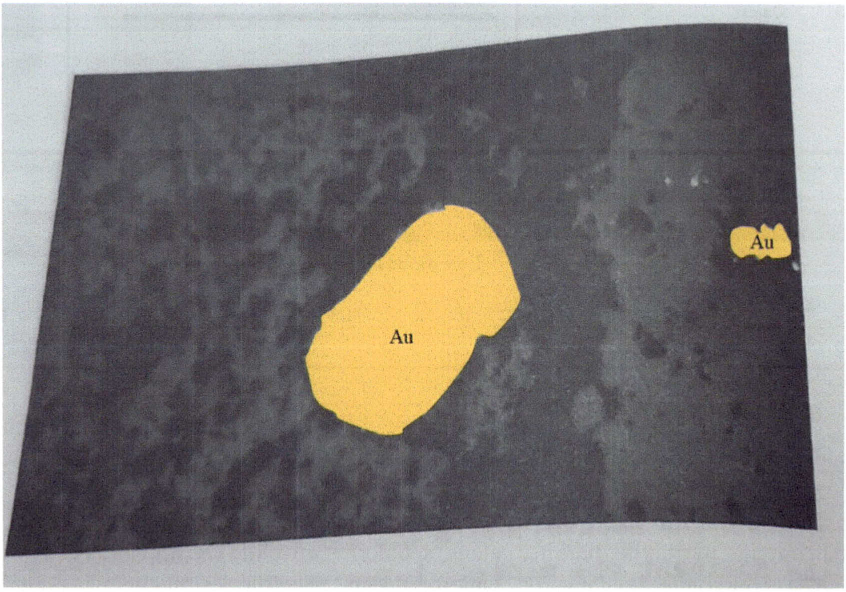

Fig. 7 Grains of gold in a quartz vein. Sample 14/08/293/3. Scale 1 cm: 0.5 mm

Fig. 8 A crack in a quartz vein completely filled with arsenopyrite. Note also the early stage of gold replacing the arsenopyrite at its edges (Sample 14/08/293/9; Scale 1 cm: 0.5 mm; Aspy = arsenopyrite, Au = gold)

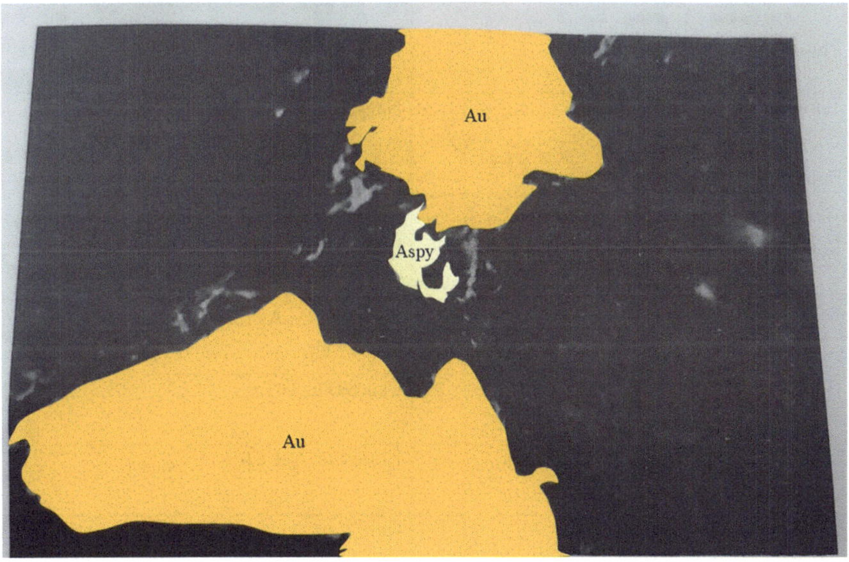

Fig. 9 An advanced stage of arsenopyrite displacement by gold in a quartz vein (Sample 14/OB/293/9; Scale 1 cm: 0.5 mm; Aspy = arsenopyrite, Au = gold)

The sulphides are the dominant ore group followed by the sulphosalts and among the sulphides, pyrite is the dominant ore mineral as against the arsenopyrite which is preponderant in the country rocks. Tellurides are rare.

In the quartz veins, the gold-sulphides-sulphosalts-tellurides parageneses suggest that there are different generations of gold formation as can be inferred from the distinct intergrowth features of gold in phases shown below (Fig. 10).

Phase I		**Phase II**
Bornite, Cu,FeS		Arsenopyrite, FeAsS
Covellite, CuS		Gold, Au
Pyrrhotite, FeS		-
Pyrite, FeS_2		-
Galena, PbS		-
Chalcopyrite, $CuFeS_2$		-
Gold, Au		-
Phase III		**Phase IV**
Tetrahedrite, $(Cu,Fe)_{12}Sb_4S_{13}$		Nagyagite, $Pb_5Au\ (Te, Sb)_4S_{5\text{-}8}$
Bournonite, $PbCuSbS_3$		Altaite, PbTe
Boulangerite, $Pb_5Sb_4S_{11}$		Petzite, Ag_3AuTe_2
Jamesonite, $Pb_4FeSb_6S_{14}$		Sylvanite, $AuAgTe_4$
Gold, Au		Calaverite, $AuTe_2$
-		Kostovite, $AuCuTe_4$
Sphalerite, ZnS		Gold, Au
Gold, Au		-

Fig. 10 Distinct phases (I, II, III and IV) deduced for the gold deposition [6, 36, 38, 41]

For example, the pyrrhotite-pyrite-galena-chalcopyrite-gold phase (I) is deduced from the intimate intergrowth of gold with pyrrhotite, pyrite, galena and chalcopyrite in the quartz vein. Phase II is the gold-arsenopyrite assemblage, and the third phase is the gold-sphalerite occurrence with the sulphosalts. The distinct gold-telluride association forms phase IV. Phases I and II represent the first generation of gold deposition in the quartz veins. The sulphosalts commonly replace the sulphides and fill their fractures [38] (i.e. Phase III being younger than I and II).

Bowell et al. [41] also report replacement of a gold-sulphide assemblage by the gold-telluride mineralization (Phase IV younger than I and II). [36] reported nagyagite ($Pb_5Au (Te, Sb)_4 S_{5-8}$) intergrown with bournonite and tetrahedrite and also found altaite (PbTe) as inclusions in bournonite or marginally intergrown with it.

The low fluid temperatures and pressures (= 170 °C, 0.8 Kbar) [41] for the telluride mineralization also suggests it to be the last gold depositing event in the quartz vein. Hence, the gold-sulphosalts mineralization (III) is deduced to have occurred after the gold-sulphide (I and II) but before the gold-telluride event (IV).

Often, the gold in phases I, II and III replaces the sulphides and the sulphosalts, which suggests that most of the gold was introduced at the end of each episode (Furthermore, it has not been observed that the sulphosalts replace the gold). In each episode, there could be some minerals from the preceding one. For example, in stage II gold and arsenopyrite dominate but there could be some galena, chalcopyrite, pyrrhotite, pyrite and gold from the previous stage I.

In the quartz veins, gold deposited in four different stages (Figs. 10 and 11) has the gold-sulphides-sulphosalts-tellurides paragenesis and there is a positive correlation for Cu, SiO_2 CO_3, Ca and Mg and less for As, Fe, S, Pb, K, and little for Zn and Te.

6 Alteration Mineralogy

Knowledge about alteration processes is deduced from (a) field (b) petrographic and (c) geochemical data and [39] gave field evidence of alterations in the Birimian as formation of pyrite, arsenopyrite, sericite, chlorite, quartz veinlets and ankerite both in the wall rocks and in the mineralized quartz reefs which suggests a close temporal relationship between the conditions of alteration and gold introduction.

The Birimian metasediments comprise of phyllites, meta-greywackes and siliceous rock, therefore the minerals in the unaltered metasediments were quartz, illitic and chloritic clay minerals, feldspar, carbonates and organic matter. Manu [6] concluded that the following minerals are consistently formed in the metamorphic plane of the metasediments (i.e. their formation which occurred at the peak of the metamorphism): Mg-rich chlorite, graphite, sericite, quartz, tourmaline, sphene, (siderite and ankerite?) and also minerals formed as a result of a hydrothermal activity (i.e. post-metamorphism) as (Fe- and Mg-rich) chlorite, sulphides, sericite, siderite, ankerite, epidote and quartz.

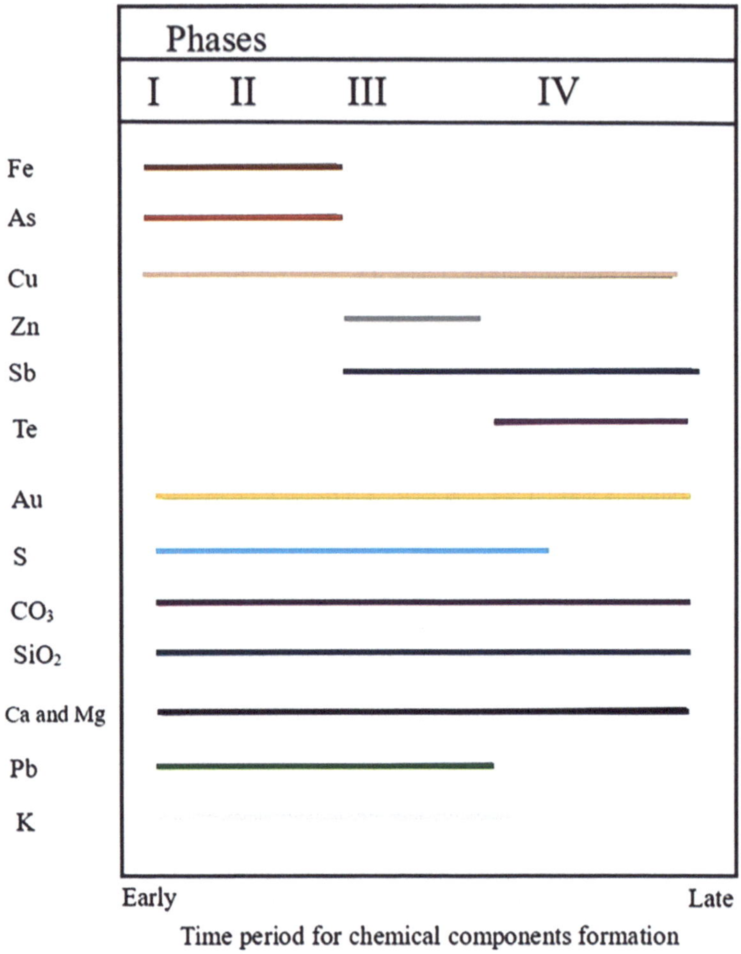

Fig. 11 Paragenetic diagram of the major chemical components in the quartz veins

The bulk composition of the Birimian metavolcanics is dacite to rhyodacite [4] and the original mineral assemblage is expected to be quartz, feldspars and biotite. According to [6] minerals produced consistently in the metamorphic plane of the metavolcanics are Mg- and Fe-rich chlorite, siderite and ankerite (?), sericite and titanite and the post-metamorphic minerals are sulphides, tourmaline, carbonates and chlorites.

A distinction between the alterations related to metamorphism and hydrothermal activity can be made if one compares the consistent minerals in metamorphic and post-metamorphic planes in both the metasediments and metavolcanics. The comparison shows that Mg-rich chlorite, carbonates, titanite and sericite are regularly formed in metamorphic planes in metasediments and metavolcanics and post-metamorphic minerals (i.e. minerals due to hydrothermal activity) are sulphides, carbonates and Mg- and Fe-rich chlorites. Therefore, the hydrothermal activity is inferred to be related to the gold deposition event because the sulphides which contain the gold are in the post-metamorphic deformation planes.

The graphite phyllite is sometimes also found together with auriferous quartz veins in shear zones [35], and since the shear zones were initiated after metamorphism, it is inferred that changes in rocks in this shear zones should come from the latter hydrothermal activity. Hence during the latter hydrothermal activity, trace elements were mainly added by the "altering hydrothermal fluid" to the metasediments (and probably generally to the country rocks). Mass-balance calculation diagrams for the metavolcanics [6] indicate that the major and minor elements show the same trend as in the metasediments (i.e. altered phyllites and the graphite phyllite which is in the shear zone) and that there is enrichment in the trace elements (except for Ba, Sr, Pb) which were lost to the alteration fluid from the country rocks.

The similarity between the trace elements distribution pattern in the altered metavolcanics and the graphite phyllite (which is in the shear zone) suggests a close link between the gold mineralization in the metavolcanics and in the quartz veins [6]. Since the quartz vein is the major ore body, the correlation of trace elements and gold in the metavolcanics and the quartz veins implies that metavolcanics could be exploration targets.

During metamorphism and hydrothermal activity, alterations which involved K, Na, Fe, Mg and Ca were due to these components moving as constituents to the minerals in the rocks (or perhaps some of the field or petrographic evidence for these alterations are only apparent and not real).

Clearly, there were gains in V, Cr, Co, Ni, Cu and Zn from the fluid alteration to the country rocks (due to the hydrothermal alteration) so whilst the metasediments were only slightly enriched, the metavolcanics recorded significant gains.

7 Fluid Inclusions

Thermometric Sample Preparation

Under the microscope, samples which showed extreme tectonic fabrics such as undulose extinction, recrystallization and deformation were rejected and not considered for the thermometric studies and only suitable samples were prepared as follows for the thermometric studies. The sample was cut as a wafer and one of the surfaces hand-ground on a glass plate using progressively finer grades of silicon carbide and further on a rotary polishing machine using MgO (or sometimes diamond grains (0.25 μm)).

This sample was then mounted with the polished face down to a metal block using Vosschemie Polyester-Gießharz MEKP liquid drops and then trimmed off to 2 to 3 mm thickness. This cut surface was then ground and polished as was done for the first surface and the sample was freed from the mounted block by slight warming on a hot plate and finally washed in acetone.

Optical examination of the samples showed three distribution patterns (Figs. 12 and 13) that the inclusions occurred as (a) isolated, most often of relatively bigger size (≈ 6 μm), (b) in two-dimensional trails, sometimes of relatively smaller size (≈ 4 μm) and (c) in cluster groups also of smaller size (≈ 2 μm). The basic principles about how fluids are trapped are known (e.g. [42], pp. 12–36; and based on these principles, inclusions are generally classified into three groups depending on the mechanism of trapping as primary (P), secondary (S) and pseudo-secondary (PS) [7] pp. 1–10) and based on these principles, inclusions are generally classified into three groups depending on the mechanism of trapping as primary (P), secondary (S) and pseudo-secondary (PS).

Primary inclusions (P) are formed during the growth of the host crystal whilst the secondary (S) ones form after the growth of the host crystal, however if during the formation of the host crystal, fractures develop due to mechanical or thermal stress, these cracks can be filled by surrounding fluids and fluids trapped in this manner are termed pseudosecondary (PS) inclusions. Due to the absence of crystal zoning and grain boundaries, secondary inclusions could not always be confidently identified. Hence, only those inclusions which were convincingly isolated were denoted as primary and those undoubtedly found in two-dimensional trails or in cluster groups were assigned as pseudosecondary.

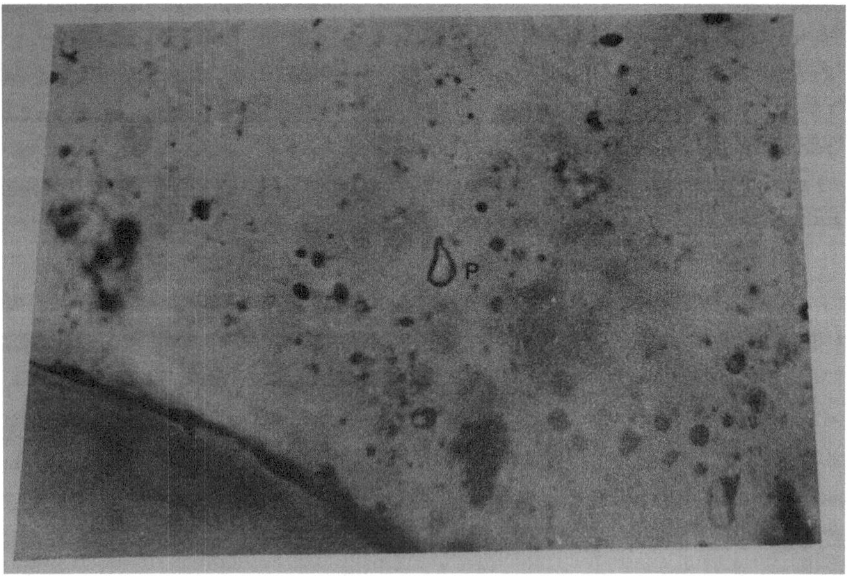

Fig. 12 Photomicrograph of a primary inclusion in an auriferous quartz vein. Sample BM 4. Scale 1 cm: 10 μm

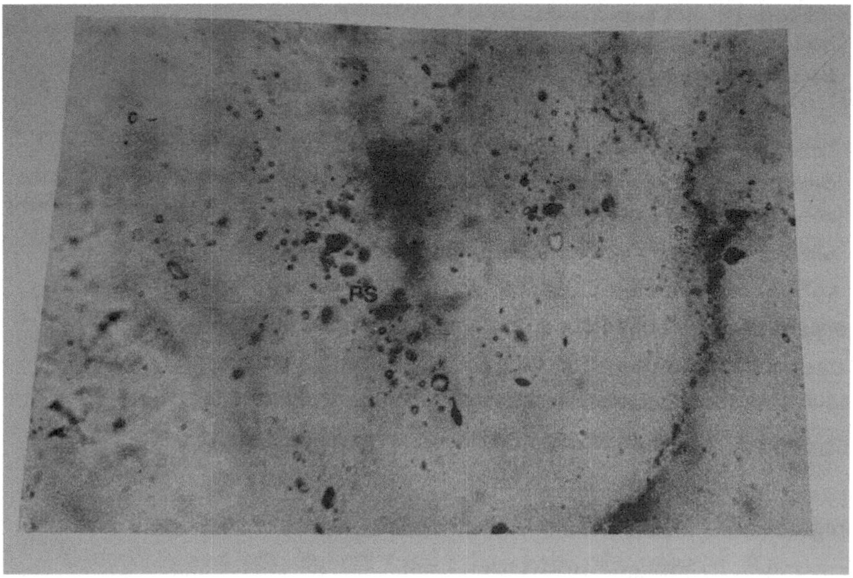

Fig. 13 Primary (P) and pseudosecondary (PS) inclusions in an auriferous quartz vein. Sample MPS 23. Scale 1 cm: 5 μm

Table 1 Sizes of fluid inclusions in μm (average values bracketed) from quartz veins in various localities in the Birimian

Deposit	Host lithology	Fluid inclusion size (μm)
Akoko	Metavolcanics	1–5 (2)
Kuberom	Metasediments	2–5 (3)
Mpesatia	Metasediments	2–10 (6)
Obuom	Metasediments/Metavolcanics	3–10 (4)
Obuasi	Metasediments/Metavolcanics	2–8 (5)
Prestea	Metasediments/Metavolcanics	2–8 (5)

For the sake of finding those inclusions which most likely represent the fluid which deposited the gold found in the quartz mineral, only the primary inclusions (Fig. 12) were chosen for the microthermometric studies (Table 1) and primary or pseudosecondary inclusions (Fig. 13) which showed evidence of necking or leaking were rejected. In one sample (14/293/OB/9), visible gold was seen as isolated grains or intergrown with sulphides in fractures in the quartz sample and primary inclusions are assumed to correspond to the gold-depositing fluid in the quartz.

8 Analytical Methods

Microthermometry

Measurements were carried out on the LINKAM THM 600 stage with the temperature range from – 196 to + 600 °C and calibration of the stage was made against the melting points of high purity chemicals. Sketches of primary inclusions were made at room temperature.

For microthermometric measurements, the sample was cooled to about – 100 °C and by gradual controlled warming brought to room temperature. During the heating, the temperatures at which phase changes occurred in the inclusion or when the inclusion had a single phase were recorded. The abbreviations which were used at various temperatures when phase changes occurred are explained below.

DF =Degree of fill; It is the volume fraction occupied by H_2O-liquid phase in the aqueous and CO_2-H_2O inclusions. The degree of fill was visually estimated at room temperature.

T_mCO_2 = Temperature of final CO_2 melting, i.e. the temperature at which all the solid CO_2 melted.
T_{fm} = Temperature of first melting, which corresponds to the eutectic temperature of an aqueous (H_2O)—salt system.
T_m = Temperature of final (or last ice) melting for the aqueous-salt system.
T_{mclath} = Temperature of final clathrate melting; corresponds to the final temperature at which a gas hydrate which formed melted.
T_hCO_2 = Temperature of partial homogenization of CO_2; i.e. temperature at which two CO_2-rich phases (+ H_2O-phase) homogenise or disappear into a single CO_2-rich phase (+ H_2O-phase). ThCO_2 (L) means homogenization into the liquid state, ThCO_2 (V) means homogenization into the vapour phase.
T_hTOT = Temperature of total homogenization: i.e. temperature at which a two-phase (CO_2 or H_2O) inclusion becomes a single phase. Homogenisation into the liquid state is denoted b (L) whilst into the vapour state is denoted by (V).
Td = Temperature of decrepitation. Temperature at which cavities or fractures around the inclusion rupture (decrepitate).

Raman Spectroscopy Analysis

To get confirmation that in the non-aqueous inclusions, besides CO_2, CH_4 and N_2 are likely to be present and perhaps some other gases too, some suitable inclusions were selected for Raman Spectroscopy whose measurements confirmed peak heights characteristics of CH_4 and N_2.

Crush-Leach Analysis

To supplement information in understanding the chemistry of the gold-forming process, micro-chemical study was done on the fluids, and this involved the determination of K, Na, Ca and Mg.

9 Results and Discussion

Fluid Formation State

Interpretations for analytical data are treated collectively rather than individually because of the similarity of the inclusions at the different mines/prospects. Two main inclusion types occur: aqueous and non-aqueous [6] and their distributions, phases and dominances are illustrated in Fig. 14. The microthermometric data is represented in the form of a scatter plot of total homogenization temperature (ThTOT) versus temperature of final clathrate melting (Tmclath) (Fig. 15), histogram for $TmCO_2$, $ThCO_2$ at Prestea mine (Prestea) (Figs. 16 and 17); at Ashanti mine (Obuasi), (Figs. 18 and 19); $TmCO_2$ at Oboum prospect (Oboum) (Fig. 20); $TmCO_2$, $ThCO_2$, Tmclath—(CO_2-H_2O) at Mpesatia prospect (Mpesatia) (Figs. 21, 22 and 23).

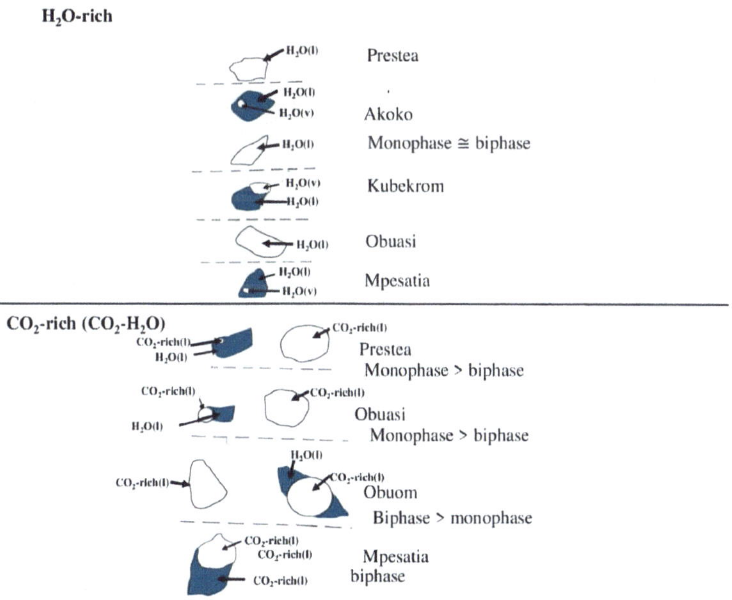

Fig. 14 Sketches of various inclusion types and their distributions in the study areas. Scale 1 cm: 10 μm

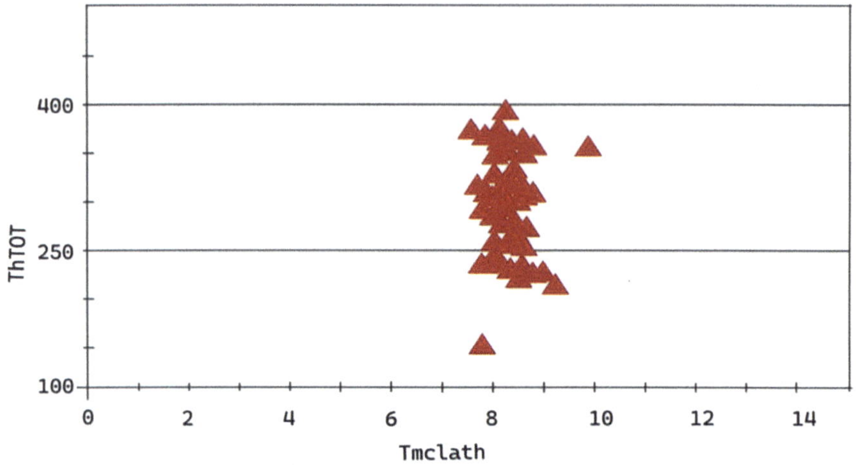

Fig. 15 Total homogenization temperature (ThTOT) versus Temperature of final clathrate melting (Tmclath) showing cooling

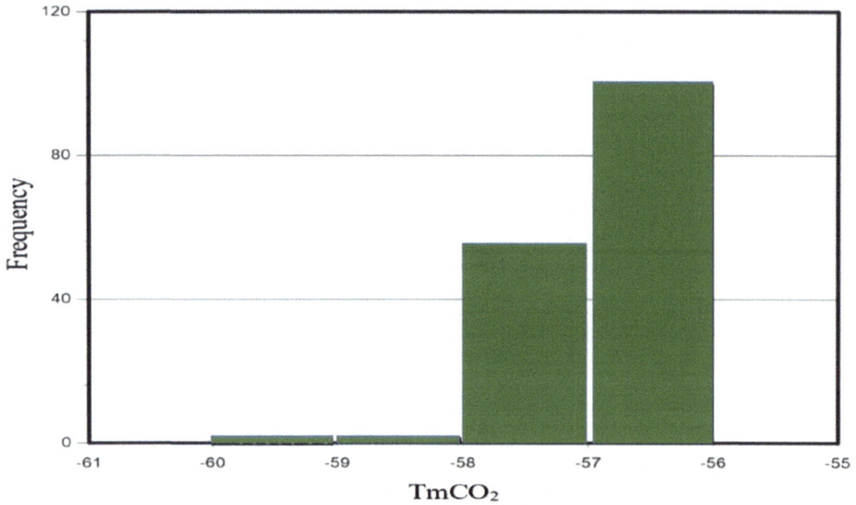

Fig. 16 Frequency histogram of $TmCO_2$ of CO_2-rich inclusions from Prestea mine

All these graphical presentations suggest that there are several populations of inclusions. Microthermometric data on primary inclusions in the quartz veinlets from the host rocks (from Ashanti and Prestea mines) also compared to those in the major quartz veins suggesting that there is an intimate association between the quartz veins and the veinlets in the host rocks.

An obvious question is the relationship between these populations and the fluid state at the time of trapping. That is, what was the nature of the fluid state from which

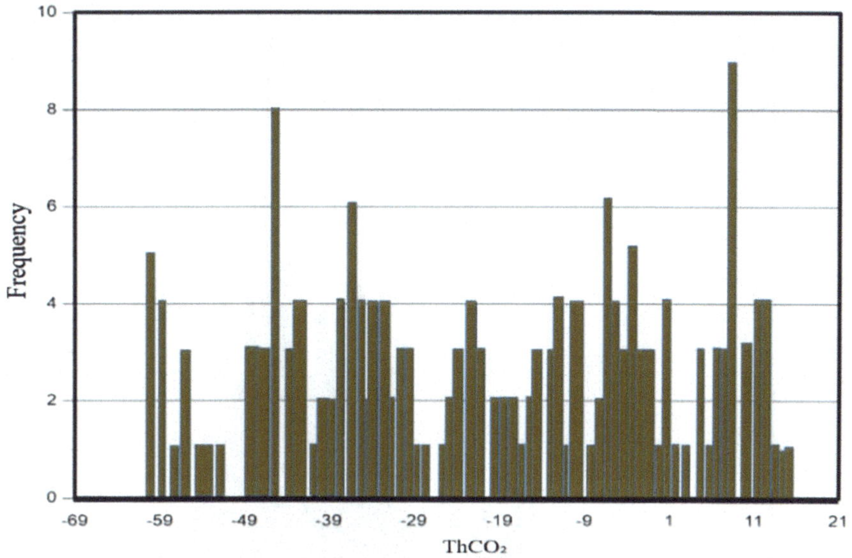

Fig. 17 Frequency histogram of $ThCO_2$ of CO_2-rich inclusions from Prestea mine

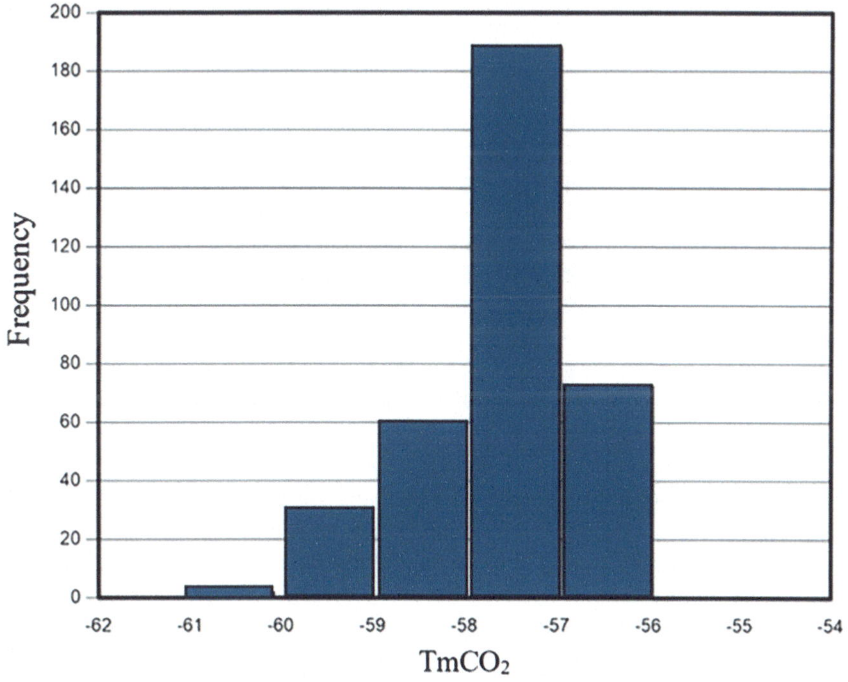

Fig. 18 Frequency histogram of $TmCO_2$ of CO_2-rich inclusions from Ashanti mine

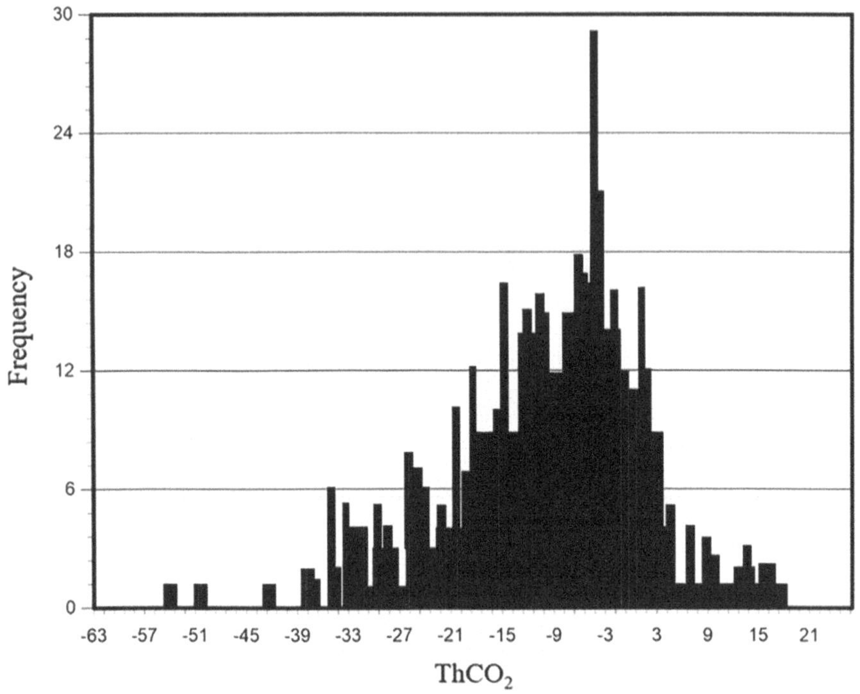

Fig. 19 Frequency histogram of $ThCO_2$ of CO_2-rich inclusions from Ashanti mine

the fluid inclusions were trapped. Was it a homogeneous fluid phase (i.e. a single phase) or a heterogeneous fluid phase (i.e. unmixing of two immiscible phases)? [35] suggested a heterogeneous trapping for the fluid because of variations in H_2O/CO_2 ratios, densities and "gas impurities" of CO_2-rich phases but did not give values to indicate the extent of variations in these parameters, particularly for the former which is a very useful criterion in support of heterogeneous trapping. Due to lack of enough fluid inclusion data base, Schwartz et al. [8] suggested that the original fluid at Ashanti Mine, Obuasi was a CO_2-H_2O mixture due to fO_2 constraints derived from fluid-mineral equilibria.

In view of misinterpretation of fluid data in relation to homogeneous and heterogeneous systems, the consensus generally adopted and recommended by several writers (e.g. [42], pp. 12–36; [44], pp. 8–10); Ramboz et al. [45] is used to interpret the fluid state of the Birimian auriferous fluid at the time of trapping. The criteria involved are (a) nature of the inclusion types and (b) degree of fill. Trapped fluid state identification based on such fundamentally unequivocal criteria are more likely to give the correct interpretation and moreover, it would provide a check for interpretations by other considerations e.g. as one based on constraints [8].

That the same type of inclusions found as primary (early) also occur as pseudosecondary (late) in both quartz veins and quartz veinlets (in the country rocks) suggests that the original parent fluid is likely to have been homogeneous. The degree of fill as

Table 2 Results of crush-leach analyses and calculated temperatures based on chemical geothermometers

Sample	Mg (ppm)	Ca (ppm)	Na (ppm)	K (ppm)	Na/K at. ratio	Ca/Na at. ratio	T(Na–K-Ca)(F)	T(Na–K)(F)	T(Na–K)(T)
BM3	0.06	0.3	3.59	1.09	5.57	0.05	180	546	261
BM8	0.12	0.34	2.1	2.05	1.73	0.1		706	507
BM5	0.01	0.11	0.74	–	–	0.09			
25/319/1	0.05	0.17	0.29	–	–	0.35			
23/301/12	0.05	0.14	2.25	–	–	0.04			
23/308/7	0.01	0.12	1.73	0.35	8.33	0.04	208	506	208
12/226/2	1.25	2.59	28.1	8.85	5.39	0.05	180	549	265
44/161/08/1	0.05	0.19	0.98	0.36	4.7	0.12	182	564	286
BWQ/08/1	0.83	0.07	0.98	0.67	2.47	0.05	229	649	411
MPS 11	–	0.06	3.23	1.72	3.18	0.01	224	613	356
MPS 23	0.01	0.02	0.8	–	–	0.01	–	–	–
K/AA7/21	–	0.44	3.65	5.76	1.07	0.07	286	804	690

given for the biphase aqueous and CO_2-H_2O inclusions fall within the limited range of 0.5 to 0.7 for all areas except at Akoko where the degree of fill is 0.95 for the aqueous biphase. That is, within the same type of inclusions (either biphase aqueous or biphase CO_2-H_2O inclusions) very limited variations of degree of fill exist and this is used to infer a homogeneous trapping.

Having established the nature of the fluid at the time of trapping, a bivariate plot of two independent parameters is used to find the likely natural mechanism which brought about the several populations of aqueous and CO_2-H_2O fluids and the plot, T_{TOT} (Temperature of total homogenization) versus Tclath (Temperature of final clathrate melting) is used. However, due to the difficulty in having enough data (mainly because of finding these independent parameters in the primary inclusions) from all the study areas, the bivariate plot is limited to Mpesatia where enough data was accumulated.

A cooling mechanism is interpreted from the T_{TOT} versus Tclath curve (Fig. 15) whereby a H_2O-rich phase and a CO_2-rich phase separated. That is, during cooling (and subsequent decrease in pressure) separation of immiscible fluids occurred in the parent homogeneous fluid.

Fluid Composition, Density and Salinity

The likely dissolved components of the fluid can be inferred from the measurements of T_{fm} and $T_m$ice values on the aqueous inclusions. The T_{fm} ranges from – 48 to – 32° C at Obuasi; – 53.5 to 47.7 °C at Mpesatia. The observation of eutectic temperatures below – 20.8 °C in the aqueous inclusions indicates the presence of salts beyond NaCl, suggesting a more complex fluid composition. The inferred solutes likely include KCl, $FeCl_2$, $MgCl_2$, and $CaCl_2$, while the presence of water–sodium carbonate or water–sodium sulphate systems is considered highly improbable based on the observed freezing behavior. The $T_m$ice values are from – 21.8 to 6.8 °C at Obuasi; – 19.9 to – 14.2 °C at Mpesatia; – 19 to – 11.9 °C at Prestea [6].

Though the possibility of other salts besides NaCl in the aqueous inclusion is mentioned earlier, the salinity is expressed in wt.% NaCl equiv., as is done conventionally. That is, the expression of the salinity in terms of the amount of NaCl that would produce the same salinity if the other salt components were also considered and using the ice melting curve of the H_2O-NaCl system [44], pp. 123, the salinities (in wt.% NaCl equiv.) for the aqueous phases are as follows: 25 (by extrapolation) to 10 at Obuasi; 23 to 18 at Mpesatia: 22 to 16 at Prestea (Figs. 17, 19, 20, 21, 22 and 23).

Two varieties of CO_2-rich (i.e. the CO_2-H_2O) inclusions occur: monophase and biphase (Fig. 14). Measurements on the biphase variety help to deduce the chemical components as well as the salinity of the CO_2-H_2O inclusions. Though the T_m for the biphase at Mpesatia which is – 25.2 °C and – 18 to – 27 °C at Obuom, does not exclude other salts besides NaCl, $T_m$clath measurements are used preferably to calculate the salinity of the CO_2–H_2O phase rather than the $T_m$ice because more values could be obtained for this phase change.

With the help of a FORTRAN program [46], the bulk inclusion compositions (expressed in terms of mole & of H_2O, CO_2 and NaCl), salinity (as wt.% NaCl equiv.), aqueous density (given as Aqd), CO_2 density (given as CO_2d) and bulk density (given as Bd) are calculated using $T_m$clath and T_hCO_2 values. Tables 3, 4, 5, 6, 7 and 8 give a summarization of the important parameters for the various study areas.

To find out which gas (or gases) caused the lowering of the formation temperature of liquid CO_2, suitable inclusions were analysed for CH_4, N_2 and H_2S using the Raman spectroscopy but only traces of CH_4, and N_2 were found in CO_2-H_2O (biphase and monophase) inclusions which suggest that the proportion of trace gases tend to be higher in the biphase than in the monophase inclusions.

Results of the crush-leach analysis confirmed the presence of Mg, Ca, K and Na cations and the variations in the alkali ratios suggest that there may be differences in the fluid which is a function of regional characteristic due to wall-rock contamination.

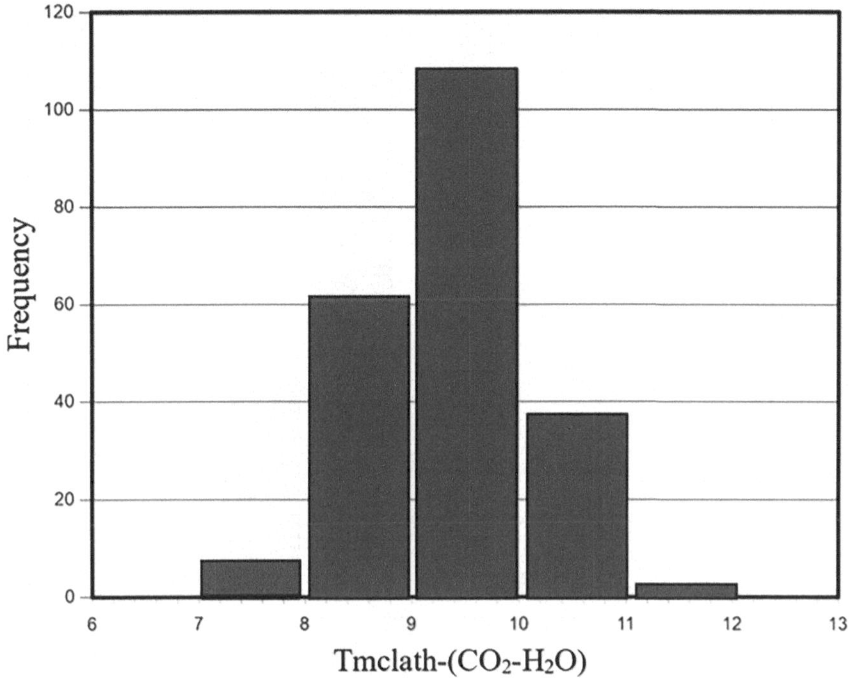

Fig. 20 Frequency histogram of $TmCO_2$ of CO_2-rich inclusions from Obuom prospect

Fluid Temperature and Pressure

Till now, formation temperatures for the quartz veins have either been estimated [25] or deduced from fluid mineral equilibria [8, 35]. Schwartz et al. [8] combined the siderite-graphite assemblage in the quartz veins with muscovite-calcite-quartz in the host rock to bracket the maximum formation temperature at between 370 and 470 °C for the quartz veins in the Ashanti mine and estimates are higher than the range proposed [35], who used the siderite-graphite association in the quartz vein to get a value of 300–440 °C. These data are consistent but do not give precise values. Moreover, these temperatures have been based on the possible assumption that the vein-mineral assemblage can only be expressed in the Fe–C–O system.

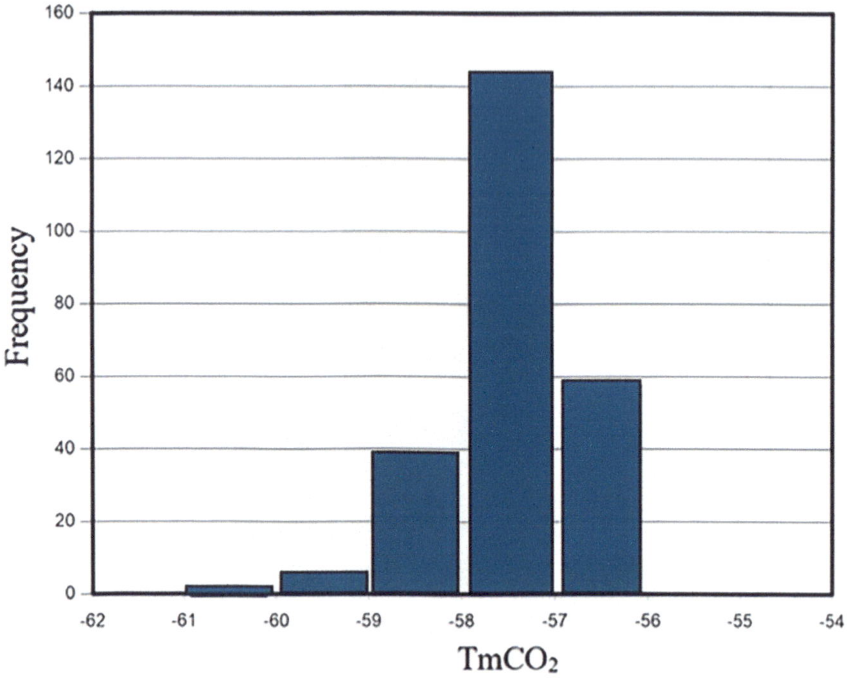

Fig. 21 Frequency histogram of $TmCO_2$ of CO_2-rich inclusions from Mpesatia prospect

Here, the alkali ratios obtained by crush-leach analysis are tentatively used to calculate the trapping temperature of the fluid based on Eqs. (1), (2) and (3).

(a) Truesdell geothermometer (Na–K) (T) [107]

$$T = \frac{855.6}{\log(\mathrm{Na}/\mathrm{K}) + 0.8573} - 273 \tag{1}$$

(b) Fournier geothermometer (Na–K) (F) [48, 81]

$$T = \frac{1217}{\log(\mathrm{Na}/\mathrm{K}) + 1.483} \tag{2}$$

(c) Fournier and Truesdell geothermometer (Na–K–Ca) (F) [26, 51]

$$T = \frac{1647}{\log(\mathrm{Na}/\mathrm{K}) + 1/3\left[\log\left(\sqrt{\mathrm{Ca}}/\sqrt{\mathrm{Na}}\right) + 2.06\right] + 2.47} - 273.15 \tag{3}$$

Results of the crush-leach analysis confirmed the presence of Mg, Ca, K and Na cations and the variations in the alkali ratios suggested that there may be differences in the fluid which is a function of regional characteristic due to wall-rock contamination.

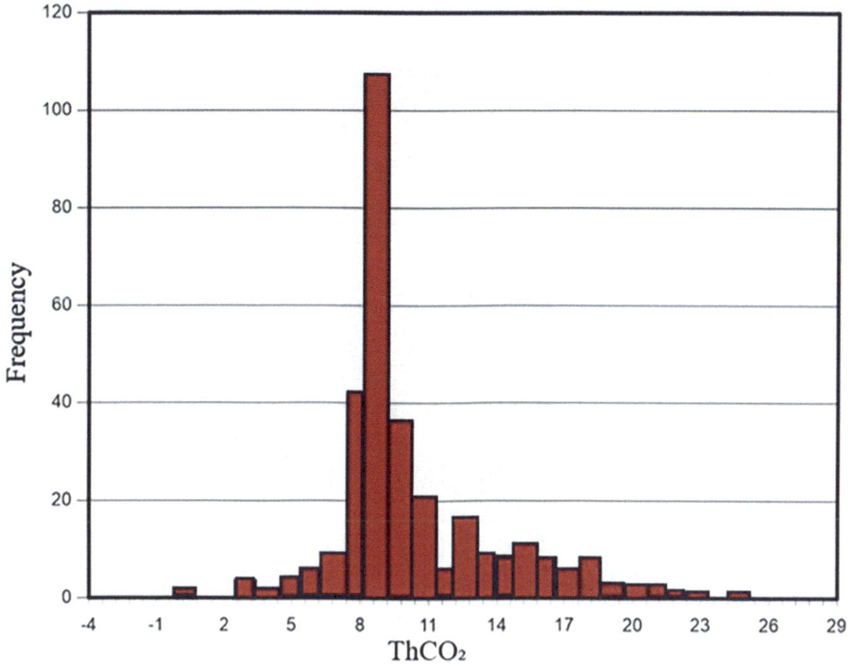

Fig. 22 Frequency histogram of $ThCO_2$ of CO_2-rich inclusions from Mpesatia prospect

The temperatures obtained (Table 2) are either too low or too high implying that the alkali ratios probably do not reflect the fluid equilibria at the site of deposition or that the crush-leach results are wrong or even as according to [49] chemical geothermometers tend to give erroneous results when applied to low-salinity fluid inclusions as it is in the Birimian gold deposits.

It was shown earlier that the different groups of inclusions resulted from cooling of a trapped homogeneous system, and this means that the biphase aqueous and CO_2–H_2O inclusions which were heated in the laboratory followed the reversed (cooling) path until it became a single phase at the total homogenization temperature, T_hTOT. This is the minimum temperature at which the single fluid phase (upon heating) exists, hence the minimum temperature of trapping, (T_t). Knowing the difference between T_hTOT and T_t. (commonly called the pressure correction) enables one to calculate the trapping (depositional) temperature.

The pressure correction given as $P_t - P_h/(AP/\Delta T)$ ([42], pp. 255), where P_t = formation pressure, P_h pressure at the homogenization temperature and AP/AT = the slope of the isochore beginning at T_h and P_h, cannot be used to find pressure corrections since the pressure during the quartz vein formation is not constrained, even though the maximum pressure for the geological setting of the host rocks is between 2 and 4 kbars based on the metamorphism of the geological setting [6].

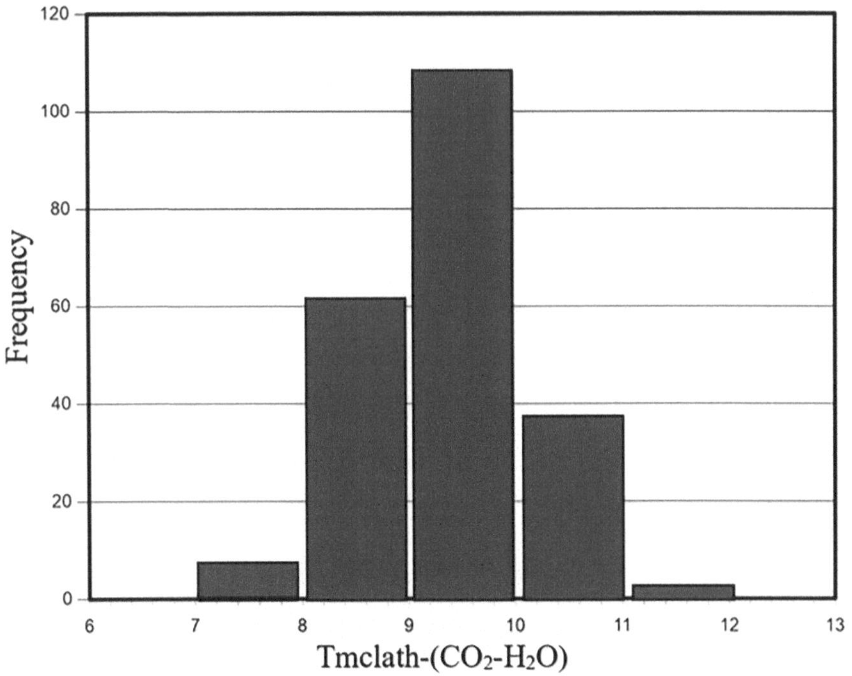

Fig. 23 Frequency histogram of Tmclath of CO_2-rich inclusions from Mpesatia prospect

Table 3 Fluid data at Prestea mine, Prestea

Inclusion type	Fluid composition	Salinity (wt.% NaCl)	ThTOT (°C)
Aqueous monophase	H_2O	22–16	–
CO_2–H_2O Biphase	XH_2O: 0.73–0.9 XCO_2: 0.082–0.255 XNaCl: 0.007–0.015 Aqd: 1.028–1.031 CO_2d: 0.829–0.868 Bd: 0.922–0.999	2.81–5.05	> 297
CO_2 Rich monophase	CO_2 CO_2d: 1.74 -0.847	–	–

Table 4 Fluid data at Kubekrom prospect, Kubekrom

Inclusion type	Fluid composition	Salinity (wt.% NaCl)	ThTOT (°C)
Aqueous biphase	H_2O	n.d	370

Table 5 Fluid data at Akoko prospect, Akoko

Inclusion type	Fluid composition	Salinity (wt.% NaCl)	ThTOT (°C)
Aqueous monophase, biphase	H_2O	n.d	310

Table 6 Fluid data at Ashanti mine, Obuasi

Inclusion type	Fluid composition	Salinity (wt.% NaCl)	ThTOT (°C)
Aqueous monophase	H_2O	10–25	–
CO_2–H_2O biphase	XH_2O:0.746–0.81 XCO_2: 0.183–0.249 XNaCl: 0.005–0.007 Aqd:1.007–1.012 CO_2d:0.806–0.819 Bd: 0.907–0.935	2.03–2.62	400
CO_2 Rich monophase	CO_2 CO_2d: 1.133–0.786	–	–

Table 7 Fluid data at Obuom prospect, Obuom

Inclusion type	Fluid composition	Salinity (wt.% NaCl)	ThTOT (°C)
Aqueous monophase	H_2O	n.d	–
CO_2–H_2O Biphase	XH_2O:0.521–0.923 XCO_2:0.249–0.446 XNaCl:0.005–0.021 Aqd:1.012–1.084 CO_2d:0.181–0.887 Bd:0.597–0.947	2.62–11.70	350
CO_2 Rich monophase	CO_2 CO_2d: 0.97–0.828	–	–

Table 8 Fluid data at Mpesatia prospect, Mpesatia

Inclusion type	Fluid composition	Salinity (wt.% NaCl)	ThTOT (°C)
Aqueous monophase	H_2O	23–18	–
CO_2–H_2O biphase	XH_2O:0.858–0.403 XCO_2:0.755–0.187 XNaCl:0.016–0 CO_2d:0.887–0.031 Bd:0.956–0.849	6.81–1.03	380

Potter [50] published pressure correction for inclusions containing NaCl solutions at various homogenization temperatures and fluid pressures, however, the corrections are only exact if the formation pressure of the fluid is correctly known which can be (a) minimum and is called hydrostatic, if the fluid is opened to the surface or (b) maximum, referred to as lithostatic which indicates the pressure due to the overlying rocks or (c) an intermediate value between (a) and (b).

Since the fluid inclusions showed no evidence of boiling (absence of homogenization into both liquid and vapour phases), it can be assumed that the pressure of the fluids at the time of trapping was greater than the minimum pressure (i.e. the hydrothermal pressure). Therefore, pressure corrections given by [50] are used as

Table 9 Minimum trapping temperatures calculated from T_hTOT on fluids using pressure correction values from [50] (Temperatures are in °C)

Locality	T_hTOT	Press. Corr	Min. Trapping temperature
Obuasi	400	10	410
Mpesatia	380	20	400
Obuom	350	20	370
Prestea	> 297	20	> 317

the minimum values for the total homogenization temperatures which then gives the minimum trapping temperatures of the fluids. Here, the pressure-corrections [50] for 5 wt.% NaCl equiv. solution for the temperatures 350 and 400°C are used. Since the different types of inclusions originated from a homogeneous fluid the trapping temperature for the CO_2-H_2O-NaCl inclusions should be representative (Table 9).

Fluid Source

Carbonaceous matter which consists of graphitic material (graphite) and carbonates (ankerite/siderite) accompany the Au-quartz vein and preliminary studies [4] suggest an admixed (organic + magmatic) source disputed by the biogenic origin [25]. Both propositions are not supported by any δ^{13}C data and therefore difficult to accept the assertion that graphite was of a direct magmatic source. However, for the carbonates, the limited data of δ^{13}C (– 9.9 to – 12.9‰, $_{PDB}$; [25] and – 9.9 to – 12.2‰, $_{PDB}$ [6] are agreeable and suggest a homogeneous carbon for the carbonates and most likely of a metamorphic (reworked carbon) origin (Table 10).

There is no significant δ^{18}O variation in individual quartz veins from Prestea or Obuom and also veins from Prestea and Obuom have relatively uniform δ^{18}O values (Table 9) which is attributable to very similar temperatures of the vein formation. This narrow range of δ^{18}O values here (Table 9) contrasts with the large variations in δ^{18}O (+8 to – 16‰) on quartz veins for epithermal and other gold deposits [20, 51, 84, 106] which reflect mixing of fluids and/or unstable P/T conditions during mineralizations.

Table 10 The values of δ^{18}O and δD of the fluid

Locality	Sample	Temp (°C)	δ^{18}O (‰)	δD (‰)
Prestea	23/321/3	410	+ 13.1	– 40
	12/226/2	410	+ 12.1	– 29
	25/319/1	410	+ 14.7	– 25
Obuom	BM 3	370	+ 10.8	– 22
	BM 4	370	+ 11.3	– 25
	BM 5	370	+ 11.4	– 34
	BM 6	370	+ 11.5	– 31

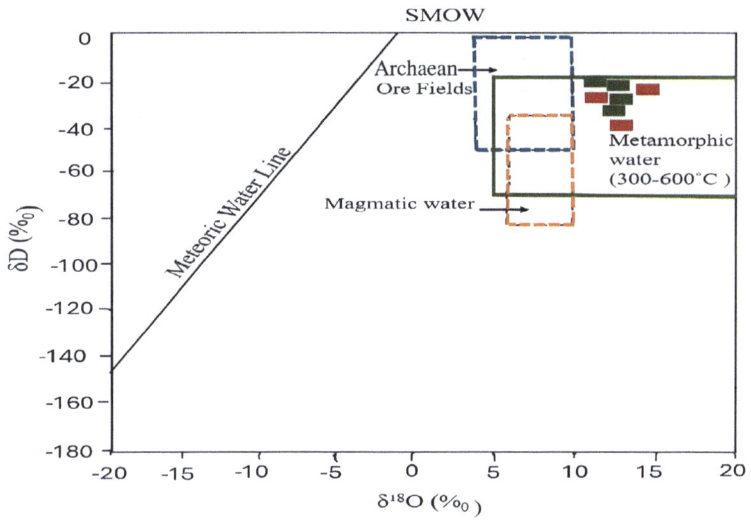

Fig. 24 Hydrogen versus oxygen isotope diagram showing hydrothermal fluid composition at Prestea mine (*red*) and Obuom (*green*) prospect. Range of Archaean ore field [52]

The δD values are used together with the $\delta^{18}O$ to illustrate the fluid composition which is in the metamorphic field (Fig. 24), though alternatively, this can also indicate low water/rock ratio as a result of equilibration of water with metamorphic rocks. As shown in Fig. 24, the $\delta^{18}O$ values for the fluids of the Prestea and Obuom gold deposits are heavier than the fluids from the Archaean gold deposits, however the δD values of the Birimian and Archaean ore fluids are in good agreement.

Since quartz veins from Prestea and Obuom (on the prominent Ashanti belt) show narrow and uniform $\delta^{18}O$ values, this can be attributed to a uniform oxygen reservoir for the fluid and a great depth during mineralization. Furthermore, the oxygen and hydrogen isotopic data together with fluid inclusion studies suggest that the mineralizing fluid was metamorphic.

Since water is the dominant constituent in ore-transporting fluids, its source is used to characterize the origin of the hydrothermal fluid and in an attempt to find the origin, one compares the hydrogen and oxygen isotopic compositions of the fluid with those of natural waters. [53] termed the natural waters: seawater, meteoric water and juvenile water as reference waters due to the exceptional definition of their H and O isotopic values and considered all other sub-surface natural waters such as connate-, geothermal-, magmatic- and metamorphic water to have been recycled from one or more of the reference waters.

The geological environment of the Birimian auriferous veins (depth of formation) excludes the consideration of geothermal fluids and the remaining likely natural waters for the Birimian are then connate-, magmatic- and metamorphic waters.

Generally, evidence for the most probable water can be made based on the salinity of the fluid and the hydrogen and oxygen isotopic compositions.

According to [54] the salinity of the fluids will increase as burial progresses, because of water consumption in the hydration minerals formed from the anhydrous minerals in the rocks, leaving more concentrated salt solution (formation waters/ brines) and it is only when these concentrated salt fluids have been removed that subsequent evolved fluids (due to rise of pressure and/or temperature) will have low salinity. Consequently, metamorphic fluids have low salinities.

The deduced low salinity of the Birimian fluids, ± 5 wt.% NaCl equiv., makes the involvement of sea- or formation-water very unlikely as the source of the auriferous fluid and points to metamorphic fluid and moreover, the isotopic composition of the fluids indicates a metamorphic origin.

Fluid Transport and Deposition of Gold

The parameters for the Birimian auriferous fluid are helpful in restraining the transport as well as deposition mechanism. Though gold can be present in solution as Au^+ or Au^{3+}, [55] concluded that at high temperatures, the Au^+ is more dominant and therefore gold in hydrothermal systems is more likely to be transported as Au^+ complexes. According to [55] the dominant species in near-neutral pH fluids is $Au(HS)_2^-$ which would therefore be expected to be predominant in the Birimian auriferous fluid of pH 5.8.

Gold deposition will be discussed for the two ore body types: disseminated ore type (i.e. mineralization in the wall rocks) and the quartz vein ore type.

An efficient mechanism for gold deposition particularly in iron-rich wall rocks is the destabilization of the $Au(HS)_2^-$ complex through sulphidation of iron in the wall rocks [47, 56] and by this process, gold is deposited accompanied by growth of pyrite, pyrrhotite or arsenopyrite. Though such rocks as high-Fe tholeiitic basalts and banded iron-formation have not been reported in the Birimian in Ghana, the presence of (a) lots of ankerite/siderite and (b) gold-rich zones in the tuffaceous bodies (i.e. the second mined ore body next to the quartz veins) which is characterized by lots of gold-rich arsenopyrite does not exclude any active role through this sulphidation mechanism for the gold deposition in the wall rocks.

Furthermore, if gold is transported as $Au\,(HS)_2^-$, then precipitation (or deposition) of the gold can occur as a result of (a) a decrease in pH, (b) a decrease in temperature or (c) a decrease or increase in fO_2 [57, 58]. The fO_2 is calculated for two depths at Ashanti Mine, approximately 300 m depth (sample 18/OB/310/9) and 900 m depth (samples 38/OB/155/10 and 38/OB/155/12). It showed virtually no changes, and the pH values remained constant, suggesting that the gold would be precipitated due to falling temperature.

Another adequate gold deposition mechanism is phase separation [59, 60] and this mechanism would be very efficient in view of the phase separation which occurred for the Birimian fluid and led to CO_2-rich and H_2O-rich fluids. The separation may have caused destabilization of the gold complexes and hence the gold deposition.

This mechanism is interesting for the Birimian fluids if one considers that CO_2-rich fluid type is dominant in auriferous quartz veins in prominent mines (e.g. Ashanti and Prestea mines) whilst H_2O-rich fluid type dominates in quartz veins in gold prospects such as for example at Kubekrom) (see Tables 3, 4, 5, 6, 7 and 8). The fH_2S values ($10^{1.5}$ to $10^{2.0}$ bars) showed the importance of aqueous reduced sulphur species and the reducing nature of the deposition environment (also shown by a large amount of carbonaceous matter).

The origin of the Birimian gold deposits has been debatable since the classical magmatic model proposed [1] suggesting that the auriferous fluid was derived from Dixcove (i.e. Na-rich) granitoids and related porphyries.

Ntiamoah-Agyakwa [3] disputed the magmatic model and suggested a volcanogenic process involving shale-chert-manganese chemical assemblage in a submarine volcanic environment for the gold whilst a metamorphic dehydration model for the Birimian gold [35] favoured auriferous fluid produced by devolatilization and dehydration reactions due to prograde metamorphism.

The above models were postulated based on the source of the fluids which had been derived from the geological observations and on prevailing ideas on metallogeny with few other constraints. However, the most important source of data for deducing the origin of ore-forming fluids are (1) fluid inclusion studies and (2) isotopic analysis of the fluids.

Fluid inclusion microthermometric data indicating that there are two main types of fluids: partly aqueous (CO_2-rich); (liquid, monophase + biphase) and aqueous (H_2O-rich); (monophase + biphase) and the limited range of the degree of fill (0.5 to 0.7) for the CO_2-rich and H_2O-rich inclusions suggests an original homogeneous fluid supporting a scheme of the fluid evolution represented below.

The original homogeneous fluid of salinity 5 wt.% NaCl equiv. separated upon cooling into the CO_2-rich type (CO_2-H_2O), which homogenize at a higher temperature (350–400 °C) and the H_2O-rich type (lower homogenizing temperature; 310–370 °C); CO_2-rich fluid salinity $\approx$ 0 wt.% NaCl equiv., H_2O-rich fluid salinity $\approx$ 16 wt.% NaCl equiv. and dominant trace gases in the fluid are CH_4 and N_2 in decreasing order (Fig. 25).

Since the auriferous fluid did not boil, because the CO_2-rich or the H_2O-rich fluids homogenized upon heating into the liquid phase, this means that the homogenization temperatures measured in the laboratory have to be corrected in order to get the trapping temperature for the fluid but unfortunately, the significant amount of $CH_4 \pm N_2$ in CO_2 do not give a reasonable intersection value of the CO_2 and H_2O isochores for determination of the true trapping temperature of the fluid as inclusions homogenizing into the liquid phase are trapped at high pressures and require pressure corrections (given in °C) for their homogenization temperatures [42].

Nevertheless, from [17, 50] a temperature correction of 10–20 °C would be necessary for the Birimian fluids of salinity 5 wt.% NaCl equiv. and using the pressure correction of 10–20°C puts the minimum formation temperature of the fluids at 370–410 °C in case of a hydrostatic system, but higher if approaching lithostatic.

The Birimian assemblage is 10–15 km thick [3], so assuming a geothermal gradient of 25 °C/km ([61] p.240) and using a deformation/depth profile [62] a

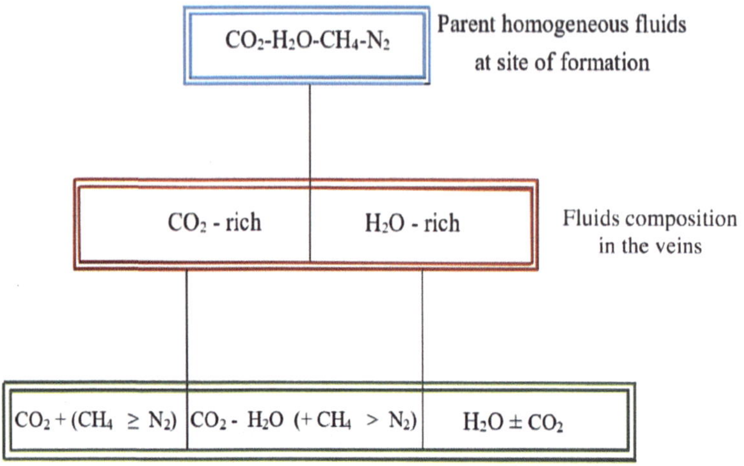

Fig. 25 A scheme showing evolution of the Birimian auriferous fluid types

lithostatic load of 2.5–4 kbar would be expected. Combining this with the geological setting of the Birimian (greenschist facies, 2 kbar) would give a lithostatic pressure range of 2–4 kbar for fluids at the base of the Birimian, but of course smaller at the level of mineralization.

Although the auriferous fluid comprises H_2O, CO_2, CH_4, and N_2, the dominant constituents are H_2O and CO_2. These components are interpreted to be the products of hydration and dehydration reactions involving silicate minerals within the host rocks, reflecting metamorphic processes that contributed to fluid generation. Decomposition of organic matter in the Birimian pile gave simple products like H_2O, CO_2, CH_4 and N_2, therefore reactions such as hydrolysis of organic matter (or graphite) or reduction of magnetite linked with oxidation of carbon also may have added more organic volatiles to the fluid as shown below in Eqs. (4) and (5). Furthermore, replacement of K^+ in silicate minerals by $NH_4{}^+$ may have contributed N_2 constituents to the fluid.

$$2C + 2H_2CO_2 + CH \quad \text{hydrolysis of carbon} \tag{4}$$

$$3FeS_2 + Fe_3O_4 + 2C6FeS + 2CO_2 \quad \text{reduction of magnetite and oxidation of carbon} \tag{5}$$

Though the H_2O-rich fluid has a higher salinity (≈16 wt.% NaCl equiv.) than the CO_2-rich fluid (≈0 wt.% NaCl equiv.), the salinity of the original homogeneous fluid is taken as ± 5 wt.% NaCl equiv. This is because (a) the absence of salt daughter crystals indicates low salinity for the parent fluid and (b) auriferous fluids in geological environments elsewhere (e.g. Archaean gold deposits in Western Australia) have low salinities. Changes which occurred after trapping of the parent fluid namely, cooling and decrease in pressure, resulted in CO_2-rich and H_2O-rich fluids separation and subsequently caused the H_2O-rich fluid to have a high salinity.

The metallogenic history of the Birimian ore deposits in West Africa has most currently been discussed by Milési et al. [23] and assigned the lithological, structural and metamorphic controls on gold and base metal deposits in West Africa to three tectonic episodes of which the Birimian gold (disseminated + quartz vein) deposits in Ghana were ascribed to the late-orogenic mineralization phase which formed after peak metamorphism, during a brittle deformation phase.

Hirdes and Leube [35] favoured a syn- to post-metamorphic model for the gold deposits. Even though these two currents models' emphasis the role of metamorphism, the discrepancy between them still exists on the interpretative timing and setting of the gold deposits. The evolution, transport and deposition of the gold can be modeled as follows in the current geological model for the gold deposits in Ghana as follows.

Following deformation and metamorphism of the volcano-sedimentary pile, large quantities of fluids released from reacting minerals were produced. The complex chemistry of the fluid made it possible that it could contain gold as complex species. Compositional characteristics of the low-saline fluid were besides water, the gases CO_2, CH_4, and N_2 in decreasing order.

The highly reduced fluid was formed at above 370–410 °C and below 2–4 kbar after peak metamorphism of the rock pile (supported by petrographic evidence) and the low salinity content and the isotopic composition indicates metamorphic fluids. The flow paths of the fluids were lithological contacts, bedding planes, shear zones and fractures created by burial of the rocks at depth and the contemporaneous emplacement of the Dixcove/Cape Coast granitoid bodies.

Gold deposition took place in two stages: first through interaction of the fluid with the country rock and later as auriferous quartz veins. Ore bodies of increasing economic importance in the Birimian are the gold-rich zones in tuffaceous rocks, which are characterized by gold-bearing arsenopyrite whose precipitation mechanism is envisaged as due to sulphidation of the wall-rocks. Destabilization of Au $(HS)_2^-$ complex led to deposition of gold and arsenopyrite.

The major ore bodies occur as auriferous quartz veins in shear zones and considering these shear zone spacings, released auriferous fluids streamed up in them and deposition of the gold occurred when phase separation due to cooling and decrease in pressure took place. Though most of the gold was transported as Au $(HS)_2^-$ complex, it is also very likely that some of the gold was carried as carbonate complex, (Au $(HCO_3)_2^-$) as this is inferred from the large amount of carbonates in the auriferous quartz veins. Since reduction of CO_2 pressure can cause carbonate formation and give rise to gold precipitation [54], this could account for the bias towards CO_2-rich fluids in the auriferous quartz veins in the Ashanti and Prestea mines.

Based on field evidence and extensive laboratory work [6] it can be concluded using the criteria given by [63] that the Birimian gold deposits fit into a prograde devolatilization model of a metamorphogenic deposit. The ore-fluid composition and depositional conditions are summarized in Table 11.

Table 11 Summary of experimentally determined and calculated formation conditions on the fluid for the Birimian gold mineralization in Ghana

Parameter	Value
Temperature, °C	370–410 (minimum)
Pressure, kbar	2–4 kbar (maximum)
Fluid composition	XH_2O: 0.7 (range 0.4–0.9)
	XCO_2: 0.3 (range 0.1–0.5)
	X_{NACL}: 0.01 (range 0–0.02)
	Aqueous density: 1.01 (range 1.00–1.08)
	CO_2 density: 0.87 (range 0.81–0.88)
	Bulk density: 0.94 (range 0.89–0.96)
Salinity, wt.% NaCl equiv	± 5
pH	5.8
Log fO_2	− 32 to − 33
Log fH_2S	1.5 to 2.0
δ^{13}Ccarbonate, ‰	− 12 to − 9
δ^{18}O carbonate, ‰	+ 13 to + 16
δ^{18}C quartz, ‰	+ 14 to + 19
δ^{18}C fluid, ‰	+ 12 to + 15
δD fluid, ‰	− 22 to − 40

10 Conclusion

The following conclusions can be drawn from this study:

1. The Birimian auriferous parent fluid was formed above 370–410 °C and below 2–4 kbar. The chemical composition of the fluid was H_2O, CO_2, CH_4 and N_2 in decreasing order. The salinity was ± 5 wt.% NaCl equiv.
2. The low salinity and hydrogen and oxygen isotopic characteristics of the fluid indicate metamorphic water as the source of the fluid.
3. The gold was transported mainly as $Au(HS)_2^-$ complex and deposited in a reducing environment at a pH of about 5.8.
4. Minor deposition of gold in the country rocks was the result of sulphidation of the corresponding rock type.
5. Alterations in the country rocks related to gold mineralization are more pronounced in the metavolcanics than in the metasediments.
6. Precipitation of the gold occurred mainly in shear zones as auriferous quartz veins due to cooling and decrease in pressure of the original homogeneous fluid by unmixing into two phases: CO_2-rich and H_2O-rich fluids.

11 Recommendations

The metamorphogenic model presented above needs more stable isotope and radiogenic isotope studies for a better understanding of the Birimian auriferous fluids because the stable isotopic data in this study could imply a multiple fluid source.

Furthermore, more studies on the timing of gold mineralization and metamorphism are very important in the acceptance of the metamorphic model. Though the data above strongly suggest a metamorphic ore fluid and that the gold deposits were formed at syn- to post-peak metamorphism (and not retrograde metamorphism), however in major fault zones (as typical in the Birimian), convective fluid circulation can also give rise to possible retrograde metamorphic gold-related alteration assemblages similar to syn- to post-metamorphic patterns.

Acknowledgements The author is very grateful to the Deutsche Akademischer Austausch Dienst (DAAD) for the full financial support during the data collection. The Editors are duly acknowledged for their critical constructive comments and suggestions on this chapter.

References

1. Junner NR (1932) The geology of the Obuasi goldfield. Gold Coast Geol Surv Memoir (2) Accra
2. Junner NR (1935) Gold in the Gold Coast. Ghana Geol Surv Memoir (4) Accra
3. Ntiamoah-Agyakwa Y (1979) Relationship between gold and manganese mineralizations in the Birimian of Ghana, West Africa. Geol Magazine 116(5):345–352. https://doi.org/10.1017/S0016756800043983
4. Leube A, Hirdes W (1986) The Birimian Supergroup of Ghana: depositional environment, structural development and conceptual model of an early Proterozoic suite. Bundesanstalt für Geowissenschaften und Rohstoffe unpbl report Hannover, Germany
5. Lüders V, Klemd R, Oberthür T, Plessen B (2015) Different carbon reservoirs of auriferous fluids in African Archean and Proterozoic gold deposits? Constraints from stable carbon isotopic compositions of quartz-hosted CO_2-rich fluid inclusions. Mineral Deposita 50:449–454. https://doi.org/10.1007/s00126-015-0588-x
6. Manu J (1993) Gold deposits of Birimian greenstone belt in Ghana: hydrothermal alteration and thermodynamics. Dissertation, TU Braunschweig, Germany
7. Oberthür T, Vetter U, Mumm AS, Weiser T, Amanor JA, Gyapong WA, Kumi R, Blenkinsop TG (1994) The Ashanti gold mine at Obuasi, Ghana: mineralogical, geochemical, stable isotope and fluid inclusion studies on the metallogenesis of the deposit. Geol Jahrb D100:31–129
8. Schwartz MO, Oberthür T, Ammanor J, Gyapong W (1992) Fluid inclusion re-equilibration and P-T-X constraints on fluid evolution in the Ashanti gold deposit, Ghana. European J Mineralogy 4:1017–1033. https://doi.org/10.1127/ejm/4/5/1017
9. Allibone A, McCuaig TC, Harris D, Etheridge MA, Munroe S, Byrne D, Amanor J, Gyapong W (2002) Structural controls on gold mineralisation at the Ashanti gold deposit, Obuasi, Ghana. Econ Geol Spec Pub 9:65–93
10. Allibone A, Teasdale J, Cameron G, Etheridge M, Uttley P, Soboh A, Appiah-Kubi J, Adanu A, Arthur R, Mamphey J, Odoom B, Zuta J, Tsikata A, Pataye F, Lamb E (2002) Timing and structural controls on gold mineralization at the Bogosu gold mine, Ghana, West Africa. Econ Geol 97:949–969. https://doi.org/10.2113/gsecongeo.97.5.949

11. Augustin J, Gaboury D, Crevier M (2017) Structural and gold mineralizing evolution of the world-class orogenic Mana district, Burkina Faso: multiple mineralizing events over 150 Ma. Ore Geol Rev 91:981–1012. https://doi.org/10.1016/j.oregeorev.2017.08.007
12. Blenkinsop T, Mumm AS, Kumi R, Sangmor S (1994) Structural geology of the Ashanti mine. Geol Jahrb D100:131–153
13. Feybesse JL, Billa M, Guerrot C, Duguey E, Lescuyer JL, Milesi JP, Bouchot V (2006) The paleoproterozoic Ghanaian province: geodynamic model and ore controls, including including regional stress modeling. Precambrian Res 149:149–196. https://doi.org/10.1016/j.precamres.2006.06.003
14. Ledru P, Pons J, Milesi JP, Feybesse JL, Johan V (1991) Transcurrent tectonics and polycyclic evolution in the Lower Proterozoic of Senegal-Mali. Precamb Res 50(4):337–435
15. Masurel Q, Thébaud N, Miller J, Ulrich S (2017) The tectono-magmatic framework to gold mineralisation in the Sadiola-Yatela gold camp and implications for the paleotectonic setting of the Kédougou-Kénieba inlier, West Africa. Precambrian Res 292:35–56. https://doi.org/10.1016/j.precamres.2017.01.017
16. McFarlene CRM, Mavrogenes J, Lentz D, King K, Allibone A, Holcombe R (2011) Geology and intrusion-related affinity of the Morila Gold Mine, Southeast Mali. Econ Geol 106:727–750. https://doi.org/10.2113/econgeo.106.5.727
17. McFarlene HB, Aillères L, Betts P, Ganne J, Baratoux L, Jessell M, Block S (2019) Episodic collisional orogenesis and lower crust exhumation during the Paleoproterozoic Eburnean orogeny—evidence from the Sefwi greenstone belt, West African Craton. Precambrian Res 325:88–110. https://doi.org/10.1016/j.precamres.2019.02.012
18. Milési JP, Ledru P, Ankrah PT, Marcoux E, Vinchon C, Johan V (1988) Geological and Structural setting of the Ghana gold deposits hosted by Birimian formations. In: Abstract, international conference and workshop on the geology of Ghana with special emphasis on gold; geology and exploration in Ghana and in selected other Precambrian terrains, 75th Anniversary Ghana Geological Surv Dept Accra, Ghana, p 31. https://doi.org/10.1007/BF00209263
19. Perrouty S, Ailleres L, Jessel M, Baratoux L, Bourassa Y, Crawford B (2012) Revised Eburnean geodynamic evolution of the gold-rich southern Ashanti belt, with new field and geophysical evidence of pre-Tarkwaian deformations. Precambr Res 204:12–39. https://doi.org/10.1016/j.precamres.2012.01.003
20. Perrouty S, Jessell M, Bourassa Y, Miller J, Apau D, Siebenaller L, Velásquez G, Baratoux L, Aillères L, Béziat D, Salvi S (2015) The Wassa deposit: a poly-deformed orogenic gold system in southwest Ghana—implications for regional exploration. J Afr Earth Sci 112:536–547. https://doi.org/10.1016/j.jafrearsci.2015.03.003
21. Allibone A, Hayden P, Cameron G, Duku F (2004) Paleoproterozoic gold deposits hosted by albite- and carbonate-altered tonalite in the Chirano district, Ghana, West Africa. Econ Geol 99:479–497. https://doi.org/10.2113/gsecongeo.99.3.479
22. Masurel Q, Thébaud N, Miller J, Ulrich S, Hein KAA, Cameron G, Béziat D, Bruguier O, Davis JA (2017) Sadiola Hill: a world-class carbonate-hosted gold deposit in Mali, West Africa. Econ Geol 112:23–47. https://doi.org/10.2113/econgeo.112.1.23
23. Milési JP, Ledru P, Feybesse JI, Donmanget A, Marcoux E (1992) Early Proterozoic ore deposits and tectonics of the Birimian orogenic belt, West Africa. Precamb Res 5:305–344. https://doi.org/10.1016/0301-9268(92)90123-6
24. Mücke A, Dzigbodi-Adjimah K (1991) Mineralogie und Genese der an Störungszonen gebundenen Goldvererzungen Ghanas. Attached copy (Beiheft) to European J Miner (3):189
25. Oberthür I, Vetter U, Schwartz MO, Weiser T, Amanor J, Gyapong WA (1991) Gold mineralization at the Ashanti mine, Obuasi, Ghana: preliminary mineralogical and geochemical data. In: Ladeira EA (ed) Brazil '91: the economics, geology, geochemistry and genesis of gold deposits. Balkema, Rotterdam, pp 533–537
26. Feybesse JL Billa M, Milési JP, Lerouge C, Le Goff E (2000), Relationships between metamorphism-deformation-plutonism in the Archean-Paleoproterozoic contact zone of West Africa. 31st international geological congress, Rio, Brazil

27. Klemd R, Hünken U, Olesch M (2002) Metamorphism of the country rocks hosting gold–sulfide-bearing quartz veins in the Paleoproterozoic southern kibi-winneba belt (SE-Ghana). J Afr Earth Sci 35(2):199–211. https://doi.org/10.1016/S0899-5362(02)00122-7
28. Lawrence DM, Treloar PJ, Rankin AH, Harbidge P, Holliday J (2013) The geology and mineralogy of the Loulo mining district, Mali, West Africa: evidence for two distinct styles of orogenic gold mineralisation. Econ Geol 108:199–227. https://doi.org/10.2113/econgeo.108.2.199
29. McCuaig TC, Fougerouse D, Salvi S, Siebenaller L, Parra-Avila LA, Seed R, Béziat D, André-Mayer AS (2016) The Inata deposit, Belahouro district, northern Burkina Faso. Ore Geol Rev 78:639–644. https://doi.org/10.1016/j.oregeorev.2015.11.014
30. Lambert-Smith JS, Allibone AH, Treloar PJ, Lawrence DM, Boyce AJ, Fanning M (2020) Stable C, O, and S isotopic record of magmatic-hydrothermal interactions between Falémé Fe-skarn and Loulo Au systems in western Mali. Econ Geol 115:1537–1558. https://doi.org/10.5382/econgeo.4759
31. Lawrence DM, Treloar PJ, Rankin AH, Boyce A, Harbidge P (2013) A fluid inclusion and stable isotope study at the Loulo mining district, Mali, West Africa: implications for multifluid sources in the generation of orogenic gold deposits. Econ Geol 108:229–257. https://doi.org/10.2113/econgeo.108.2.229
32. Abouchami W, Boher M, Michard A, Albarède F (1991) A major 2.1 Ga old event of mafic magmatism in West Africa: an early stage of crustal accretion. J Geophys Res 95 (B11):17605–17629. https://doi.org/10.1029/JB095iB11p17605
33. Kesse GO (1985) The mineral and rock resources of Ghana. Balkema, Rotterdam
34. Mensah MK, Iddirisu Y, Gaisie JS, Manu J, Attoh K (1988) The gold belts of Ghana. Unpublished report prepared for the Government of Ghana and United Nations, project GHA/86/005 technical assistance to the Minerals Commission, Accra, Ghana
35. Hirdes W, Leube A (1989) On gold mineralization of the Proterozoic Birimian Supergroup in Ghana/West Africa. Unpublished report Bundesanstalt für Geowissenschaften und Rohstoffe, Hannover, Germany
36. Homrighausen P (1983) Gold- und Buntmetall Prospektion in der-Nord-Bibiani Range, SW-Ghana. Unpublished report. Bundesanstalt fur Geowissenschaften und Rohstoffe, Hannover, Germany
37. Gyapong WA (1980) Factors affecting ore localization at Ashanti mine. Dissertation, Univ London
38. Adjimah CL (1988) Petrologische Untersuchungen zur Genese der Goldlagerstätte Prestea (Ghana). Dissertation, Univ Berlin
39. Appiah H (1991) Geology and mine exploration trends of Prestea Goldfields, Ghana. J African Earth Sci 2(13):235–241. https://doi.org/10.1016/0899-5362(91)90008-M
40. Baako BA (1972) Mining geology of the Prestea Goldfields. Dissertation, Univ Cagliari, Italy
41. Bowell RJ, Foster RP, Stanley CJ (1990) Telluride mineralization at Ashanti gold mine. Mineral Mag 54:617–627. https://doi.org/10.1180/minmag.1990.054.377.15
42. Roedder E (1984) Fluid inclusions. Min Soc Amer Rev Min (12) Washington, D.C.
43. Milési JP, Ledru P, Dommanget A, Johan V, Diallo M (1989) Lower Proterozoic succession in Senegal and Mali (West Africa): position of sediment-hosted Au and Fe deposits of Loulo area and significance in terms of crustal evolution. Abstr 28th Int Geol Cong Washington D.C. USA (2–3):433–434
44. Shepherd T, Rankin AH, Alderton DHM (1985) A practical guide to fluid inclusion studies. Blackie, Glasgow. https://doi.org/10.1180/minmag.1986.050.356.32
45. Ramboz C, Pichavani M, Weisbrod A (1982) Fluid immiscibility in natural processes: use and misuse of fluid inclusion data. Part II: Interpretation of Fluid inclusion data in terms of immiscibility. Chem Geol 37:29–48. https://doi.org/10.1016/0009-2541(82)90065-1
46. Nicholls J, Crawford ML (1985) Fortran programs for calculating fluid properties from microthermometric data on fluid inclusions. Comp Geosci 11(5):619–645. https://doi.org/10.1016/0098-3004(85)90090-1

47. Philips GN, Groves DI, Martyn JE (1984) An epigenetic origin for Archaean banded-iron-formation-hosted gold deposits. Econ Geol 79:162–171. https://doi.org/10.2113/gsecongeo.79.1.162
48. Faure G (1977) Principles of isotope geology. J. Wiley and Sons, New York
49. Christie AB, Goguel RL, Robinson BW (1989) Problems of crush-leach analyses of low-salinity inclusions-poor material. Chem Geol 78:35–49. https://doi.org/10.1016/000925 41(89)90050-8
50. Potter RW (1977) Pressure corrections for fluid inclusion homogenization temperatures based on the volumetric properties of the system $NaCl\text{-}H_2O$. US Geol Surv J Res 5:603–607
51. Fournier RO, Truesdell AH (1973) An empirical Na-K-Ca geothermometer for natural waters. Geochimica Cosmochimica Acta 37:1255–1275. https://doi.org/10.1016/0016-7037(73)900 60-4
52. Nesbitt BC, Muehlenbachs K (1989) Geology, geochemistry, and genesis of mesothermal lode gold deposits of the Canadian cordillera: evidence for ore formations from evolved meteoric water. Econ Geol monograph 6:553–563
53. Ohmoto H (1986) Systematics of metal ratios and sulfur isotopic ratios in low temperature basemetal deposits. Terra Cognita (6):134–135
54. Fyfe WS, Kerrich R (1984) Gold: Natural concentration processes. In: Foster RP (ed) Proceedings of Gold 82: the geology, geochemistry, and genesis of gold deposits, Geol Surv Zimbabwe Spec Publ Balkema, Rotterdam, pp 99–127
55. Seward TM (1973) Thio complexes of gold in hydrothermal ore solutions. Geochimica Cosmochimica Acta 37:379–399. https://doi.org/10.1016/0016-7037(73)90207-X
56. Gilligan JM, Foster RP (1987) Gold mineralization in iron-formation: the importance of contrasting modes of deformation at the Lennox mine, Zimbabwe. In: Africa mining; Inst of Mining and Metallurgy, London, pp 127–138
57. Foster RP (1989) Archaean gold mineralization in Zimbabwe: implications for metallogenesis and exploration. In: Keays RR, Ramsay WRH, Groves DI (eds) The geology of gold deposits: The perspective in 1988. Econ Geol Monograph 6:54–70. https://doi.org/10.5382/Mono.06.04
58. Seward TM (1984) The transport and deposition of gold in hydrothermal systems. In: Foster RP (ed) Proceedings of Gold '82: the geology, geochemistry, and genesis of gold deposits, Geol. Surv. Zimbabwe, Spec Publ Balkema, Rotterdam, pp 165–181
59. Cole DR, Drummond SE (1986) The effect of transport and boiling on Ag-Au ratios in hydrothermal solutions: a preliminary assessment and implications for the formation of epithermal precious-metal ore deposits. J Geochem Exploration 25:45–80. https://doi.org/10.1016/0375-6742(86)90007-5
60. Seward TM (1989) The hydrothermal chemistry of gold and its implications for ore formation: boiling and conductive cooling as examples. Econ Geol Monograph 6:398–404. https://doi.org/10.5382/MONO.06.31
61. Condie KC (1981) Archaean greenstone belts. Elsevier, Amsterdam
62. Colvine AC, Fyon JA, Heather KB, Marmont S, Smith PM, Troop DG (1988) Archaean lode gold deposits in Ontario. Ontario Geol Surv Misc Paper 139:137
63. Pohl W (1992) Defining metamorphogenic mineral deposits - an introduction. Min Petrol (45):45–52. https://doi.org/10.1007/BF01163109
64. Afenya PM (1981) The carbonaceous material in the Birimian gold deposits of Ghana. Bull Fac Sci Univ Ghana 1:4
65. Afenya PM (1976) Studies in the processing of a carbonaceous gold ore. Dissertation, Univ London. https://doi.org/10.1016/0892-6875(91)90082-7
66. Annovitz LM, Essene EJ (1987) Phase equilibria in the system $CaCO_3\text{-}MgCO_3\text{-}FeCO_3$. J Petrol Pt 2(28):389–414. https://doi.org/10.1093/petrology/28.2.389
67. Asprey LB (1976) The preparation of very pure fluorine gas. J flourine Chem 7:359361. https://doi.org/10.1016/S0022-1139(00)84009-9
68. Audéoud D (1982) Les mineralisations uranifères et leur environnement à Kamoto, Kambove et Shinkolobwe (Shaba, Zaire) petrographie geochemie et inclusions fluides. Dissertation, Univ Lyon, France. https://doi.org/10.3749/9780921294726.ch02

69. Beran A (1977) Die Kluftankerite des Steirischen Erzberges und ihre mögliche Verwendung als Geothermometer. Miner Deposita 12:90–95. https://doi.org/10.1007/BF00204507
70. Berman RG (1991) Thermobarometry using multi-equilibrium calculations: a new technique with petrological applications. Canadian Min 29:833–855
71. Bryndzia LT, Scott SD (1984) Experimentally determined activity model, $\Delta H°_f$, $\Delta S°_f$ and $\Delta G°_f$ of iron chlorite [$Fe_5Al_2Si_3O_{10}(OH)_8$] Geol Soc America Abstracts with Programs (16):458
72. Burrows DR, Wood PC, Spooner ETC (1986) Carbon isotope evidence for a magmatic origin for Archaean gold-quartz vein ore deposits. Nature 1:851–854. https://doi.org/10.1038/321851a0
73. Cathelineau M, Nieva D (1985) A chlorite solid solution geothermometer. The Los Azufres geothermal system (Mexico). Contrib Mineral Petrol 91:225–244. https://doi.org/10.1007/BF00413350
74. Cooper WGC (1934) The geology of Prestea goldfields. Gold Coast Geol Surv Memoir 3:20
75. Cozens B (1988) Some aspects of the regional geology of Ghana and its relationship in gold mineralization. In: Abstracts of international conference workshop on the geology of Ghana with special emphasis on gold; Geology and Exploration in Ghana and in selected other Precambrian terrains. 75th Anniversary Ghana Geol Surv Dept Accra, Ghana, p 10
76. Dachs E (1990) Geothermobarometry in metasediments of the southern Grossvenediger area (Tauern Window, Austria). J Metamor Geol 8:217–230. https://doi.org/10.1111/j.15251314.1990.tb00467.x
77. Deer WA, Howie RA, Zussman J. (1963) Rock-forming minerals, vol 5. Non-silicate. Longman, London. https://doi.org/10.1180/minmag.1996.060.400.20
78. Eisenlohr BN (1992) Conflicting evidence on the timing of mesothermal and paleoplacer gold mineralization in early Proterozoic rocks from southwest Ghana, West Africa. Miner Deposita 27:23–29. https://doi.org/10.1007/BF00196077
79. Essene EJ (1983) Solid solutions and solvi among metamorphic carbonates with applications to geologic thermometry. In: Reeder RJ (ed) Carbonates: mineralogy and chemistry. Reviews in Mineralogy, vol 11. pp 77–96
80. Fiadonu EB, Hammond NQ, Dzigbodi-Adjimah K, Adomako-Ansah K (2022) Ore fluid characteristics and depositional processes of the hydrothermal gold mineralisation at Damang, southwestern Ghana. J Afr Earth Sc 197:104741. https://doi.org/10.1016/j.jafrearsci.2022.104741
81. Fournier RO (1981) Application of water geochemistry to geothermal exploration and reservoir engineering. In: Rybach, L Muffler LJP (eds) Geothermal systems: principles and case histories. J Wiley and Sons, London, pp 109–143
82. Fyon JA, Crocket JH (1982) Gold exploration in the Timmins district using field and lithogeochemical characteristics of Carbonate alteration zones. Canadian Inst Mining Metallurgy Spec 24:113–129
83. Gary N, McAfee Jr, R Wolf CL (eds) (1973) Glossary of geology. Amer Geol Inst Washington, D.C. https://doi.org/10.1016/0012-8252(73)90103-7
84. Golding SD, Groves DI, Monaughton NJ, Barley ME, Rock NMS (1987) Carbon isotopic composition of carbonates from contrasting alteration styles in supracrustal rocks of the Norseman-Wiluna belt, Yilgan block, Western Australia: Their significance to the source of Archaean auriferous fluids. In: Ho, SE, Groves DI (eds) Recent advances in understanding Precambrian gold deposits. Geology Dept. and University Extension, The Univ. of Western Australia, Publ (11):215–238
85. Gresens RL (1967) Composition-volume relationships of metasomatism. Chem Geol 2:47–65. https://doi.org/10.1016/0009-2541(67)90004-6
86. Helgeson HC, Kirkham DH (1974) Theoretical prediction of the thermodynamics behaviour of aqueous electrolytes at high pressures and temperatures. II. Debye-Hückel parameters for activity coefficients and relative partial molar properties. Amer J Sci 274:1199–1261. https://doi.org/10.2475/ajs.274.10.1089

87. Helgeson HC, Delany JM, Nesbitt HW, Bird DK (1978) Summary and critique of the thermodynamic properties of rock-forming minerals. Amer J Sci 278A:229
88. Hoefs J (1973) Stable isotope geochemistry. Springler-Verlag, New York, USA. https://doi.org/10.1007/978-3-662-09998-8
89. Kerrich R, Fyfe WS (1981) The gold-carbonate association: source of CO_2 fixation reactions in Archaean Lode deposits. Chem Geol 33:265–294. https://doi.org/10.1016/0009-2541(81)90104-2
90. Kerrich R (1986) Archaean lode gold deposits of Canada, Part 11: characteristics of the hydrothermal systems and models of origin. Univ Witwatersrand Johannesburg, Econ Geol Res Unit Inf Circular (183)
91. Kielland J (1937) Individual activity coefficients of ions in aqueous solutions. J Amer Chem Soc 59:1675–1678. https://doi.org/10.1021/ja01288a032
92. Laird J (1988) Chlorites: metamorphic petrology. In: Bailey SW (ed) Hydrous phyllosilicates. Rev Min (19):405–453
93. Manu J (1991) Fluid inclusions in quartz veins in the Birimian gold deposits of Ghana. In: Pagel M, Leroy JD (eds) Source, transport and deposition of metals, proceedings of the 25 yrs society for geology applied to mineral deposits (SGA) Ann Meeting Nancy, France, pp 673–675
94. Manu J, Agyei EK (1993) Minimum Au mineralisation age of the Birimian in Ghana. In: Peters JW, Kesse GO (eds) Regional trends in African Geology, Proceedings of the 9th international conference Geological Society of Africa Accra, Ghana, pp 234–242
95. Matsuhisa Y, Goldsmith JR, Clayton RN (1979) Oxygen: isotopic fractionation in the system quartz-albite-anorthite-water. Geochimica Cosmochimica Acta 42:173–182. https://doi.org/10.1016/0016-7037(79)90099-1
96. McFarlene HB, Thébaud N, Parra-Avila LA, Armit R, Spencer C, Ganne J, Aillères L, Baratoux L, Betts PG, Jessel MW (2019) Onset of the supercontinent cycle: evidence for multiple oceanic arc accretion events in the Paleoproterozoic Sefwi Greenstone Belt of the West African Craton. Precambr Res 335. https://doi.org/10.1016/j.precamres.2019.105450
97. Neall FB, Philips GN (1987) Fluid-wall rock interaction in an Archaean hydrothermal gold deposit: a thermodynamic for the Hunt mine, Kambalda. Econ Geol 82:1679–1694. https://doi.org/10.2113/gsecongeo.82.7.1679
98. Philips GN (1987) Metamorphism of the Witwatersrand goldfields: conditions during peak metamorphism. J Metamor Geol 5:307–322. https://doi.org/10.1016/0169-1368(94)90017-5
99. Powell R, Condliffe DM, Condliffe E (1984) Calcite-dolomite geothermometry in the system $CaCO_3$-$MgCO_3$-$FeCO_3$: an experimental study. J Metamor Geol 2:33–42. https://doi.org/10.1111/j.1525-1314.1984.tb00283.x
100. Reeder RJ (1983) Crystal chemistry and rhombohedral carbonates. In: Reeder RJ (ed) Carbonates: mineralogy and chemistry. Min Soc Amer Rev Mineral (11):1–47
101. Rosenberg PE (1967) Subsolidus relations in the system $CaCO_3$-$MgCO_3$-$FeCO_3$, between 350 and 550 °C. Amer Min 52:787–796
102. Rumble D (1982) The role of perfectly mobile components in metamorphism. Annual Review Earth Planetary Sci 10:221–233. https://doi.org/10.1146/annurev.ea.10.050182.001253
103. Scott SD (1973) Experimental calibration of the sphalerite geobarometer. Econ Geol 68:466–474. https://doi.org/10.1016/0016-7037(83)90094-7
104. Sibson RH (1977) Faults and fault rock mechanisms. J Geol Soc London (133):191–213. https://doi.org/10.1144/gsjgs.133.3.0191
105. Sibson RH (1983) Continental fault structures and the shallow earthquake source. J Geol Soc London (140):741–767. https://doi.org/10.1144/gsjgs.140.5.0741
106. Taylor Jr. HP (1979) Oxygen and hydrogen isotope relationships in hydrothermal mineral deposits. In: Barnes H1 (ed) Geochemistry of hydrothermal ore deposits, 2nd ed. J. Wiley and Sons, New York, pp 236–277
107. Truesdell AH (1984) Chemical geothermometers for geothermal exploration. In: Henley RW, Truesdell AH, Barton Jr. PB, Whitney JA (eds) Rev Econ Geol (1):31–44

108. Valley JL, O'neil JR (1981) $^{13}C/^{12}C$ exchange between calcite and graphite: a possible thermometer in Greenville marbles. Geochimica Cosmochimica Acta (45):411–419. https://doi.org/10.1016/0016-7037(81)90249-0
109. Walshe JL, Solomon M (1981) An investigation into the environment of formation of the volcanic-hosted Mt. Lyell copper deposits using geology, mineralogy, stable isotopes, and a six-component chlorite solid solution model. Econ Geol 76:246–284. https://doi.org/10.2113/gsecongeo.76.2.246
110. Wright JB, Hastings DA, Jones WB (1985) Geology and mineral resources of West Africa. G. Allen and Unwin, London. https://doi.org/10.1007/978-94-015-3932-6
111. Yacé I (1984) Le Prècambrien de l'Afrique de l' Ouest et ses corrélations avec le Brèsil oriental. Rap. final du projet PICG, CIFEG, publication occasionelle, (2):28

Application of Machine Learning in Predicting Gold-Bearing Lithologies in Ghana

Application of Machine Learning Algorithms to Predict Rock Types Using Geochemical Data: A Case Study from the Obuasi Gold District, Ghana

Emmanuel Daanoba Sunkari, Abdullah Bello Muhammed, and Abdul Wahab Basit

1 Introduction

Rock identification procedures require meticulous measures to produce unbiased, reliable and objective results. In different mining and exploration companies, production and exploration geologists encounter difficulties recording the lithology types during manual logging. The practical and accurate classification of rock types has been an area of concern for the past decade. Rock-type classification uses a more subjective, time-consuming, and cost-unfriendly approach [1]. The major issues that mining and exploration companies usually strive to avoid when it comes to lithological classification are the drawbacks mentioned above. Exploration and production geologists are exposed to conventional manual logging, which is highly subjective and may not yield the required results [2]. It has been indicated that metamorphism, alteration, and deformation impede the identification of a rock since they alter its primary features [3]. According to our review, the lithology of the Obuasi Gold District is complex to identify by manual inspection since structures control it and have undergone extensive metamorphism and alteration. The rock units found in the

E. D. Sunkari (✉) · A. B. Muhammed
Department of Geological Engineering, Faculty of Geosciences and Environmental Studies, University of Mines and Technology, Tarkwa, Ghana
e-mail: sunkariemmanuel@rocketmail.com

E. D. Sunkari
Department of Chemical Sciences, Faculty of Science, University of Johannesburg, Johannesburg, South Africa

Mining Engineering, Faculty of Integrated and Advanced Technology, Sir Padampat Singhania University, Bhatewar, Udaipur, Rajasthan, India

A. W. Basit
Department of Computer Science and Engineering, Faculty of Computing and Mathematical Sciences, University of Mines and Technology, Tarkwa, Ghana

E. D. Sunkari and S. Ali (eds.), *Geology and Gold Mineralization in Ghana*, Earth and Environmental Sciences Library, https://doi.org/10.1007/978-3-031-97163-1_8

Obuasi Gold District are often distinguished by several modifications that make it difficult to classify the rock types.

ML techniques are currently used to curb the issue of biases and exploratory errors associated with deriving mineral prospectivity maps. The traditional statistical methods such as the index overlay, a knowledge-driven approach, rely on the judgement of the expert [4]. This leads to biases in classifying discrete classes of a given continuous spatial value. ML techniques such as linear regression and neural networking have been reported to adequately hint at the weights of evidence. These significant data-driven methods employed comparative quantitative analysis of spatial relationships between a new feature class and a similar feature class that has been detected [5, 6]. Moreover, ML techniques provide the avenue to use numerous variables concurrently, thereby resulting in high precision [2, 7]. Numerous researchers indicated that the good performance of ML techniques that are integrated with geochemical XRF data or well-log data shows good results [3, 8–10].

Moreover, it is crucial to recognise the substantial threat posed by waste rocks to the environment. Many mining communities commonly employ waste rocks to construct tailings embankments, fill waterlogged regions, and surface roads [11, 12]. However, some of these rocks harbour hazardous heavy (non)metals such as arsenic, iron, lead, and other chemical compounds [12]. These substances can potentially trigger acid mine drainage, endangering the water sources and ecosystems [12, 13]. Prolonged contact with these contaminants through human activities like drinking, bathing, swimming, or irrigation, significantly heightens health risks for nearby communities [13]. Furthermore, the mineralogical complexity of the ore in high-sulphide alteration zones can pose challenges for gold recovery technologies, requiring innovative approaches and specialised processing techniques [14]. Understanding the ore body's mineralogical composition and geochemical characteristics is essential for optimising gold recovery strategies in such environments [14].

ML is changing the landscape of geosciences by focusing on predicting lithofacies and classifying ore/waste materials as exploring minerals and processing geochemical data efficiently using ML techniques, despite challenges such as data reliance and geological variations that emphasize the significance of blending domain expertise with computational approaches.

Peng et al. [15] developed a Stacking Machine Learning Model to improve lithofacies prediction in shale reservoirs. The model, which used classifiers like random forest, gradient boosting, and categorical boost, achieved an accuracy rate of 95.87% [15]. This improved reservoir assessment and drilling location identification, but its practical applicability may be limited to similar shale formations [15].

Goldstein et al. [16] used Measure While Drilling (MWD) data and machine learning models like Decision Trees, Random Forests, Gaussian Processes, and Neural Networks to predict geophysical attributes like density, gamma radiation, magnetic susceptibility, and resistivity. These models showed high performance, with R^2 values reaching 0.98. This study provides insights into the effective use of geological modelling for orebody analysis [16]. However, a significant limitation is that data-driven machine learning strategies may inaccurately represent non-existent geological connections, underscoring the importance of balancing computational models with geological principles, as discussed in [16].

Boiger et al. [17] used ML to predict mineral content from drill core images, achieving an impressive 96% accuracy rate. Convolutional neural networks with transfer learning techniques outperformed multidimensional log analysis methods, offering a cost-effective alternative to traditional laboratory-based analyses [17]. However, the study was limited to a single set of borehole data, raising questions about the model's applicability to other environments. Additionally, the precision of core image processing affects the model's accuracy, underscoring the importance of improving imaging procedures [17].

Zhou et al. [18] tackled the issue of missing values in geochemical datasets by using a Random Forest model. They found that optimising model parameters and features significantly improved the accuracy of predictions for major and trace element values, enhancing data comprehensiveness and applicability in mineral exploration and environmental studies [18]. Notwithstanding, the study cautions that machine-learning-generated geochemical data should be interpreted in conjunction with domain knowledge to adequately account for complex geochemical interactions, rather than relying solely on algorithmic methods [18].

When using ML for rock identification, geochemical data is crucial. ML deals with feeding computers and machines with data and making meaningful information from this data. Geochemical data such as XRF data is vital in geoscience, mainly in mining and mineral exploration [19]. Geochemical data such as element concentrations are the basis for identifying rocks, lithology and ore grades [20, 21]. For example, the genesis of metamorphic rocks was defined in the ML environment using geochemical data, explicitly using the nine principal oxides [22]. Recent research has demonstrated the prowess of the integration of ML with XRF and MWD data in differentiating between ore and waste in mining operations, a challenge due to the reliance on time-intensive laboratory tests and subjective manual sampling methods [23]. Numerous successful applications of ML techniques for lithological classification using XRF data have been conducted, as enumerated in Table 1. These applications have highlighted the significance of ML techniques to identify rock types in terms of efficiency and accuracy.

In a nutshell, it can be deduced from the literature that the amalgamation of various ML algorithms produces the best results. Moreover, deep learning methods are more effective and accurate than traditional ML algorithms as indicated by Lawley et al. [24]. However, conventional ML algorithms could be boosted for better results. Again, some datasets, particularly geochemical data, may possess imbalanced samples. Thus, the inculcation of sampling techniques such as SMOTE becomes significant in this sense [2]. Fernández et al. [23] also recommend the use of oversampling methods to improve accuracy in classifying ores versus waste materials. Despite the role, as mentioned earlier, ML played in the application of rock type identification, Siman et al. [25] and Ibrahim et al. [2] indicated that complex geologic terrain and imbalanced datasets could hinder the application of the tool. Houshmand et al. [1] further stated that the worth of the information gathered immediately influences the output of machine learning models since employing inconsistent, skewed,

Table 1 Compilation of scientific works conducted on rock-type classification using geochemical XRF dataset

Authors	Geological setting/Country	Developed models	Highlights
Caté et al. [9]	Lalor volcanogenic massive sulphide deposit, Snow Lake, Manitoba, Canada	KNN, GNB, SVM, RF, GTB	SVM resulted in higher accuracy in predicting volcanic units, whilst SVM, GTB, and RF scored the best in the prediction of alteration types. KNN yielded lower scores in both. They further mentioned that the elemental concentrations of the data perform better when untransformed
Lawley et al. [24]	Golden Triangle, NW British Columbia, Canada	Deep Learning (Autoencoder + Artificial Neural Networks)	The authors used LA-ICP-MS pyrite geochemistry to classify four deposit types (porphyry Cu–Au, epithermal Au–Ag, magmatic Ni–Cu, and VMS Cu–Pb–Zn). This resulted in 99% classification accuracy and identified new exploration targets. ML identified previously unrecognized exploration targets. Demonstrated that pyrite morphology and trace elements are effective indicators of deposit type
Hood et al. [3]	Minto Cu–Au–Ag mine, Whitehorse, Yukon, Canada	K-means++, RF	Data were normalised and clustered into various groups representing protolith lithologies. This was established using the K-means++ unsupervised ML technique. Subsequently, RF was used to establish the classification of the various rock types. The RF successfully classified the lithology labels with AUC, sensitivity and precision of 0.94, 0.84 and 0.84 respectively. The classification scores from the RF assist geologists in auditing the quality of rock associations
Fernández et al. [23]	Narrow vein underground mine, Spain	K-means Clustering + Ensemble ML Model + MWD drilling data	In this study, XRF data was integrated with MWD drilling parameters for ore/waste classification. It achieved an F1 score of 80% + for ore classification and 86% for waste, reducing operator bias through continuous sampling. It also improved performance with continuous drill chip sampling, demonstrating strong correlations between rock mineralogy and drilling response

(continued)

Table 1 (continued)

Authors	Geological setting/Country	Developed models	Highlights
Siman et al. [21]	Zinkgruvan Zn–Pb–Ag Deposit, Bergslagen, Sweden	Classification and Regression Trees (CART), K-means clustering- Self Organising Map (SOM)	The results indicated SOM coupled with K-means clustering as the most suitable algorithm with an accuracy of 68.02%, while the CART yielded an accuracy score of 62.79. The authors indicated that the complex nature of the study area affected the test scores, thus leading to low accuracy scores
Ibrahim et al. [2]	Tarkwaian Paleoplacer, Ghana	XGB, DT and RF coupled with SMOTE	The authors needed to address the issue of oversampling in the dataset, thus implementing the Synthetic Minority Oversampling Technique (SMOTE) in the XRF dataset to create a balance in the dataset. Ultimately, SMOTE and extreme gradient boosting (XGB) hybridisation outperform other hybrid approaches in classifying rock types within the area
Houshmand et al. [1]	Kinross Gold, Canada	XGBoost, RF, SVM, DT, kNN, Naïve Bayes, LR and CNN (deep learning)	In their study, various methods were utilised to classify rock types. However, among them were different ML algorithms coupled with the geochemical data used to classify the types of rocks, where XGBoost outperformed all other algorithms, constituting the study's first approach. A CNN approach was used to analyse core sample images, forming the study's second approach. Ultimately, it was deduced that a third approach, hybridising the first and second approaches, yielded the most accurate results compared to the individual methods

and insufficient data will not yield appropriate results. The reliance on diverse types of data can improve the efficiency and efficacy of ML in rock-type identification [16].

The Obuasi Gold District is found within the prolific Ashanti Belt, rich in gold mineralization in the Paleoproterozoic Birimian Supergroup. It is notable for various mine and exploration operators, notably the AngloGold Ashanti Obuasi Mine. [26, 27]. The area is severely marked by alterations and metamorphism which may hinder accurate rock type classification. The application of ML techniques will assist geologists in auditing the results of their classification, thus saving time and cost.

This study seeks to address the existing knowledge gap by evaluating the efficacy of machine learning (ML) algorithms in predicting rock types using X-ray fluorescence (XRF) data. Specifically, four ML models—Gradient Boosting, AdaBoost, Support Vector Machines (SVM), and Random Forest (RF)—were assessed. These models were employed to categorize lithologies based on a collection of XRF analyses derived from rock samples in the Obuasi Gold District. In this study, 205 datasets from whole-rock geochemical analysis were expanded to 10,215 to reduce the risk of model overfitting and generalisation. The dataset encompasses a range of units that reflect the complex geological history of the region. The models were trained and validated using a structured dataset and evaluated according to key classification metrics, including accuracy, precision, recall, and F1 score. By comparing these models, the study provides insights into the benefits and limitations of machine-learning-based classification in high-quality, structurally controlled gold deposits. Ultimately, the findings support the advancement of automated, data-driven classification processes that can enhance decision-making, exploration, and resource assessment.

2 Materials and Methods

Figure 1 shows the general methodology adopted to achieve the objectives of this research. The methodology illustrated in Fig. 1 is explained in the following sections.

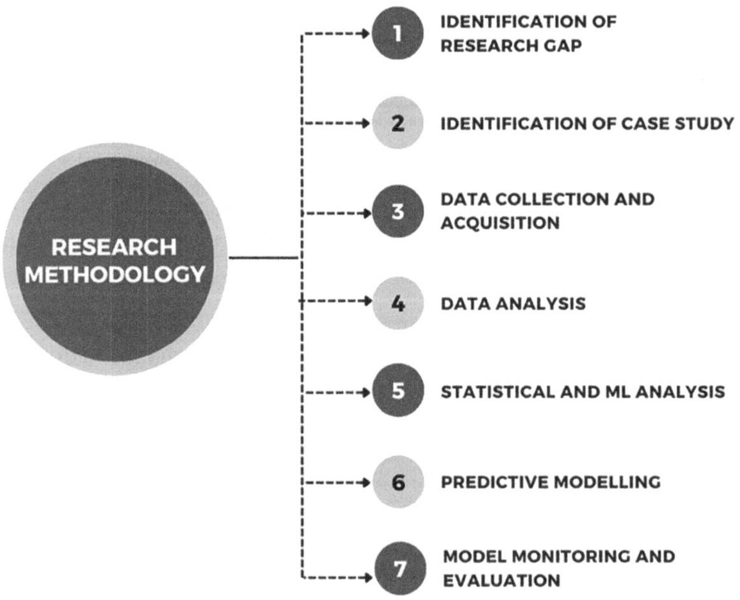

Fig. 1 Flowchart of the research methodology

Identification of Research Gap

In applying the best algorithms for the predictive model, there was a need to conduct a thorough desk study to ascertain the various ML types and their associated algorithms. The essence of the literature review was to understand the different ML techniques and their integration with geochemical and other datasets, their pros, cons, and numerous applications, thus serving as a foundation to identify the research gap.

Moreover, the study involved scrutinising various related research and relevant information about the study area. The Obuasi mining district is one of Ghana's most prolific mining communities. The region is heavily altered and metamorphosed, which could make it difficult to classify the rocks accurately. Geologists will save time and money by using modern techniques to audit the outcomes of their classification. Thus, the problems identified, and the study's aim will improve the area's mining and exploration activities.

In conclusion, the desk study indicates that supervised classification algorithms will be the best algorithms to apply in the predictive model analysis of this study. Various related studies using geochemical data for rock-type classification indicated accurate results using supervised classification algorithms.

Identification of the Case Study

The Obuasi Gold District is located within the Obuasi Municipal in the Ashanti Region of Ghana (Fig. 2a, b). It is approximately 240 km northwest of the capital city, Accra, and about 60 km south of Kumasi [27, 28].

The Obuasi Gold District is located within the north-eastern striking Ashanti Belt, a prolific belt, rich in gold mineralization in the Paleoproterozoic Birimian Supergroup (Fig. 1b). Therefore, the Obuasi gold deposit has been the most significant gold discovery [26]. The Ashanti belt mainly comprises the Birimian metasedimentary and the Birimian metavolcanic and, thus, the most prominent among the other Birimian belt units [27, 31]. The rocks within the study area are basin volcanic-sedimentary (metasedimentary) and are in contact with the greenstone belt of the Sefwi Group [32]. The contact accounts for the area being situated near an obscured connection between two primary faults, the Ashanti and Akropong thrust faults (Fig. 2b, c). The primary lithologies within the area are greywackes, phyllites, schist and carbonaceous schist [29, 32]. The area also consists of metavolcanic but possesses minimal linear structures. The rocks, as mentioned earlier, are intruded by dolerite dykes (Fig. 2c), which are usually unmineralized and unaltered [29, 32].

The rocks in the area are associated with various structural sequences [33]. The location possesses an early weak-developed S0 cleavage. The S0 cleavage is acute to the angle of the bedding [33]. The existence of an S2 cleavage that trends northeast and sharply descends to the northwest is the most dominant [33]. Moreover, the rocks have been subjected to a series of deformations. According to [26] and [32], the rocks

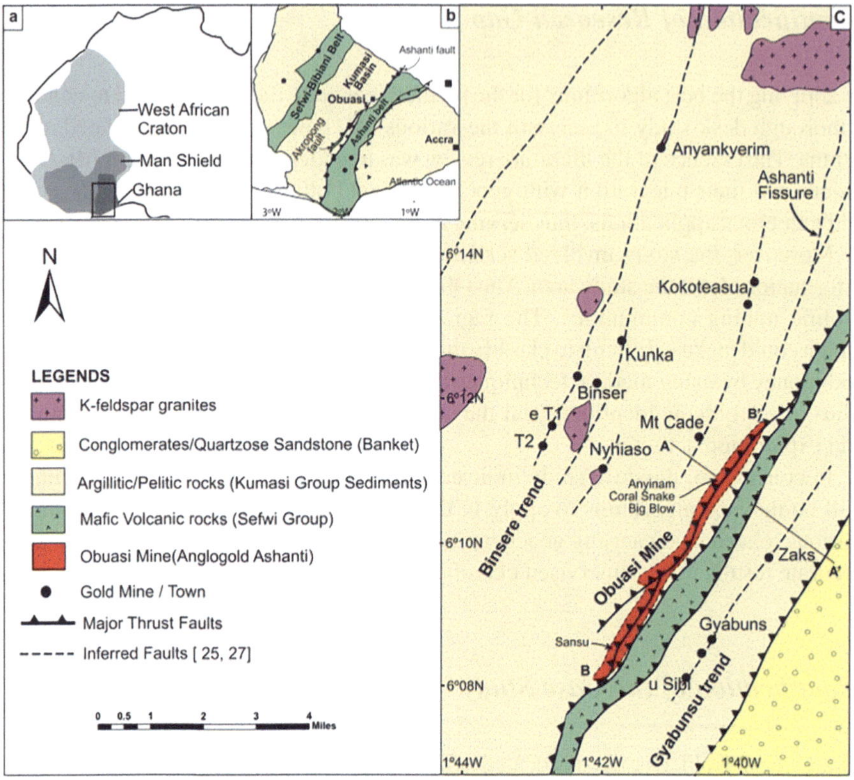

Fig. 2 **a** The position of Ghana in the Man Shield of the West African Craton. **b** The geology of south-western Ghana is mainly made up of the Birimian Supergroup (Kumasi Group and Sefwi Group). The Obuasi Gold District is situated at the contact of the two Birimian Groups. **c** The geology of Obuasi Mining District illustrating various mines along fault zones (modified from [29, 30])

of Obuasi have undergone three main deformation events: (i) these deformations are related to folding events. D1, the initial deformation event, consists of a sequence of parallel shear beds. They remain in the hinges of late folds as early foliations [30, 32], (ii) the next deformation event is D2, associated with the regional NW–SE shortening [32, 33]. Their associated folds are termed F2 folds. The F2 folds give rise to the significant initiation of gold mineralization [32], and (iii) Opong et al. [32] further explained that the latest deformation event is the D3, identified by NW–SE shortening and associated with cross-cutting F3 folds. The F3 folds give rise to late crenulation cleavages with an axial plane. The late crenulations are termed S3. High-grade ore shoots known as the "visible gold" are located in the F3 fold hinges at the AngloGold Ashanti Obuasi mine. These folds overprint the quartz veins during the D3 deformation event [32, 33].

According to Griffis et al. [34], the mineralization style in Obuasi is characterised by graphite-chlorite-sericite fault zones, as displayed in Fig. 3. These fault zones are identified with mineralization, usually associated with quartz or sulphides [27, 34]. Opong et al. [32] recognised five major shear zones. These shear zones are typically identified by extensive hydrothermal alteration of silica, carbonate, and sulphur, which is characterised by critical folding from the different deformation events. The shear amidst the folding occurs within the schists, phyllites, metagreywackes and tuffs of the Kumasi Group [27]. The metasedimentary rocks mentioned host the bulk of the minerals and are more vulnerable to ductile and brittle deformation [32]. The central and most prominent shear system associated with mineralization is characterised by anastomosing structural forms.

Furthermore, the persistent, lenticular-shaped gold mineralization extends over 8 km in strike length and descends more than 1700 m below the surface, both laterally and vertically [26, 29]. The gold mineralization in the study area possesses a structural setting that is in concordance with the regional structure of south-western Ghana. The mineralization strikes the northeast and dips steeply to about 75–85° to the northwest [36].

The Obuasi gold deposit is the largest gold deposit in West Africa, with more than 60 MOz [29, 35]. The gold deposit is located in the contact between the Sefwi Group's metavolcanic rocks and the Kumasi Group's metasedimentary rocks, containing more

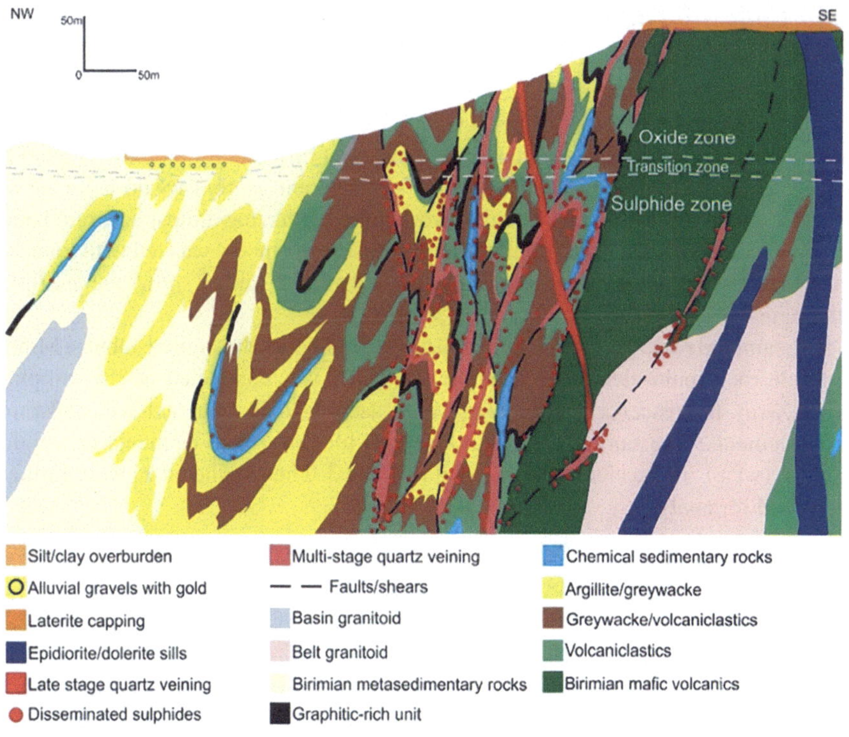

Fig. 3 Schematic cross-section of a typical Ashanti-type mineralization (adapted from [34, 35])

Table 2 Various rock types sampled

Lithology	Identification code
Quartz	MLO-001
Granitoid	MLO-002
Phyllite	MLO-003
Greywacke	MLO-004
Felsic Dyke	MLO-005

than 20 individual ore shoots [32]. Figure 3 displays two major mineralization types in the Obuasi Gold District: the quartz vein type and sulphide ore [35]. Quartz with free-milling coarse gold and trace levels of metal sulphides like iron, zinc, lead, and copper make up the quartz vein type [34]. The presence of gold within the crystal structure of arsenopyrite identifies the sulphide ore [27]. The quartz vein type is associated with very high grade, whilst the refractory sulphide ore constitutes more than 50% of the ore and has a grade of about 4 to 7 g/t. Higher grades of sulphide ore are associated with finer grains of arsenopyrite [27, 32].

Data Collection and Acquisition

The actualisation of the study was initiated with fieldwork, where various rock samples were collected for further analysis. Various localities within the Obuasi Gold District were visited to collect rock samples of different lithologies. The fieldwork was supervised and accompanied by a competent geologist who is aware of the geology of the study area. The fieldwork aimed at having almost accurate hand specimen identification of the rock samples to ensure the model was fed with reliable data, which would serve as target variables for the model. Table 2 shows the multiple lithologies identified.

The sampled rock types were delivered to the SGS Soluserv Limited, Obuasi Mine. A whole rock multi-element geochemical analysis was conducted on the samples using X-ray fluorescence (XRF). The study accounted for major, trace, and rare earth elements. The samples were all analysed for forty-three elements in weight per cent (wt%). These elements are listed in Table 3. The obtained data were further processed for analysis.

Data Analysis

At this stage, the data acquired from the laboratory were explored, inspected, visualised, and screened to obtain a thorough knowledge of the data and its relationships. The data were cleaned, and some unrequired data points were removed using Microsoft Excel. Before the ML modelling, the complete dataset was pre-processed.

Table 3 Statistical distribution of the geochemical elements used for the analysis

Element (wt%)	Mean	Mode	Median	Dispersion	Min	Max
MgO	1.49406	0.00	1.6698	0.522437	0.0	2.4193
Al_2O_3	10.8074	13.9128	13.2714	0.508644	0.0	15.6891
SiO_2	63.0815	100	54.4341	0.29382	51.0255	100
P	0.0504327	0.00	0.0597	0.637114	0.00	0.1037
S	0.0136121	0.00	0.0001	1.53296	0.00	0.0675
K_2O	1.95165	0.0636	2.1957	0.578123	0.0568	3.3588
Ca	1.2339	0.0169	1.095	0.649987	0.0137	2.4533
Ti	0.193744	0.0318	0.1856	0.716674	0.0269	0.3975
V	0.00437756	0.00	0.0044	0.832626	0.00	0.0116
Cr	0.01754	0.0166	0.0164	0.276316	0.0105	0.0288
Mn	0.0547654	0.0326	0.0353	0.955495	0.007	0.1602
Fe	2.55084	1.1059	2.3671	0.662694	0.3439	4.9122
Co	0.00106244	0.002	0.0009	0.647461	0.001	0.0021
Ni	0.00272341	0.0000	0.0025	0.728089	0.00	0.0062
Cu	0.00189561	0.0006	0.0014	0.718788	0.00	0.0052
Zn	0.00435756	0.0001	0.696885	0.00	0.0092	0.0092
As	0.0148761	0.004	1.72161	0.00	0.0737	0.0737
Se	0.000214634	0.0002	0.311455	0.001	0.0004	0.0004
Rb	0.00463707	0.00	0.601008	0.00	0.0087	0.0087
Sr	0.0463707	0.0006	0.955967	0.0006	0.1259	0.1259
Y	0.000995122	0.0014	0.675806	0.00	0.0019	0.0019
Zr	0.0111215	0.0003	0.684431	0.0002	0.0234	0.0234
Nb	0.0235834	0.0235	0.0822818	0.0192	0.0288	0.0288
Mo	0.000130732	0.00	0.0001	1.08129	0.00	0.0006
Ag	0.00157902	0.0017	0.0015	0.303043	0.0007	0.0047
Sn	0.00146244	0.0016	0.0014	0.258161	0.0006	0.0033
Sb	0.0011278	0.001	0.0011	0.450213	0.0001	0.0024
Ba	0.0661107	0.0841	0.0663	0.586492	0.0017	0.132
La	0.00150976	0.00	0.0013	0.73344	0.00	0.0057
Ca	0.00336488	0.00	0.0028	0.747501	0.00	0.0136
Pb	0.000516098	0.003	0.0003	0.986668	0.00	0.0019
Bi	8.87805e-05	0.00	0.0001	1.12478	0.00	0.0004
Th	0.000294146	0.0002	0.0003	0.628797	0.00	0.0008
U	0.000533659	0.00	0.0004	1.01273	0.00	0.02

Data points without identification numbers or omitted values were sorted and deleted, giving rise to 35 elements and 205 samples.

Statistical and Machine Learning Analyses

Kumar and Garg [37] indicated the robustness of ML algorithms over conventional statistical analysis for formulating a predictive model. This study employed classification algorithms due to the context of our research and data. From the desk study, four classification algorithms were selected for the analysis of the data, namely, Gradient Boosting (GBoost), Support Vector Machine (SVM), Random Forest (RF) and Adaptive Boosting (AdaBoost).

Gradient Boosting (GBoost)

Gradient boosting (GBoost) is a boosting ML technique that usually builds on ensemble models like decision trees [38]. Ali et al. [38] further mentioned that, in boosting, the trees are built sequentially, with each succeeding tree attempting to reduce the faults of the preceding tree. Each tree's construction is based on the knowledge derived from its precursors and updates the remaining errors. By minimising the anticipated value of a particular loss function, GBoost seeks to identify an estimate of the function that links occurrences to their output values [39]. GBoost has been to possess high predictivity; however, its employment was minimal [38]. The algorithm should produce one tree at a time to minimise errors from earlier tree trees in the model. Thus, the processing time is very long, even for small models [38].

Support Vector Machine (SVM)

Support vector machine (SVM) is a supervised learning algorithm that analyses data for classification and regression analysis using an n-dimensional classification space. The SVM (developed by Cortes and Vapnik) algorithm seeks out the hyperplane with the most significant margin between classes as the ideal one [40]. In its simplest version, SVM is a linear twofold classifier that differentiates a single border between two classes [41]. However, Boateng et al. [40] indicated that the algorithm solves problems that are not linearly separated. As a result, for such a task, an SVM with a non-linear root, such as the Radial Basis Function (RBF), would suffice. Using the training data, SVMs attempt to determine the best plane or line to split the dataset into predefined classes [41]. SVMs, as support vectors, maximise the separation or margin by selecting a subset of the training sample that is proximal to the ideal determination boundary in the attribute space [41].

Random Forest (RF)

Random forest (RF) is a classification technique (ensemble method) that uses several classifiers with tree structures and independently dispersed, identically distributed random vectors to generate a single vote at a selection for the most popular class [42]. A tree is constructed using the training test and a random vector unrelated to the previous random vectors of the same distribution to determine an upper border for RF and the generalisation error concerning two parameters. A tree is built using the training test and a random vector independent of the prior random vectors of the same allocation to obtain an upper bound for RF and the generalisation error for two parameters [43]. RF usually uses the bagging method to reduce massive variance within the dataset [38]. In other words, combining multiple classifiers can reduce variation and produce more accurate results when using unstable classifiers. A large number of decision trees make up RF. The independence of each decision tree must be ensured by increasing classification accuracy and enhancing generalisation capacity. The random operation is added to the build process, along with selecting sample and feature subsets [44]. According to [41], RF frequently employs the voting strategy (majority voting), which selects the label for each unlabelled sample that receives the most votes from different classifiers—the simplicity and efficiency of the majority voting technique account for its popularity.

Adaptive Boosting (AdaBoost)

An adaptive boosting (AdaBoost) classifier starts by fitting a classifier into the original data set. Next, it provides a copy of that classifier on the revised dataset, balancing the inaccurate and imprecise points so that the subsequent classifiers can focus on the instances that result in increased inaccuracy [45]. AdaBoost was proposed in 1995 by Freund and Schapire, and it operates by changing the weight without requiring a prior understanding of how learners learn [46]. AdaBoost is a type of iterative calculation [45]. The basic idea behind the analysis is to prepare a variety of classifiers (powerless classifiers) with a preparation set and then use various methods to coordinate them to create a more robust classifier. The computation is carried out by altering the information flow, indicated by the order amendment of the preparation set test, just as the final accuracy of the general arrangement determines the weight of each example [45].

Predictive Modelling

Predictive modelling will follow the pre-work for predictive analytics [47]. The user will demonstrate the robustness of ML algorithms over conventional statistical analysis for formulation using predictive modelling tools to produce accurate predictions. In this research, the previously stated ML techniques were applied as the basis for predictive modelling in a comparative manner, as indicated in the methodology flowchart in Fig. 4.

Firstly, the limited data samples posed a significant risk of overfitting in machine learning models. Through interpolation between maximum and minimum observed values and finding the mean, there was a successful expansion from 205 to 10,215 samples. This helped to reduce the risk of model overfitting and generalisation. The presented results and findings are based on the augmented dataset. The data were split with 80% of the data used for training and 20% for testing in the data sampler widget of the software. In addition, 15-fold cross-validation was applied to further evaluate the model's performance with 10 subsets used for training and 5 subsets used for testing. It has been emphasised that it is essential to keep in mind that the dataset's random division could lead to an uneven or variable interpretation of a particular rock type in the training and testing samples [2]. Stratified sampling was employed to guarantee that the classes were distributed proportionately.

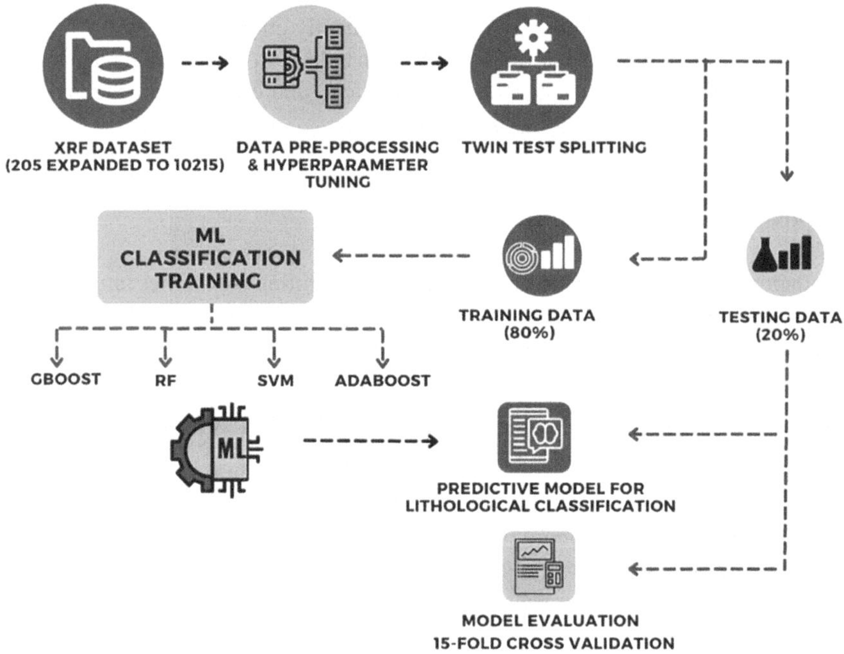

Fig. 4 Flowchart of the predictive analysis process

Model Monitoring and Evaluation

It is critical that after establishing a model for the prediction, the model ought to be monitored and evaluated using specific evaluation parameters. The Google Collaboratory platform provides various sampling techniques to exploit in the model's performance assessment. In the case of this study and the nature of the data, 15-fold cross-validation was used. 15-fold cross-validation helps overcome the limitation of the hold-out method by ensuring the model is tested on all samples [48]. It offers a more thorough assessment of the model's functionality and aids in determining how well it can be generalised [48]. The various performance parameters were used.

Cross-Validation Accuracy (CVA)

Cross-validation accuracy is a vital metric for evaluating the performance of machine learning models (Eq. 1). It involves partitioning the dataset into multiple subsets, or "folds," to ensure that each data point is used for both training and testing. The most common method is k-fold cross-validation, where the dataset is divided into k equal parts. The model is trained on $k - 1$ folds and validated on the remaining fold, repeating this process k times and averaging the results to compute the overall accuracy [49].

$$\text{Cross Validation Accuracy} = \frac{1}{2}\sum_{i=1}^{k} Accuracy_i \tag{1}$$

where $Accuracy_i$ represents the accuracy of the model on the ith fold.

Classification Accuracy (CA)

The CA statistic is used to evaluate the performance of a classification model. It represents the proportion of accurate forecasts to all input samples [50]. The formula for calculating accuracy is shown as Eq. 2.

$$\text{Accuracy} = \frac{Correct\ predictions}{Total\ predictions} \tag{2}$$

Although accuracy is a regularly used statistic for rating classification models, it has several drawbacks, mainly when dealing with unbalanced class distributions. In these situations, accuracy can be deceptive and not give a whole picture of the model's performance [51]. Additional measuring parameters such as area under the curve, F1 score, precision, and recall can be used to evaluate the model's performance in this scenario.

Confusion Matrix

A table that shows a machine learning model's performance with a test data sample is called a confusion matrix. It is frequently used to assess how well categorisation models work. For each input event, these models attempt to forecast a categorical label [52]. The matrix displays the number of true positives (TP), true negatives (TN), false positives (FP), and false negatives (FN) that the model produced using the test data [52]. The matrix will be a 2 × 2 table for binary classification and an n × n table for n classes or multi-class classification [53]. Since accuracy might produce false results when the number of observations in different categories varies significantly, the matrix is conducive when the dataset is unbalanced [53].

Precision

Precision is the percentage of correctly true values in expected positive circumstances [2]. It is represented mathematically as Eq. 3:

$$\text{Precision} = \frac{TP}{TP + FP} \tag{3}$$

where TP = True Positive and FP = False Positive.

Recall

Alférez et al. [54] define recall as displaying the percentage of genuine positives accurately categorised out of all positive samples (true positives plus false negatives. The following formula (Eq. 4) defines it:

$$\text{Recall} = \frac{TP}{TP + FN} \tag{4}$$

F1 Score

It is the harmonic mean of precision and recall [54]. The following definition (Eq. 5) serves as its basis:

$$\text{F1 score} = 2 \times \frac{Precision \times Recall}{Precision + Recall} \tag{5}$$

3 Results and Discussion

Classification Performance of Various Models

Table 3 shows the statistical distribution of the elements (data attributes). The lithologies were almost equal, as shown in Fig. 5; thus, the data was balanced and, there was no need for sampling techniques to make the data balanced. The target class of the models, as indicated earlier, are felsic dyke, granitoid, greywacke, phyllite and quartz of the Obuasi Gold District. For easy comprehension, the confusion matrix was employed to display the performance of the various models considering the discussed performance evaluation parameters: CVA, CA, F1-score, Precision, and Recall.

Performance Evaluation of Gradient Boosting

The confusion matrix for the classification performance for GBoost is indicated in Fig. 6. Table 4 also shows the detailed performance evaluation by the model. The GBoost algorithm was excellent in predicting the felsic dyke, greywacke and quartz. However, the model misclassified phyllite as greywacke and misclassified greywacke as phyllite. As [38] mentioned, GBoost is a boosting ML technique that

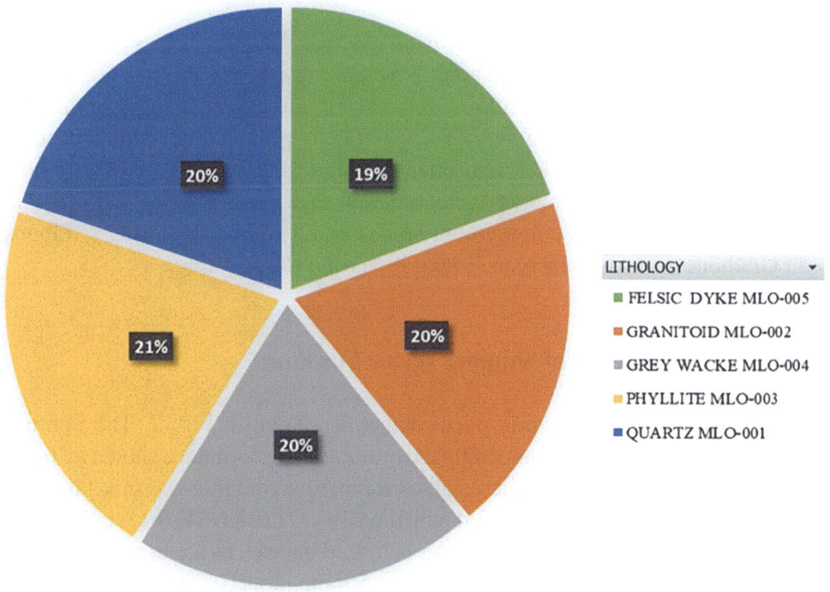

Fig. 5 Distribution of lithological classes

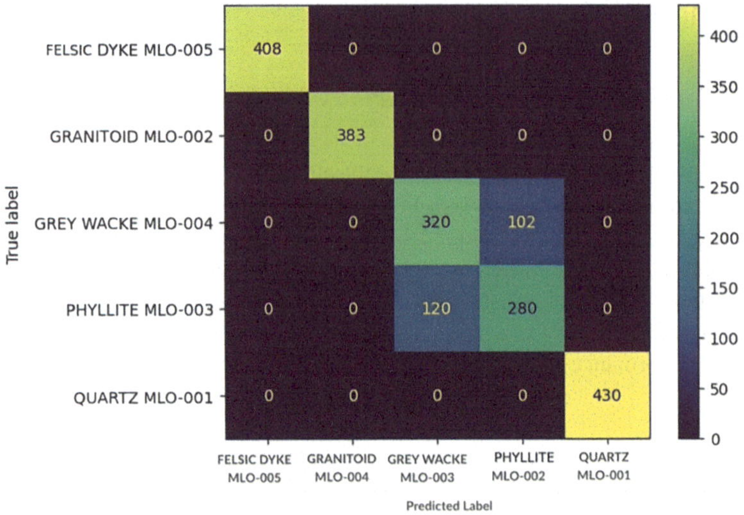

Fig. 6 Confusion matrix of the classification of rock types using GBoost

Table 4 Performance evaluation of the GBoost model

Model	CVA	CA	Precision	F1	Recall
GBoost	0.899	0.891	0.892	0.892	0.892

usually builds on ensemble models like decision trees. Various studies have indicated ensemble models as suitable for supervised classification problems. For instance, different machine learning algorithms were used to identify rock types in the Ordos Basin, China, where ensemble methods, specifically gradient boosting and random forest, emerged the best [8]. Dev and Eden [55] indicated in their study that gradient boosting models could further be increased, and their predictability and scalability could be enhanced by the employment of Extreme Gradient Boosting (XGboost), Light Gradient Boosting Machine (LightGBM) and CatBoosting.

Performance Evaluation of Support Vector Machine

The SVM model's classification performance is shown in Fig. 7. The confusion matrix in Fig. 7 is further expounded by the evaluation performance shown in Table 5. The SVM model accurately predicted all rock types, but phyllite was misclassified as greywacke and vice versa, as observed with the GBoost model. SVM's relatively good performance in this study depends on the balanced dataset, as it can adequately split the different classes linearly [40]. Deng et al. [56] mentioned that SVM performance was ranked very poor when the dataset used was imbalanced for their prediction of crystalline rocks from the Chinese Continental Scientific Drilling Main Hole

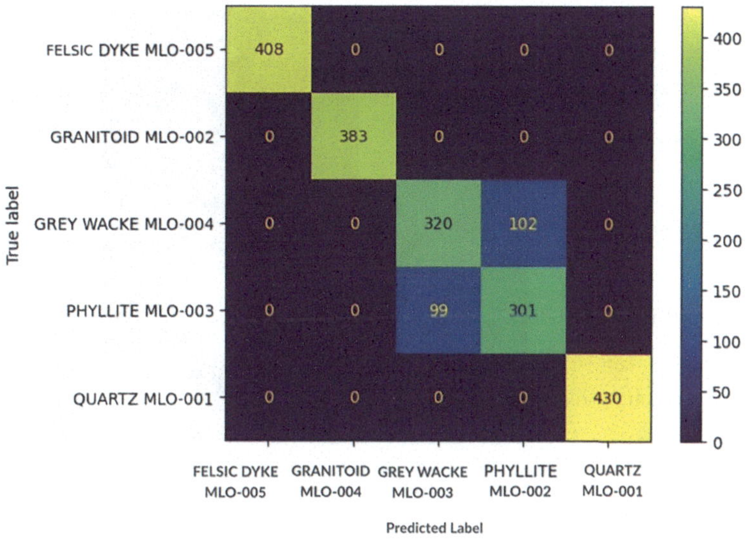

Fig. 7 Confusion matrix of the classification of rock types using SVM

Table 5 Performance evaluation of the SVM model

Model	CVA	CA	Precision	F1	Recall
SVM	0.903	0.902	0.902	0.902	0.902

(CCSD-MH) data. They further discussed SVM could be an excellent performer for imbalanced datasets when introduced to the Synthetic Minority Over-Sampling Technique (SMOTE). Therefore, considering the nature of the data based on its linear segregation and balance is vital for implementing the SVM model for litho-classification.

Performance Evaluation of Random Forest

The confusion matrix displayed in Fig. 8 summarises the classification performance by RF. The performance evaluation is further elucidated in Table 6. The RF model indicated strong classification accuracies in all lithologies. The model possessed weakness in classifying greywacke and phyllite samples, just like GBoost, which also belongs to the ensemble methods family. A study by Trott et al. [57] in the Late Miocene to early Pliocene porphyry belt in Chile indicated the robustness of random forest in rock type classification. The robustness of RF lies in its ability to reduce variance and errors in the dataset as it can determine the particular input features that are more significant upon the final classification, providing a helpful response to the user [38, 57]. More so, using k-fold cross-validation to evaluate the performance

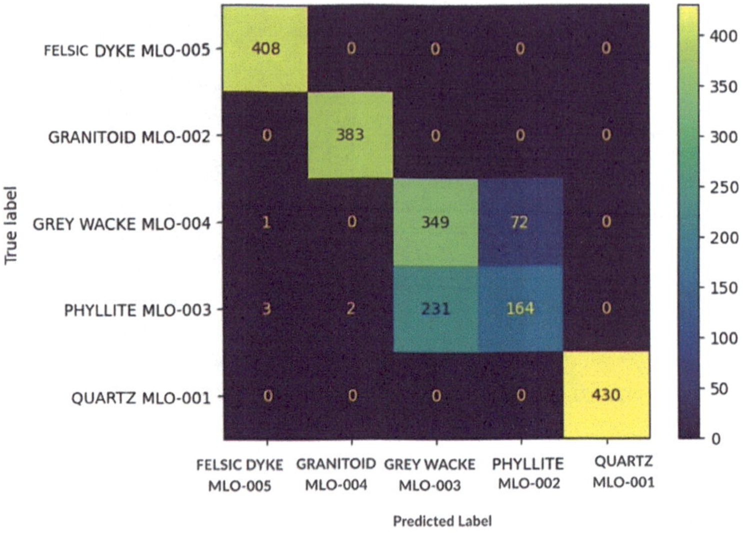

Fig. 8 Confusion matrix of the classification of rock types using RF

Table 6 Performance evaluation of the RF model

Model	CVA	CA	Precision	F1	Recall
RF	0.856	0.849	0.840	0.810	

of RF has proven to be an excellent step to minimise errors; as such, threefold cross-validation was implanted in this study, increasing the efficacy of RF [57]. Trott et al. [57] further described using k-fold cross-validation in RF models to assess the standard deviation between metrics produced by multiple folds to understand the degree of representation in any given subset.

Performance Evaluation of Adaptive Boosting

Figure 9 indicates the classification performance of the AdaBoost model. The confusion matrix in Fig. 9 is further explained by the evaluation performance shown in Table 7. The AdaBoost model accurately predicted only the quartz sample. On the other hand, the model performed relatively poorly in predicting the other lithologies. AdaBoost is one of the most utilised boosting approaches [58]. AdaBoost successively trains models such that misclassified examples are identified at the end of each repetition, and their weight is improved in a new training set [55].

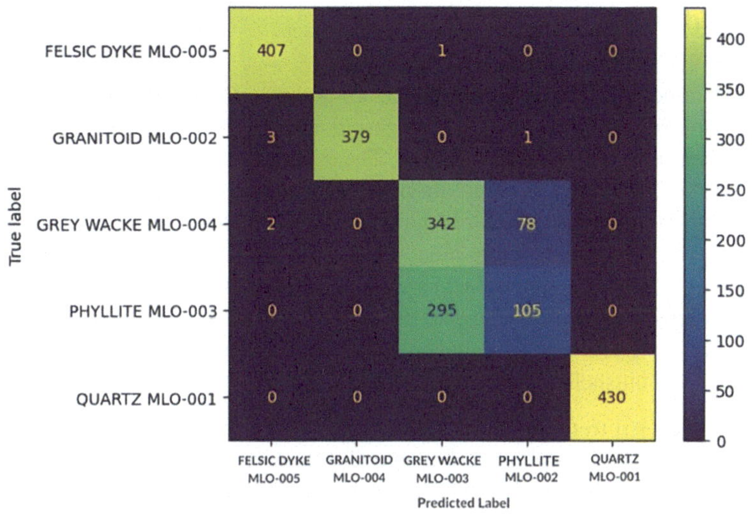

Fig. 9 Confusion matrix of the classification of rock types using AdaBoost

Table 7 Performance evaluation of the AdaBoost model

Model	CVA	CA	Precision	F1	Recall
AdaBoost	0.806	0.841	0.820	0.812	0.798

Overall Model Performance

The various models were subjected to a ranking framework to understand the best algorithm for the predictive model. Table 8 shows the ranking of the models, and SVM outperformed all models from the ranking. GBoost seconded with the best performance, followed by SVM and AdaBoost, as shown in Table 8.

Table 8 Overall performance evaluation of the various models

Model	CVA (Cross-validation accuracy)	CA (Classification accuracy)	Precision	F1 score	Recall	Total rankings
GBoost	0.899 **(2)**	0.891 **(2)**	0.892 **(2)**	0.892 **(2)**	0.892 **(2)**	10
SVM	0.903 **(1)**	0.902 **(1)**	0.902 **(1)**	0.902 **(1)**	0.902 **(1)**	5
RF	0.856 **(3)**	0.849 **(3)**	0.840 **(3)**	0.810 **(4)**	0.805 **(3)**	16
AdaBoost	0.806 **(4)**	0.841 **(4)**	0.820 **(4)**	0.812 **(3)**	0.798 **(4)**	19

4 Conclusion

The study aimed to establish a framework to understand various machine learning (ML) algorithms and their vital role in geosciences, particularly general rock-type and gold-bearing host rock identification. The establishment of the machine learning model can aid mining and exploration firms in understanding the scope of gold recovery fully in a high-sulphide ore deposit. Production geologists can also conceptualise an effective model for ore blending and ore grading since the rock types are now fully understood and identified objectively. Desk study has unravelled different ML algorithms, their characteristics, usefulness and limitations. The comparative analysis of the models indicated that SVM was the best model for deployment in the case study. In a nutshell, the key findings from the study are:

1. Desk study led to the identification of the characteristics of various ML algorithms. ML algorithms were grouped under supervised, unsupervised, reinforcement and semi-supervised. The multiple attributes of these ML types were reviewed and elaborated under the literature review. Furthermore, the desk study established that supervised learning had demonstrated robust performance and reliability in lithology classification where the geology is well-based and there is a distinctive feature between the rock types.
2. The case study indicates that the output was known (various rock classes), therefore the employment of supervised ML algorithms. Furthermore, classification algorithms were regarded as the best since the input data (XRF) corresponds to a definite target class (rock types).
3. GBoost, SVM, RF and AdaBoost were selected as the algorithms for the multi-class classification modelling. Desk study and literature review indicate their robustness, particularly for multi-class classification problems in this case study.
4. SVM indicated the best algorithm for deployment with a CA of 0.902, followed by GBoost and RF with a CA of 0.891 and 0.849 respectively. The ranking results showed that GBoost outperformed RF in the overall performance evaluation, whilst AdaBoost lags with a CA of 0.841.
5. As shown in the confusion matrices, the models indicated strong effectiveness in classifying the rock types with a few misclassifications.
6. It can be deduced that SVM can be deployed for best results in similar cases like the Obuasi Gold District.

5 Recommendations

This study may not account for all case studies, and thus, the following are recommended:

1. In deploying ML for general rock-type and gold-bearing host rock identification, the data samples should be almost equal to prevent oversampling or under-sampling in the datasets. Sampling techniques like Synthetic Minority Oversampling (SMOTE) could be employed in cases of unbalanced datasets.
2. Gold mining and exploration companies within the study area should employ the tool in the exploration and production stages to save time and enhance productivity. Deploying the model in their operations will save time, avoid subjective classification improve productivity during logging and boost overall operational success.
3. Geochemical data should be integrated with other datasets, such as geophysical data and drill core images, to unveil the actual characteristics of any lithology.
4. Due to diverse geological terrains and the nature of the dataset, a rigorous desk study should be conducted before employing ML algorithms in rock classification. Some datasets and the area's geology may require USL algorithms due to the rigorous labelling of input data (rock classes).

Acknowledgements This research is part of the second author's BSc Project Work/Thesis under the supervision of the first author at the University of Mines and Technology, Tarkwa, Ghana. SGS Soluserv Limited, Obuasi Mine, is duly acknowledged for its immense contributions towards the geochemical analysis of the rock samples in their laboratory. Mr Nathaniel Ansah Bruno formerly of AngloGold Ashanti Obuasi Mine, Geology Department, and now a Technical Specialist at Chrysos Corporation is highly appreciated for his endless guidance and direction towards the research from its inception. Immeasurable gratitude also goes to Dr. Yao Yevenyo Ziggah of the Geomatic Engineering Department, University of Mines and Technology, for his knowledge contribution while formulating the predictive model. The authors declare that no funds, grants, or other support were received during the preparation of this manuscript. The Editors are duly acknowledged for their fruitful suggestions.

References

1. Houshmand N, Goodfellow S, Esmaeili K, Ordóñez Calderón JC (2022) Rock type classification based on petrophysical, geochemical, and core imaging data using machine and deep learning techniques. Appl Comput Geosci 16:1–13
2. Ibrahim B, Ahenkorah I, Ewusi A, Majeed F (2023) A novel XRF-based lithological classification in the Tarkwaian paleo placer formation using SMOTE-XGBoost. J Geochem Explor 245:1–14. https://doi.org/10.1016/j.gexplo.2022.107147
3. Hood SB, Cracknell MJ, Gazley MF (2018) Linking protolith rocks to altered equivalents by combining unsupervised and supervised machine learning. J Geochem Explor 186:270–280. https://doi.org/10.1016/j.gexplo.2018.01.002
4. Sabbaghi H, Tabatabaei SH (2023) Data-driven logistic function for weighting of geophysical evidence layers in mineral prospectivity mapping. J Appl Geophys 212. https://doi.org/10.1016/j.jappgeo.2023.104986
5. Sabbaghi H, Tabatabaei SH (2020) A combinative knowledge-driven integration method for integrating geophysical layers with geological and geochemical datasets. J Appl Geophys 172. https://doi.org/10.1016/j.jappgeo.2019.103915
6. Sabbaghi H, Tabatabaei SH (2022) Application of the most competent knowledge-driven integration method for deposit-scale studies. Arab J Geosci 15(11):1057. https://doi.org/10.1007/s12517-022-10217-z

7. Verma SP, Agrawal S (2011) New tectonic discrimination diagrams for basic and ultrabasic volcanic rocks through log-transformed ratios of high field strength elements and implications for petrogenetic processes. Revista Mexicana de Ciencias Geologicas 28:24–44
8. Xie Y, Zhu C, Zhou W, Li Z, Liu X, Tu M (2018) Evaluation of machine learning methods for formation lithology identification: a comparison of tuning processes and model performances. J Petrol Sci Eng 160:182–193. https://doi.org/10.1016/j.petrol.2017.10.028
9. Caté A, Schetselaar E, Mercier-Langevin P, Ross PS (2018) Classification of lithostratigraphic and alteration units from drillhole lithogeochemical data using machine learning: a case study from the lalor volcanogenic massive sulphide deposit, Snow Lake, Manitoba, Canada. J Geochem Explor 188:216–228. https://doi.org/10.1016/j.gexplo.2018.01.019
10. Lui TCC, Gregory DD, Anderson M, Lee WS, Cowling SA (2022) Applying machine learning methods to predict geology using soil sample geochemistry. Appl Comput Geosci 16:1–13. https://doi.org/10.1016/j.acags.2022.100094
11. Kossoff D, Dubbin WE, Alfredsson M, Edwards SJ, Macklin MG, Hudson-Edwards KA (2014) Mine tailings dams: characteristics, failure, environmental impacts, and remediation. Appl Geochem 51:229–245. https://doi.org/10.1016/j.apgeochem.2014.09.010
12. Maruthupandian S, Chaliasou NA, Kanellopoulos A (2021) Recycling mine tailings for a sustainable future built environment. In: Gewerbestrasse Mporas I, Kourtessis P, Al-Habaibeh A, Asthana A, Vukovic V, Senior J (eds) Energy and sustainable futures proceedings of the 2nd ICESF 2020, Springer proceedings in energy, pp 163–169
13. Simate GS (2021) Environmental and health effects of acid mine drainage. In: Simate GS, Ndlovu S (eds) Acid mine drainage from waste to resources, 1st ed. CRC Press, Boca Raton, pp 18
14. Li J, Yang H, Zhao R, Tong L, Chen Q (2023) Mineralogical characteristics and recovery process optimisation analysis of a refractory gold ore with gold particles mainly encapsulated in pyrite and arsenopyrite. Geochemistry 83(1):125941. https://doi.org/10.1016/j.chemer.2022.125941
15. Peng S, Feng C, Qiu Z, Zhang Q, Liu W, Feng J, Hu Z (2025) Prediction of lithofacies in heterogeneous shale reservoirs based on a robust stacking machine learning model. Minerals 15:240. https://doi.org/10.3390/min15030240
16. Goldstein D, Aldrich C, Shao Q, O'Connor L (2025) Unlocking subsurface geology: a case study with measure-while-drilling data and machine learning. Minerals 15:241. https://doi.org/10.3390/min15030241
17. Boiger R, Churakov SV, Llagaria IB, Kosakowski1 G, Wüst R, Prasianakis NI (2024) Direct mineral content prediction from drill core images via transfer learning. Swiss J Geosci 117:8. https://doi.org/10.1186/s00015-024-00458-3
18. Zhou S, Cheng Z, Wang J, Li N, Jiang G (2025) Uncover implicit associations among geochemical elements using machine learning. Ore Geol Rev 179(2025):10650. https://doi.org/10.1016/j.oregeorev.2025.106506
19. Acosta ICC, Khodadadzadeh M, Tolosana-Delgado R, Gloaguen R (2020) Drill-core hyperspectral and geochemical data integration in a superpixel-based machine learning framework. IEEE J Sel Top Appl Earth Obs Remote Sens 13:4214–4228. https://doi.org/10.1109/JSTARS.2020.3011221
20. Kaplan UE, Topal E (2020) A new ore grade estimation using combine machine learning algorithms. Minerals 10(10):1–17. https://doi.org/10.3390/min10100847
21. Hussain M, Liu S, Ashraf U, Ali M, Hussain W, Ali N, Anees A (2022) Application of machine learning for lithofacies prediction and cluster analysis approach to identify rock type. Energies 15(12):1–15. https://doi.org/10.3390/en15124501
22. Hasterok D, Gard M, Bishop CMB, Kelsey D (2019) Chemical identification of metamorphic protoliths using machine learning methods. Comput Geosci 132:56–68. https://doi.org/10.1016/j.cageo.2019.07.004
23. Fernández A, Segarra P, Sanchidrián JA, Navarro R (2024) Ore/waste identification in underground mining through geochemical calibration of drilling data using machine learning techniques. Ore Geol Rev 168(2024):106045. https://doi.org/10.1016/j.oregeorev.2024.106045

24. Lawley CJM, Petts DC, Lee WS, Brueckner S (2025) Pyrite geochemistry for deposit type prediction and exploration in the golden triangle, Northwest British Columbia, Canada. Ore Geology Rev 177(2025):106447. https://doi.org/10.1016/j.oregeorev.2025.106447
25. Siman F, Jansson N, Kampmann T, Liwicki F (2021) Rock classification with machine learning: a case study from the Zinkgruvan Zn-Pb-Ag deposit, Bergslagen, Sweden. In: Swedish Artificial Intelligence society workshop. https://doi.org/10.1109/SAIS53221.2021.9483959
26. Charles Amadu C, Foli G, Abrokwa BK, Akpah S (2021) Geostatistical approach for the estimation of shear-hosted gold deposit: a case study of the Obuasi Gold deposit, Ghana. Malaysian J Geosci 5(2):76–84. https://doi.org/10.26480/mjg.02.2021.76.84
27. AngloGold Ashanti (2023) Mineral resource and reserve report. AngloGold Ashanti. https://reports.anglogoldashanti.com/22/rr/. Accessed 22 March 2023
28. Bempah CA, Ewusi A, Obiri-Yeboah S, Asabere SB, Mensah F, Boateng J, Voigt H (2013) Distribution of arsenic and heavy metals from mine tailings dams at Obuasi municipality of Ghana. Am J Eng Res 2(5):61–70
29. Fougerouse D, Micklethwaite S, Ulrich S, Miller J, Godel B, Adams DT, McCuaig TC (2017) Evidence for two stages of mineralization in West Africa's largest gold deposit: Obuasi, Ghana. Econ Geol 112(1):3–22
30. Allibone HA, Harris D, Etheridge M, Munroe S, Byrne D, Amanor J, Gyapong W (2002) Structural controls on gold mineralization at the Ashanti Gold Deposit, Obuasi, Ghana. J Soc Econ Geol 9:65–69. https://doi.org/10.5382/SP.09.04
31. Foli G, Gawu SKY (2017) Modified acid-base accounting model validation and pH buffer trend characterisation in mine drainage at the AngloGold Ashanti Obuasi Mine in Ghana, West Africa. Environ Earth Sci 76(19):1–10. https://doi.org/10.1007/s12665-017-7005-y
32. Opong JW, Amedjoe CG, Asante A, Wilson MC (2022) The importance of integrating geological mapping information with validated assay data for generating accurate geological wireframes in orebody modelling of mineral deposit in mineral resource estimation: a case study in AngloGold Ashanti, Obuasi Mine. Int J Geosci 13(6):426–437. https://doi.org/10.4236/ijg.2022.136023
33. Fougerouse D, Micklethwaite S, Miller J, McCuaig CT (2013) Multistage mineralisation of the Giant Obuasi Gold Deposit, Ghana. In: Proceedings 12th SGA biennial meeting 3, pp 1105–1108
34. Griffis RJ, Barning K, Agezo FL, Akosah FK (2002) Gold deposits of Ghana. Minerals Commission, 432
35. Smith AJB, Henry G, Frost-Killian S (2016) A review of the Birimian Supergroup- and Tarkwaian group-hosted gold deposits of Ghana. Episodes 39(2):177–197. https://doi.org/10.18814/epiiugs/2016/v39i2/95775
36. Amadu AC, Foli G, Sunkari ED (2022) A geostatistical approach to optimal drilling and sampling design: a case study of the Obuasi gold deposit in southern Ghana. J Sci Technol 4(1):1–17
37. Kumar V, Garg ML (2018) Predictive analytics: a review of trends and techniques. Int J Comp Appl 182(1):31–37
38. Ali ZA, Abduljabbar ZH, Taher HA, Sallow AB, Almufti SM (2023) Exploring the power of eXtreme gradient boosting algorithm in machine learning: a review. Academic J Nawroz Univ 12(2):320–334. https://doi.org/10.25007/ajnu.v12n2a1532
39. Bentéjac C, Csörgő A, Martínez-Muñoz G (2021) A comparative analysis of gradient boosting algorithms. Artif Intell Rev 54(3):1937–1967. https://doi.org/10.1007/s10462-020-09896-5
40. Boateng EY, Otoo J, Abaye DA (2020) Basic tenets of classification algorithms K-Nearest-Neighbor, support vector machine, random forest and neural network: a review. J Data Anal Inf Process 08(4):341–357. https://doi.org/10.4236/jdaip.2020.84020
41. Sheykhmousa M, Mahdianpari M, Ghanbari H, Mohammadimanesh F, Ghamisi P, Homayouni S (2020) Support vector machine versus random forest for remote sensing image classification: a meta-analysis and systematic review. IEEE J Sel Top Appl Earth Obs Remote Sens 13:6308–6325. https://doi.org/10.1109/JSTARS.2020.3026724

42. Reis I, Baron D, Shahaf S (2019) Probabilistic Random Forest: a machine learning algorithm for noisy data sets. Astron J 157(1):1–12. https://doi.org/10.3847/1538-3881/aaf101
43. Abdulkareem NM, Abdulazeez AM (2021) Machine learning classification based on radom forest algorithm: a review. Int J Sci Bus 5(2):128–142. https://doi.org/10.5281/zenodo.4471118
44. Parmar A, Katariya R, Patel V (2019) A review on Random Forest: an ensemble classifier. In: International conference on intelligent data communication technologies and Internet of Things 2018. ICICI, pp 758–763. https://doi.org/10.1007/978-3-030-03146-6_86
45. Azmi SS, Baliga S (2020) An overview of boosting decision tree algorithms utilising AdaBoost and XGBoost boosting strategies. Int Res J Eng Technol 7(5):6867–6870
46. Chengsheng T, Huacheng L, Bing X (2017) Adaboost typical algorithm and its application research. MATEC Web Conf 139:1–6. https://doi.org/10.1051/matecconf/20171390022
47. Lee CS, Cheang PYS, Moslehpou M (2022) Predictive analytics in business analytics: Decision Tree. Adv Decis Sci 26(1):1–30. https://doi.org/10.1016/j.bushor.2019.10.005
48. Lyashenko V, Jha A (2023) Cross-validation in machine learning: how to do it right. Neptune AI. https://www.neptune.ai/blog/cross-validation-in-machine-learning-how-to-do-it-right. Accessed 20 Aug 2023
49. Arlot S, Celisse A (2010) A survey of cross-validation procedures for model selection. Statistics Surveys 4:40–79. https://doi.org/10.1214/09-SS054
50. Mishra A (2018) Metrics to evaluate your machine learning algorithm. Towards data science. https://www.towardsdatascience.com/metrics-to-evaluate-your-machine-learning-algorithm-f10ba6e38234. Accessed 20 Aug 2023
51. Brownlee J (2021) Failure of classification accuracy for imbalanced class distributions. Machine Learning Mastery. https://machinelearningmastery.com/failure-of-accuracy-for-imbalanced-class-distributions/. Accessed 20 Aug 2023
52. Narkhede S (2018) Understanding confusion matrix. Towards data science. https://www.towardsdatascience.com/understanding-confusion-matrix-a9ad42dcfd62. Accessed 19 Aug 2023
53. Chicco D, Jurman G (2023) The Matthews Correlation Coefficient (MCC) should replace the ROC AUC as the standard metric for assessing binary. BioData Mining 16(1):1–23. https://doi.org/10.1186/s13040-023-00322-4
54. Alférez GH, Esteban OA, Clausen BL, Ardila AMM (2022) Automated machine learning pipeline for geochemical analysis. Earth Sci Inf 15(3):1683–1698. https://doi.org/10.1007/s12145-022-00821-8
55. Dev VA, Eden MR (2019) Gradient boosted decision trees for lithology classification. In: Muñoz SG, Laird C, Realff M (eds) Proceedings of the 9th international conference on foundations of computer aided process design, pp 113–118. https://doi.org/10.1016/B978-0-12-818597-1.50019-9
56. Deng C, Pan H, Fang S, Konaté AA, Qin R (2017) Support vector machine as an alternative method for lithology classification of crystalline rocks. J Geophys Eng 14(2):341–349. https://doi.org/10.1088/1742-2140/aa5b5b
57. Trott M, Leybourne M, Hall L, Layton-Matthews D (2022) Random forest rock type classification with integration of geochemical and photographic data. Appl Comp Geosci 15:1–15. https://doi.org/10.1016/j.acags.2022.100090
58. Rokach L (2016) Decision forest: twenty years of research. Inf Fusion 27:111–125. https://doi.org/10.1016/j.inffus.2015.06.005

The manufacturer's authorised representative in the EU is Springer Nature Customer Service Centre GmbH, Europaplatz 3, 69115 Heidelberg, Germany. If you have any concerns regarding our products, please contact ProductSafety@springernature.com

Printed and bound by CPI Group (UK) Ltd, Croydon, CR0 4YY
30/09/2026
02225588-0004